Sustainable Food Packaging Materials and Technology

Sustainable Food Packaging Materials and Technology

Juan Manuel Tirado-Gallegos, PhD
Juan Daniel Rivaldi Chávez, PhD
José Juan Buenrostro-Figueroa, PhD
Editors

AAP APPLE ACADEMIC PRESS

First edition published 2027

Apple Academic Press Inc.
1265 Goldenrod Circle, NE,
Palm Bay, FL 32905 USA

760 Laurentian Drive, Unit 19,
Burlington, ON L7N 0A4, Canada

CRC Press
2385 NW Executive Center Drive,
Suite 320, Boca Raton FL 33431

4 Park Square, Milton Park,
Abingdon, Oxon, OX14 4RN UK

Library and Archives Canada Cataloguing in Publication

Title: Sustainable food packaging materials and technology / Juan Manuel Tirado-Gallegos, PhD, Juan Daniel Rivaldi Chávez, PhD, José Juan Buenrostro-Figueroa, PhD, editors.
Names: Tirado-Gallegos, Juan Manuel, editor. | Rivaldi Chávez, Juan Daniel, editor. | Buenrostro-Figueroa, José Juan, 1985- editor
Description: First edition. | Includes bibliographical references and index.
Identifiers: Canadiana (print) 20260148709 | Canadiana (ebook) 20260148717 | ISBN 9781779641427 (hardcover) | ISBN 9781779641434 (ebook)
Subjects: LCSH: Food—Packaging—Technological innovations. | LCSH: Food—Packaging—Environmental aspects. | LCSH: Packaging—Materials. | LCSH: Biopolymers.
Classification: LCC TP374 .S87 2026 | DDC 664/.09—dc23

Library of Congress Cataloging-in-Publication Data

CIP data on file with US Library of Congress

ISBN: 978-1-77964-142-7 (hbk)
ISBN: 978-1-77964-143-4 (ebk)

DOI: 10.1201/9781779641434

About the Editors

Juan Manuel Tirado-Gallegos, PhD
Research Professor, School of Animal Sciences and Ecology, Autonomous University of Chihuahua, Mexico

Juan Manuel Tirado-Gallegos, PhD, is a Research Professor at the School of Animal Sciences and Ecology at the Autonomous University of Chihuahua, Mexico. Dr. Tirado-Gallegos has experience in isolating, modifying, and characterizing natural polymers and developing biodegradable packaging with application in meat products. He has published 34 scientific articles, ten book chapters, and more than 40 contributions in scientific meetings. Dr. Tirado-Gallegos has been a member of the National System of Researchers in México since 2017. He graduated as an engineer in Food Technology from Antonio Narro Agrarian Autonomous University and obtained his Master of Science in Biochemical Engineering from the Technological Institute of Veracruz, Mexico. He earned his science doctorate in Food Biopolymers from the Research Center for Food and Development, A.C., Mexico.

Juan Daniel Rivaldi Chávez, PhD
Chemical Engineer and Researcher, Department of Industrial Applications, Faculty of Chemical Sciences, Universidad Nacional de Asunción, San Lorenzo, Paraguay

Juan Daniel Rivaldi Chávez, PhD, is a Chemical Engineer and Researcher at the Department of Industrial Applications of the Faculty of Chemical Sciences of the Universidad Nacional de Asunción, San Lorenzo, Paraguay, working in bioprocess, biomaterials, and circular bioeconomy, with emphasis in the characterization and valorization of byproducts and industrial residues for obtaining valuable products. He has experience as a quality control and process manager in the paper and cardboard recycling industry, and as a technical expert and auditor of quality control laboratories. Dr. Rivaldi was a lecturer at the University of São Paulo and National University of Asuncion. As researcher, he has been categorized as a researcher level I in the National Program for Incentive to Researchers (PRONII) by the National Science and Technology Council (CONACYT) of Paraguay. Additionally, he is a member of the Paraguayan Society of Materials, Paraguayan Society of Microbiology, and Associate Programme of the Organisation for the Prohibition of Chemical Weapons. He has published 24 scientific articles in indexed journals, three book chapters, and a book. In addition, he has presented 35 full papers and 24 abstracts and has been chairman and invited speaker at 15 scientific events. Dr. Rivaldi has been visiting research at the National Institute of Advanced Industrial Science and Technology (AIST-Japan) and at the Cellulose and Paper Group of the Institute of Forest Science, Spanish National Research Council (CSIC), Spain. He graduated from the National University of Asuncion, Paraguay, and earned his master's degree in Industrial Biotechnology and doctorate in Sciences in the bioprocess area at the University of Sao Paulo, Brazil.

José Juan Buenrostro-Figueroa, PhD
Researcher, Research Center for Food and Development,
Chihuahua, Mexico

José Juan Buenrostro-Figueroa, PhD, is a titular Researcher at the Research Center for Food and Development, Chihuahua, Mexico. He is a food engineer. Dr. Buenrostro has experience in bioprocess development, including microbial processes for enzyme production and recovery of bioactive compounds, valorization of agroindustrial byproducts, and extraction and characterization of bioactive compounds from plants and agroindustrial wastes. He has published 57 papers in indexed journals, 31 book chapters, nine books as editor, and more than 117 contributions at scientific meetings, with over 1701 citations and h-index of 23 (Google Scholar database). He is a member of the Mexican Association of Food Science (AMECA), Mexican Society for Food Protection affiliate of the International Association for Food Protection, and Food Safety Thematic Network: Agroindustrial By-products Valorization and Reduction of Food Losses and Waste (PDA). Dr. Buenrostro is a member of SNII (National System of Researches, Mexico), since 2014. His MSc and PhD degrees were Food Science and Technology.

Contents

Contributors

Mónica Daniela Alvarado-Macías
Universidad Autónoma Agraria Antonio Narro. Departamento de Ciencia y Tecnología de Alimentos, Buenavista, Saltillo, Coahuila, México

Andrea Araya
LANOTEC CeNAT CONARE, San José, Costa Rica

Diego Batista
LANOTEC CeNAT CONARE, San José, Costa Rica

Juan Carlos Bustillos-Rodríguez
Tecnológico Nacional de México, Instituto Tecnológico de Cd. Cuauhtémoc, Mexico

Jose Maria Carbajo
Cellulose and Paper Group, Forest Product Department, Forest Research Centre (ICIFOR-CSIC), Madrid, Spain

Miroslava Carrera-Romero
Faculty of Zootechnics and Ecology, Autonomous University of Chihuahua, Chihuahua, México

Rebeca Corrales
LANOTEC CeNAT CONARE, San José, Costa Rica

Yendry Corrales
LANOTEC CeNAT CONARE, San José, Costa Rica

Ndahita De Dios-Avila
Faculty of Agrotechnological Sciences, Autonomous University of Chihuahua, Chihuahua, Mexico

Alexia Domínguez-Villarreal
Universidad Autónoma Agraria Antonio Narro. Departamento de Ciencia y Tecnología de Alimentos, Buenavista, Saltillo, Coahuila, México

Magdalena Espínola
Bio and Materials Laboratory, Polytechnic School, National University of Asuncion, San Lorenzo, Central, Paraguay

Omayra B. Ferreiro
Food Engineering and Technology Department, Faculty of Chemical Sciences, National University of Asuncion; San Lorenzo, Paraguay, Bio and Materials Laboratory, Polytechnic School, National University of Asuncion, San Lorenzo, Central, Paraguay

Úrsula Fillat-Latorre
Cellulose and Paper Group, Forest Product Department, Forest Research Centre (ICIFOR-CSIC), Madrid, Spain; CELBIOTECH Paper Engineering Research Group, Universitat Politècnica de Catalunya (UPC), Terrassa, Barcelona, Spain

Junior García-Candido
Food Engineering and Technology Department, Faculty of Chemical Sciences, National University of Asuncion, San Lorenzo, Paraguay

Félix García-Ochoa
Chemical and Materials Engineering Department, Faculty of Chemical Sciences, Universidad Complutense de Madrid, Madrid, Spain

Nuria Gómez-Hernández
Cellulose and Paper Group, Forest Product Department, Forest Research Centre (ICIFOR-CSIC), Madrid, Spain

Natalia González
Instituto de Química del Sur, INQUISUR (UNS-CONICET), Bahía Blanca, Argentina; Departamento de Química, Universidad Nacional del Sur (UNS), Bahía Blanca, Argentina

Luis Castillo Henríquez
LANOTEC CENAT CONARE, San José, Costa Rica; Facultad de Farmacia, Universidad de Costa Rica, San Pedro, Costa Rica

Francisco Hernández-Centeno
Universidad Autónoma Agraria Antonio Narro. Departamento de Ciencia y Tecnología de Alimentos, Buenavista, Saltillo, Coahuila, México

María Hernández-González
Universidad Autónoma Agraria Antonio Narro. Departamento de Ciencia y Tecnología de Alimentos, Buenavista, Saltillo, Coahuila, México

Ruth Lechuga-Valles
Faculty of Zootechnics and Ecology, Autonomous University of Chihuahua, Chihuahua, México

Olivia V. López
Departamento de Química, Universidad Nacional del Sur (UNS), Bahía Blanca, Argentina; Planta Piloto de Ingeniería Química, PLAPIQUI (UNS-CONICET), Bahía Blanca, Argentina

Haydeé Yajaira López-De la Peña
Universidad Autónoma Agraria Antonio Narro. Departamento de Ciencia y Tecnología de Alimentos, Buenavista, Saltillo, Coahuila, México

Mary Lopretti
UDELAR, Montevideo, Uruguay

Gisela Martínez-Mendoza
Faculty of Agrotechnological Sciences, Autonomous University of Chihuahua, Chihuahua, Mexico

Susana Mesén
LANOTEC CeNAT CONARE, San José, Costa Rica

Magna Monteiro
Bio and Materials Laboratory, Polytechnic School, National University of Asuncion, San Lorenzo, Central, Paraguay

José Alberto Núñez-Gastélum
Departamento de Ciencias Químico Biológicas, Instituto de Ciencias Biomédicas, Universidad Autónoma de Ciudad Juárez, Ciudad Juárez, Chihuahua, México

Gabriela Montes de Oca-Vásquez
Centro de Estudios Sobre Desarrollo Sostenible, Universidad Técnica Nacional, Alajuela, Costa Rica

Magali Ordóñez-García
Tecnológico Nacional de México, Instituto Tecnológico de Cd. Cuauhtémoc, Mexico

Yuridia Ortiz-Rivera
Departamento de Ciencias Químico Biológicas, Instituto de Ciencias Biomédicas, Universidad Autónoma de Ciudad Juárez, Ciudad Juárez, Chihuahua, México

Sergio A. Paniagua
LANOTEC CENAT CONARE, San José, Costa Rica

Ailen M. Poza
Instituto de Investigaciones Bioquímicas de Bahía Blanca, INIBIBB (UNS-CONICET), Bahía Blanca, Argentina; Centro I-Mar, Universidad de Los Lagos, Puerto Montt, Los Lagos, Chile

Claudio Ríos-Velasco
Centro de Investigación en Alimentación y Desarrollo- CIAD, AC, Unidad Cuauhtémoc, Chihuahua, México

Juan Daniel Rivaldi Chávez
Industrial Applications Department, Faculty of Chemical Sciences, Universidad Nacional de Asunción, San Lorenzo, Paraguay; Cellulose and Paper Group, Forest Product Department, Forest Research Centre (ICIFOR-CSIC), Madrid, España; Chemical and Materials Engineering Department, Faculty of Chemical Sciences, Universidad Complutense de Madrid, Madrid, Spain

Cinthya Gabriela Rocha-Cortés
Universidad Autónoma Agraria Antonio Narro, Departamento de Ciencia y Tecnología de Alimentos, Buenavista, Saltillo, Coahuila, México

Ana Margarita Rodríguez-Hernández
Centro de Investigación en Química Aplicada, Col. San José de los Cerritos, Saltillo, Coahuila, México

Jesús Rubén Rodríguez-Núñez
Programa de Biotecnología, División de Ciencias de la Salud e Ingenierías, Universidad de Guanajuato Campus Celaya-Salvatierra, Celaya, Guanajuato, México

María Janeth Rodríguez-Roque
Faculty of Agrotechnological Sciences, Autonomous University of Chihuahua, Chihuahua, Mexico

Nora Aideé Salas-Salazar
Faculty of Agrotechnological Sciences, Autonomous University of Chihuahua, Chihuahua, Mexico

Rogelio Sánchez-Vega
Faculty of Zootechnics and Ecology, Autonomous University of Chihuahua, Chihuahua, México

Juan Manuel Tirado-Gallegos
Faculty of Zootechny and Ecology, Autonomous University of Chihuahua, Chihuahua, Mexico

Vera Torres-Armendariz
Faculty of Agrotechnological Sciences, Autonomous University of Chihuahua, Chihuahua, Mexico

Daiana V. Trapé
Planta Piloto de Ingeniería Química, PLAPIQUI (UNS-CONICET), Bahía Blanca, Argentina; Departamento de Ingeniería Química, Universidad Nacional del Sur (UNS), Bahía Blanca, Argentina

Maynor Alberto Vargas-Vargas
Universidad Técnica Nacional, Alajuela, Costa Rica

José Roberto Vega-Baudrit
LANOTEC CeNAT CONARE, San José, Costa Rica; Escuela de Química, Universidad Nacional de Costa Rica, Heredia, Costa Rica

Priscilla Vergara
Cellulose and Paper Group, Forest Product Department, Forest Research Centre (ICIFOR-CSIC), Madrid, Spain

Juan Carlos Villar
Cellulose and Paper Group, Forest Product Department, Forest Research Centre (ICIFOR-CSIC), Madrid, Spain

Marcelo A. Villar
Planta Piloto de Ingeniería Química, PLAPIQUI (UNS-CONICET), Bahía Blanca, Argentina; Departamento de Ingeniería Química, Universidad Nacional del Sur (UNS), Bahía Blanca, Argentina

Claudia Chaves Villareal
LANOTEC CeNAT CONARE, San José, Costa Rica

Paul Baruk Zamudio-Flores
Centro de Investigación en Alimentación y Desarrollo, A.C., Unidad Cuauhtémoc.
Laboratorio de Carbohidratos, Empaques y Alimentos Funcionales, Cd. Cuauhtémoc, Cuauhtémoc, Chihuahua, México

Juan Miguel Zúñiga-Umaña
LANOTEC CeNAT CONARE, San José, Costa Rica

Abbreviations

α-TP	α-tocopherol
3R	reduce, reuse, and recycle
ACE	angiotensin I-converting enzyme
AITC	allyl isothiocyanate
ASL	alcohol-soluble lignin
BDO	1,4-butanediol
bFGF	basic fibroblast growth factor
BHA	butylated hydroxyanisole
BPA	bisphenol A
BSE	bovine spongiform encephalopathy
CAGR	compound annual growth rate
CD	carbon dot
CEPI	Confederation of European Paper Industries
CF	Cirish fructans
CMQ	carboxymethyl chitosan
CNC	cellulose nanocrystal
CNF	cellulose nanofiber
CNF	cellulose nanofibrils
COD	chemical oxygen demand
DPP-IV	dipeptidyl peptidase-IV
EAB	elongation at break
ECs	edible coatings
EDC	1-ethyl-3-(3-dimethylaminopropyl) carbodiimide
EIP	isoelectric point
EVOH	ethylene vinyl alcohol copolymer
FDA	Food and Drug Administration
FDCA	2,5-furan dicarboxylic acid
FDMC	furan-2,5-dimethylcarboxylate
GHG	greenhouse gas
GRAS	generally recognized as safe
HAp	hydroxyapatite
HDPE	high-density polyethylene
HPP	high-pressure processing method
HT	hydroxytyrosol
Hyp	hydroxyproline
IUPAC	International Union of Pure and Applied Chemistry

KL	kraft lignins
LCBW	leveraging lignocellulosic biomass waste
LDPE	low-density polyethylene
LEO	lemon essential oil
LLDPE	linear low-density polyethylene
MAAs	mycosporine-like amino acids
MC	microbial cellulose
MPM	minimally processed mangoes
MW	molecular weight
NCFs	nanocellulose fibrils
NFC	nanofibrillated cellulose
NHS	N-hydroxysuccinimide
OMW	olive mill wastewater
OP	oxygen permeability
P(HB-HV)	poly(3-hydroxybutyrate-co-3-hydroxyvalerate)
PB	polybutylene
PBAT	poly(butylene adipate-co-terephthalate)
PBF	poly(butylene 2,5-furan dicarboxylate)
PBS	polybutylene succinate
PBSA	succinate-co-adipate copolymer
PCL	polycaprolactone
PE	polyethylene
PEF	polyethylene furanoate
PET	polyethylene terephthalate
PGA	polyglycolic acid
PHA	polyhydroxyalkanoates
PHB	polyhydroxybutyrate
PLA	polylactic acid
PP	polypropylene
PPE	pomegranate peel extract
PPF	poly(propylene 2,5-furanoate)
PPTF	poly(propylene 2,5-thiophendicarboxylate
Pro	proline
PS	Pachyrhizus starch
PS	polystyrene
PVA	polyvinyl alcohol
RC	refined carrageenan
ROP	ring-opening polymerization
RPP	Rambutan peel phenoli
SCGE	spent coffee ground
SRC	semirefined carrageenan
SSP	solid-state polycondensation

TEO	thyme essential oil
TPS	thermoplastic starch
TS	tensile strength
TSC	tilapia skin collagen
US	ultrasonication
UV	ultraviolet
WG	wheat gluten
WL	wine lees
WVP	water vapor permeability
WVTR	water vapor transmission rate
XRD	X-ray diffraction
Zn-MOFs	metal-organic frameworks ZIF-8

Preface

Food packaging, often underestimated, plays a crucial role in the supply chain and our daily lives. During storage, distribution, and marketing, packaging must contain and protect food, communicate its characteristics, and present convenience to the consumer. Several packaging materials are available, most notably synthetic polymers derived from petroleum, to accomplish this. However, the growing concern about the pollution and environmental impact generated by this type of material drives us to rethink our choices, and the search for solutions has become more urgent than ever. This book, *Sustainable Food Packaging Materials and Technology,* was born out of the need to address one of the most pressing challenges of our era: how to preserve the quality, safety, and, in some cases, freshness of food with packaging based on sustainable materials that do not compromise the well-being of the planet?

Each chapter of this book provides a technical analysis of materials and their application and presents a comprehensive overview of how we can use available resources responsibly and creatively.

Some chapters reveal the transformative potential of waste resources, from using by-products from the fishing industry to materials derived from agaves. We will also explore the use of natural polymers such as proteins and polysaccharides in the design of packaging and coatings, highlighting their capacity to offer biodegradable solutions that can also be highly effective. Within polysaccharides, we will delve into the great potential offered by starch and cellulose, which, thanks to their versatility and properties, have become some of the most explored renewable materials in creating biodegradable packaging and coatings.

An innovative aspect of this book is its emphasis on active biodegradable packaging and coatings, which integrate antioxidant and antimicrobial properties by utilizing agroindustrial waste. This convergence of sustainability and functionality offers an effective solution to combat food waste, a global challenge requiring immediate attention. In addition, we will delve into two revolutionary materials: polybutylene succinate (a biodegradable biopolymer) and polyethylene furanoate (a recyclable material), which redefine expectations about packaging sustainability.

These advances promote the circular economy and offer a roadmap to a greener future.

We thank all the collaborators who have contributed their knowledge and experience to make this project possible. May these pages serve as a guide and source of inspiration in the search for more sustainable and responsible food packaging.

CHAPTER 1

Biopolymers Derived from Fish Industry By-Products for Food Packaging Applications

JUNIOR GARCÍA-CANDIDO,[1] MAGDALENA ESPÍNOLA,[2] MAGNA MONTEIRO,[2] and OMAYRA B. FERREIRO[1,2*]

[1]Department of Food Engineering and Technology, Faculty of Chemical Sciences, National University of Asuncion, San Lorenzo, Paraguay

[2]Bio and Materials Laboratory, Polytechnic School, National University of Asuncion, San Lorenzo, Central, Paraguay

*Corresponding author

ABSTRACT

Fish skins, scales, and bones account for approximately 15–23% of the fish's weight. Generally, they are considered waste by the fishing industry. However, these wastes are an essential source of raw material that can be used to produce biomaterials such as gelatin, collagen, and hydroxyapatite and have the potential to be transformed into high-value-added products for cosmetics, pharmaceutical, and food industries. This chapter presents an overview of the fishing industry, focusing on gelatin and collagen extraction processes and their applications. A critical analysis of the characteristics of these compounds for their use in food packaging is also included. Finally, the feasibility of recovering fish waste for application in food packaging is explored.

Sustainable Food Packaging Materials and Technology. Juan Manuel Tirado-Gallegos, Juan Daniel Rivaldi Chávez, and José Juan Buenrostro-Figueroa (Eds)

DOI: 10.1201/9781779641434-1

1.1 AN OVERVIEW OF THE FISHING INDUSTRY

The rapid increase in the world population is causing an increase in demand for food, which also includes the fishing and aquaculture sectors. According to estimates from studies by the Food and Agriculture Organization of the United Nations [1], aquiculture has gained significant momentum in the last decade compared to inland and marine fisheries. Between 2018 and 2020, aquaculture increased by 5 million tons, while capture or fishing decreased by 6.2 million tons in live weight equivalent. Human food consumption has represented a low level of increase in this period because it was affected by the COVID-19 pandemic due to global confinement. However, the growing global population is expected to reach approximately 10 billion by 2050 [2]. This projection is worrying and leads to thinking about "what and how" to achieve the Sustainable Development Goals (SDGs) objectives. Tremendous efforts have been made to reduce the damage caused by increased food demand, mainly related to waste production.

In the circular economy era, eco-friendliness and sustainability are fundamental terms that must be applied in developing new products and implementing improvements in production chains. In this context, the bioeconomy plays an essential role in Blue Growth, transforming products from raw fishing and aquiculture to obtain high-added value [3] and reducing fishing waste production. Bioeconomy is a recent term applied to materials and processes that are of biological origin, which include mimetic or natural mechanisms to construct products and systems with the highest efficient performance [4].

Fishing and aquaculture are among the most polluting industrial sectors and are part of the food industry. Given the large amount of biowaste generated, the aim is to increase the circular economy through the valuation of by-products to strengthen the bioeconomy. In the search for the consolidation of the bioeconomy, science is not sparing efforts, as well as community actions at the state level. Carcass, head, skin, scales, bones, and viscera are considered fish by-products, corresponding to approximately 60% of fish production [5, 6]. To break this paradigm of by-products, studies have reported the transformation into high-value products, including proteins (proteolytic enzymes, protein hydrolysates, bioactive peptides, collagen, and gelatin), fish oils, and hydroxyapatite (HAp). Within this context, this chapter intends to highlight in a significant way the importance of these high-value products for improving the quality of life and the environment, generating a positive impact on human life.

1.2 STRUCTURE AND EXTRACTION METHODS OF COMPOUNDS FROM FISH BY-PRODUCTS

Collagen and gelatin have enormous potential in the food, pharmaceutical, and medical industries; their applications can range from tissue engineering to their use as additives in food products. However, to understand the significance of collagen and gelatin at an industrial level, it is important to have prior knowledge of their structure and how it relates to their possible applications.

1.2.1 STRUCTURE AND PROPERTIES

Collagen is the most abundant structural protein in vertebrate and invertebrate organisms, representing up to 30% of the total protein in an animal. It is also the most important component in the extracellular matrix and is found in muscles, bones, and teeth [7, 8].

The collagen molecule is composed of three polypeptide chains called α chains, which contain approximately 1000 amino acids and have molecular weights (MWs) of approximately 100 kDa. Each α-chain is made up of an amino acid sequence of the form Gly-X-Y, where Gly represents glycine, X represents proline (Pro), and Y generally represents hydroxyproline (Hyp).

Pro and Hyp are important amino acids that ensure the structural integrity of collagen [9]. Solari and Córdova [10] reported a Hyp concentration of 6.5 and 52.9 mg per gram of sample for *Engraulis ringens* (anchovies) wastes and the collagen extracted from them, respectively. At the same time, Menezes et al. [11] obtained at least 6% of Hyp in the collagen extracted from Nile tilapia skins. Zhang et al. [12] reported that the type I collagen extracted from freshwater carp fish scales contained relatively high half-cysteine but lower denaturation temperature than the porcine dermis type I collagen.

Which amino acid is located at the X or Y position will depend on the type and source of collagen. Thus, the three α-chains of the collagen molecule may be identical (i.e., they are encoded by a single gene), giving rise to a homotrimer, or they may be different, giving rise to a heterotrimer [13].

Each α-chain forms a levorotatory helix on itself, and once the three α-chains are joined, the characteristic dextrorotatory triple helix structure of collagen is formed. The location of glycine in such a structure is fundamental since being a nonpolar molecule, it is strongly favorable to hide it within the triple helix, away from any external polar substance, mainly water. In addition, being the smallest amino acid, it does not present any significant steric hindrance when located in the center of the triple helix, giving great stability to the collagen molecule [13, 14].

On the other hand, collagen molecules can be laterally associated by covalent cross-linking to allow the formation and subsequent expansion of the so-called collagen fibrils, through the process of fibrillogenesis. The fibrils, which are responsible for providing mechanical strength and integrity to the body, can also adhere to each other longitudinally to give rise to the formation of collagen fibers (Figure 1.1) [15, 16].

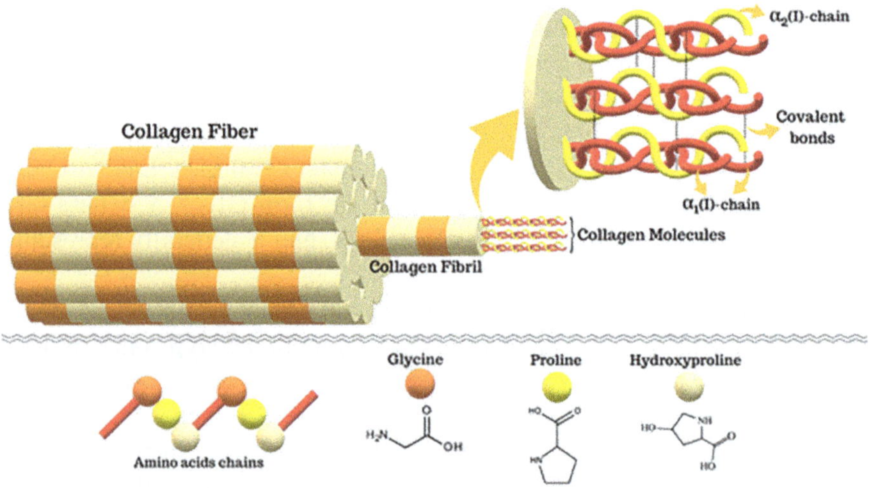

FIGURE 1.1 Collagen type I molecular structure.

To date, a total of 29 collagen types have been identified, all of which have in common the coiled triple helix of α-chains. Although the nomenclature of collagen types has not been precisely established, in general, each type of collagen is identified by a Roman numeral, being type I collagen the most abundant in connective tissues. However, a specific type of collagen may present more than one combination of α-chains. Therefore, each α-chain is identified by an Arabic and Roman numeral in parentheses. In the case that two or three of the chains are synthesized by a single gene, the number of identical chains in the molecule is specified with Arabic numerals outside of square brackets (or as a subscript after the parentheses). For example, the type I collagen molecule can have two different chains, the α1(I) chain and the α2(I) chain, which can combine in two different ways to give two different molecular species: type I collagen of the form [α1(I)]2, α2(I); or type I collagen of the form [α1(I)]3. Note that, in both cases, the total number of α-chains in the molecule is three. Finally, collagen types can be classified according to their structure and supramolecular organization into fibrillar and nonfibrillar, fibrillar being those involved in the formation of cross-linked striated fibrils [3, 17–20]. Table 1.1 presents a summary of the classification of collagen types.

On the other hand, gelatin is a heterogeneous mixture of polypeptides obtained from the partial hydrolysis or thermal degradation of collagen from animal connective tissues [26–28]. Hydrolysis can be accomplished by heating the collagen in an aqueous solution and at a temperature high enough to allow the covalent cross-links of the fibrils to break and the triple helix to unravel, forming gelatin in the form of random coils (Figure 1.2A). If the solution is cooled, the small segments of two or three polypeptide chains regroup to form a structure similar to the triple helix of collagen, forming a gelatin gel (Figure 1.2B) [29, 30].

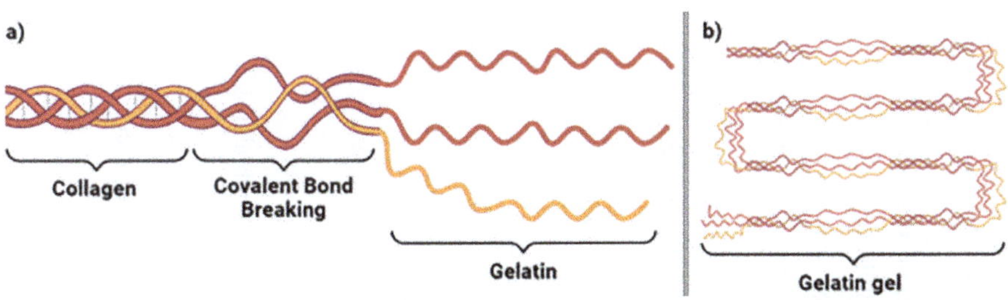

FIGURES 1.2 A and B. Gelatin structure: (A) from collagen hydrolysis and (B) in the gel form.

There are two types of gelatins depending on the treatment it receives during the extraction process: (1) type A gelatin (acid treatment), which is generally obtained from pigs, poultry, and fish and (2) type B gelatin

TABLE 1.1 Classification of Collagen Types [18, 20–25]

Type	Location	Class	α-Chains	Molecular Species
I	Skin, bones, tendons, and cornea	Fibrillar	α1(I) and α2(I)	$[\alpha1(I)]_2$, $\alpha2(I)_2$, and $[\alpha1(I)]_3$
II	Gristle and vitreous body	Fibrillar	α1(II)	$[\alpha1(II)]_3$
III	Skin, vessels, intestine, and uterus	Fibrillar	α1(III)	$[\alpha1(III)]_3$
IV	Basal membrane and capillaries	Nonfibrillar	α1(IV), α2(IV), α3(IV), α4(IV), α5(IV), and α6(IV)	$[\alpha1(IV)]_2$, α2(IV), α3(IV), α4(IV), α5(IV), $[\alpha5(IV)]_2$, and α6(IV)
V	Skin, bones, cornea, and placenta	Fibrillar	α1(V), α2(V), α3(V), and α4(V)	$[\alpha1(V)]_3$, $[\alpha1(V)]_2$, α2(V), $[\alpha1(V)]_2$, α4(V), α1(XI), α1(V), and α3(XI)
VI	Bones, vessels, skin, cornea, and gristle	Nonfibrillar	α1(VI), α2(VI), α3(VI), α4(VI), and α5(VI)	α1(VI), α2(VI), and α3(VI) α1(VI), α2(VI), and α4(VI)
VII	Mucous membranes, skin, bladder, umbilical cord, and amniotic fluid	Nonfibrillar	α1(VII) and α2(VII)	$[\alpha1(VII)]_3$, $[\alpha1(VII)]_2$, and α2(VII)
VIII	Skin, brain, heart, kidneys, vessels, bones, and gristle	Nonfibrillar	α1(VIII) and α2(VIII)	$[\alpha1(VIII)]_2$, α2(VIII), α1(VIII), $[\alpha2(VIII)]_2$, $[\alpha1(VIII)]_3$, and $[\alpha2(VIII)]_3$
IX	Cornea, vitreous body, and gristle	Nonfibrillar	α1(IX), α2(IX), and α3(IX)	α1(IX), α2(IX), and α3(IX)
X	Gristle	Nonfibrillar	α1(X)	$[\alpha1(X)]_3$
XI	Gristle and intervertebral disk	Fibrillar	α1(XI), α2(XI), and α3(XI)	α1(XI), α2(XI), α3(XI), α1(XI), α1(V), and α3(XI)
XII	Gristle, tendons, and skin	Nonfibrillar	α1(XII)	$[\alpha1(XII)]_3$
XIII	Skeletal muscles, heart, eye, skin, and endothelial cells	Nonfibrillar	α1(XIII)	$[\alpha1(XIII)]_3$

TABLE 1.1 *(Continued)*

Type	Location	Class	α-Chains	Molecular Species
XIV	Vessels, eyes, nerves, tendons, bones, skin, and gristle	Nonfibrillar	α1(XIV)	$[\alpha1(XIV)]_3$
XV	Capillary vessels, ovaries, heart, testicles, skin, placenta, and kidneys	Nonfibrillar	α1(XV)	$[\alpha1(XV)]_3$
XVI	Heart, skin, kidneys, and smooth muscle	Nonfibrillar	α1(XVI)	$[\alpha1(XVI)]_3$
XVII	Skin	Nonfibrillar	α1(XVII)	$[\alpha1(XVII)]_3$
XVIII	Kidneys, lungs, and liver	Nonfibrillar	α1(XVIII)	$[\alpha1(XVIII)]_3$
XIX	Skin, kidneys, liver, placenta, spleen, and prostate gland	Nonfibrillar	α1(XIX)	$[\alpha1(XIX)]_3$
XX	Corneal epithelium	Nonfibrillar	α1(XX)	$[\alpha1(XX)]_3$
XXI	Stomach, kidneys, vessels, heart, placenta, and skeletal muscles	Nonfibrillar	α1(XXI)	$[\alpha1(XXI)]_3$
XXII	Tissue connections	Nonfibrillar	α1(XXII)	$[\alpha1(XXII)]_3$
XXIII	Metastatic carcinogenic cells	Nonfibrillar	α1(XXIII)	$[\alpha1(XXIII)]_3$
XXIV	Bones and cornea	Fibrillar	α1(XXIV)	$[\alpha1(XXIV)]_3$
XXV	Eye, brain, heart, and testicles	Nonfibrillar	α1(XXV)	$[\alpha1(XXV)]_3$
XXVI	Testicles and ovaries	Nonfibrillar	α1(XXVI)	$[\alpha1(XXVI)]_3$
XXVII	Gristle	Fibrillar	α1(XXVII)	$[\alpha1(XXVII)]_3$
XXVIII	Nervous system cells	Nonfibrillar	α1(XXVIII)	$[\alpha1(XXVIII)]_3$
XXIX	Skin	–	α1(XXIX) or α5(VI)	$[\alpha1(XXIX)]_3$

(alkaline treatment), which is obtained from bovine sources. The main difference between the physicochemical properties of both treatments for the same gelatin source lies in the isoelectric point (EIP), since the EIP for type A gelatin is between the pH range of 8–9, while in type B gelatin, it is between 4.8 and 5.5. This is because, in type B gelatin, there is a greater amount of aspartic acid and glutamic acid residues than in type A gelatin, so the number of negative charges is greater [27, 31–33].

Fish gelatin has attracted significant attention due to the risk of human–animal infectious diseases such as bovine spongiform encephalopathy (BSE) and foot-and-mouth disease, even if no evidence of transmission has been reported. In addition, some sociocultural religions do not allow the consumption of derivative products of porcine [33, 34]. Thus, researchers and companies are still exploring safer sources of fish gelatin due to demand in many areas such as agriculture, food, cosmetic, and pharmaceutical industries [26, 35] and overcoming the problem mentioned above.

From gelatin, it is possible to obtain a hydrolysate rich in proteins, which has higher antioxidant activity as well as antimicrobial activity than the original protein. This confers on fish gelatin hydrolysates the ability to convert free radicals to more stable products before terminating the oxidative chain reaction [36].

Another compound that can be obtained from fish waste is HAp, a calcium phosphate ceramic found in the bones and teeth of all vertebrate animals and other sources, including fish scales [37–39]. HAp particles have been widely used for biomedical applications [40, 41] and for the remediation of water, soils, and air [42, 43].

1.2.2 ADDED-VALUE COMPOUNDS OBTAINED FROM FISH BY-PRODUCTS

Fish skin, scales, and spines with suitable pretreatment are important sources of collagen [44–46] and other fish by-products, such as gelatin [33, 47], hydrolysates [48], among others [39, 42].

Depending on the compound of interest, different steps are necessary to produce them. Nonetheless, in any case, it is important to separate organic (albumins, globulins, and fat) and inorganic matter (demineralization using acid). The sequential treatments could include some biotechnological process using enzymes or extraction using water or acid solutions.

In this way, it is possible to obtain collagen, gelatin, protein hydrolysates, and HAp from fish by-products [49]. The pretreatments for gelatin and collagen extraction promote an increase in the process yield and, consequently, will have an important influence on the final characteristics of the products [48, 50]. Below are some of the most common methods for extracting these products, including the pretreatment step.

1.2.2.1 COLLAGEN EXTRACTION METHODS

Fish collagen can be extracted from various anatomical parts of the fish, including skins, scales, bones, and fins [34]. Different fish species, such as tilapia [51], cod [52], salmon [53], and tuna [54, 55], have been used as a source of collagen, each with its unique properties and characteristics. Bi et al. [56] found that depending on the extraction method, the extracted collagen presented different MWs and different protein secondary structures characterized by sodium dodecyl sulfate-polyacrylamide gel electrophoresis and Fourier transform infrared spectroscopy.

A pretreatment step allows the elimination of compounds different from collagen. A defat process is commonly made using ethanol [56] or butyl alcohol [57]. A deproteinized step was also proposed using 0.1 M NaOH [57]. Song et al. [44] proposed a fermentation pretreatment using *Bacillus velezensis* FEL-BM21 to eliminate the noncollagen ingredients with similar extraction yield and collagen characteristics to those achieved using chemical pretreatments. The authors also concluded that the fermentation pretreatment is safe, environment-friendly, and time-saving, making it viable for industrial applications.

NaOH is the alkali most widely used to remove noncollagenous protein from fish by-products. It is worth noting that NaOH concentration, extraction temperature, and contact time between fish by-products and the alkali agent could cause collagen hydrolysis [45].

For the same purpose, saturated calcium hydroxide was also used or a combination of NaOH and $Ca(OH)_2$ as reported by Chinh et al. [46]. The last resulted in the successful pretreatment of fish scales for collagen extraction. Chinh et al. [46] tested different alkali solutions and concentrations [$Ca(OH)_2$ NaOH or a mixture of both] to remove fat and protein, followed by a process of demineralization performed in acid solution (HCl, H_2SO_4, H_3PO_4, or citric acid). When 0.1 M NaOH was combined with a saturated $Ca(OH)_2$ (2/1 v/v), followed by a mixture of 0.2 M HCl and 0.5 M H_2SO_4 (1/2 v/v), the largest weight of raw collagen was obtained.

Giraldo-Rios et al. [58] performed a design of experiment 33 to evaluate the effects of temperature (10–20°C), NaOH concentration (0.3–0.7 M), and size of skin Nile tilapia sample (1–5 cm) on the response variables of noncollagenous protein extraction and losses of collagen. In this chapter, the stirring speed was not considered a variable because the authors found that above 400 rpm, the concentration of noncollagenous protein did not vary. Hence, this stirring speed was used in all the experiments to minimize the external mass transfer limitations. The authors found that due to the higher surface area of smaller sizes of skin, there was a higher protein exposure, causing an inverse effect on the extraction efficiency. On the other hand, the extraction temperature and the NaOH concentration showed direct effects on the noncollagenous protein extraction. The maximum noncollagenous protein concentration (27.2%) was obtained using a skin size of 1 cm at 20°C and 0.7 M NaOH solution. In these conditions, the highest loss of collagen was also reported (6.4%). The authors performed a multiobjective optimization and found that using a 1 cm sample, 10°C, and 0.7 M NaOH, it is possible to decrease the loss of collagen to 1.76%, notwithstanding the extraction yield of noncollagenous protein (24.3%). The use of NaOH for removing noncollagenous protein and EDTA for eliminating the presence of calcium was also reported in the literature [9, 59].

The acetic acid method is the most reported method for extracting acid-soluble collagen [44]. In the presence of pepsin, it has been reported a higher extraction yield that could be due to a covalent cross-linkage resulting from the condensation of aldehyde groups of lysine and hydroxylysine in the nonhelical region, which decreases the solubility of collagen in the acid solution and, pepsin which is responsible for the cleavage of the peptide localized at the telopeptide region to readily dissolved collagen [44, 60].

When the extraction is made using acetic acid, the collagen has to be precipitated from the solution, and NaCl is widely used for this purpose due to its high ionic force [45]. It has been reported that collagen has good solubility at pH 1–4 and NaCl concentrations less than 3% (w/v) [44].

Besides acetic acid, Bhuimbar et al. [61] evaluated the extraction efficiency of other acids such as lactic, tartaric, citric, formic, sulfuric, and hydrochloric. Lactic acid showed the highest efficiency, tartaric, citric, and formic acids performed better than acetic acid, while sulfuric and hydrochloric acids presented negligible yield.

Extraction temperature has a crucial influence on extraction efficiency, and it is important to consider improving the extraction yield of the chosen method [56]. The effects of temperature, acetic acid concentration, and extraction time have been studied to optimize collagen extraction yield [11, 59]. Menezes et al. [11] found a synergistic positive interaction between time and temperature, while a high acetic acid concentration was to the extraction yield's detriment. The highest collagen yield obtained by Menezes et al. [11] was 19% using 0.35 M of acetic acid at 20°C for 65 h of reaction, while Costa et al. [59] found the highest protein concentration (80%) using 0.5 M acetic acid for 24 h.

Bi et al. [56] used the sodium hydroxide method for collagen extraction. However, this method caused a higher hydrolysis of the collagen molecules than other methods, such as acetic acid and hot water.

Kuwahara [62] recently performed collagen extraction from tilapia scales by bubbling ultrafine bubbles of gases such as oxygen, carbon dioxide, and ozone into an acetic acid solution. The highest yield (1.58%) was obtained using CO_2 ultrafine bubbles in 0.1 M acetic acid solution for 5 h. Ultrafine bubbles of O_2 were not as efficient as CO_2, while the presence of O_3 gas could have caused the collagen transformation into gelatin or collagen fragmentation into peptides. Table 1.2 summarizes collagen extraction methods found in the literature.

1.2.2.2 GELATIN EXTRACTION METHODS

The functional and technological properties of fish gelatin are influenced by two factors: (1) the characteristics of the source (species type, anatomical part, warm–cold water fish) and (2) the specific production process influenced by parameters such as temperature, time, and pH [33, 63]. These properties are generally gel strength, viscosity, thermal stability, gelation, and melting point, which are directly dependent on MW, MW distribution, Pro and Hyp imino acid compositions, as well as the ratio of α/β chains [26, 27, 33, 36, 64].

Despite glycine being one of the primary amino acids, followed by alanine and Pro presented in gelatin [65, 66], Hyp is commonly considered a gelatin purity indicator because this amino acid may lead to more desirable

TABLE 1.2 Methods for Collagen Extraction

Extraction Method	Source	Conditions	Pretreatment	Yield	Characteristics	References
Acetic acid	Tilapia skin	1000 mL 0.5 M acetic acid for 24 h at room temperature	Two volumes (v/w) of ethanol for 24 h at room temperature without stirring	Wet weight of fish skin: 11.7%	Type I collagen (280, 150, and 130 kDa)	[56]
Hot water	Tilapia skin	1000 mL 80°C ultrapure water for 6 h in an oscillatory water bath shaker at 10 rpm	Two volumes (v/w) of ethanol for 24 h at room temperature without stirring	Wet weight of fish skin: 10.7%	Type I collagen (150 and 130 kDa)	[56]
Sodium hydroxide	Tilapia skin	1000 mL 0.12 M NaOH for 24 h at room temperature	Two volumes (v/w) of ethanol for 24 h at room temperature without stirring	Wet weight of fish skin: 8.5%	Type I collagen (280, 160, 140, and 10 kDa)	[56]
Acetic acid	Ikan belida (*Notopterus lopis*) skin	0.5 M acetic acid at a sample/solvent ratio of 1:2.5 (w/v) for 24 h	0.1 M NaOH sample/solvent 1:8 (w/v) for 6 h; 10% butyl alcohol sample/solvent 1:10 (w/v) for 24 h	Dry weight basis: 0.5%		[57]
Acetic acid	Tilapia (*Oreochromis* spp.) skin.	From Tilapia (*Oreochromis* spp.) skin. 0.7 M acetic acid at 18.5°C, sample/solution 1:10 (w/v) for 5 h.	Cleaning using 250 ppm NaClO; 0.4 M NaOH at 25°C; 10% n-butanol sample/solvent 1:10 (w/v) for 12 h	Dry weight basis: 1.59%	Acid-soluble collagen	[45]
Acetic acid	Tilapia (*Oreochromis* spp.) scales	0.7 M acetic acid at 10°C, sample/solution 1:10 (w/v) for 12 h.	Cleaning using 250 ppm NaClO; 0.4 M NaOH at 25°C; 0.5 M EDTA + NaOH pH 8 sample/solution 1:10 (w/v) for 48 h	Dry weight basis: 1.56%	Acid-soluble collagen	[45]
Acetic acid	Tilapia (*Oreochromis* spp.) spines.	0.7 M acetic acid at 21.5°C, sample/solution 1:10 (w/v) for 12 h.	Cleaning using 250 ppm NaClO; 0.4 M NaOH at 25°C; 0.5 M EDTA + NaOH pH 8 sample/solution 1:10 (w/v) for 48 h	Dry weight basis: 0.54%	Acid-soluble collagen	[45]

TABLE 1.2 *(Continued)*

Extraction Method	Source	Conditions	Pretreatment	Yield	Characteristics	References
Acetic acid	Nile tilapia (*Oreochromis niloticus*) skin	0.5 M acetic acid, sample/solution 1:60 (w/v) for 48 h. Precipitated using NaCl; dissolved in 0.5 M acetic acid and dialyzed against 0.1 M acetic acid	Fermentation pretreatment: 15 g of fish skin in a mixture of 12 mL of seed fluid in 88 mL of sterile water incubated at 20°C and 200 rpm for 12 h. Seed fluid: pelleted cells from *B. velezensis* FEL-BM21	Wet weight: 4.76%	Acid-soluble collagen	[44]
Acetic acid + pepsin	Nile tilapia (*O. niloticus*) skin	0.5 M acetic acid containing 5% pepsin (w/w), sample/solution 1:80 (w/v) for 24 h.	Fermentation pretreatment: 15 g of fish skin in a mixture of 12 mL of seed fluid in 88 mL of sterile water incubated at 20°C and 200 rpm for 12 h. Seed fluid: pelleted cells from *B. velezensis* FEL-BM21	Wet weight: 8.14%	Pepsin-soluble collagen	[44]
Acetic acid	Nile tilapia (*O. niloticus*) skin	0.5 M acetic acid, sample/solution 1:60 (w/v) for 48 h. Precipitated using NaCl; dissolved in 0.5 M acetic acid and dialyzed against 0.1 M acetic acid	Chemical pretreatment: 0.1 M NaOH sample/solution 1:30 (w/v) for 24 h; 10% butyl alcohol sample/solution 1:30 (w/v) for 24 h	Wet weight: 4.27%	Acid-soluble collagen	[44]
Acetic acid + pepsin	Nile tilapia (*O. niloticus*) skin	0.5 M acetic acid containing 5% pepsin (w/w), sample/solution 1:80 (w/v) for 24 h.	Chemical pretreatment: 0.1 M NaOH sample/solution 1:30 (w/v) for 24 h; 10% butyl alcohol sample/solution 1:30 (w/v) for 24 h	Wet weight: 7.6%	Pepsin-soluble collagen	[44]
Acetic acid	Carp fish (*Cyprinus carpio*) scales	0.5 M acetic acid for 24 h.	0.1 M NaOH − $Ca(OH)_2$ saturation for 8 h; 0.2 M HCl + 0.5 M H_2SO_4 sample/solution 1:20 for 20 min	Extraction yield for raw collagen: 13.6%		[46]

TABLE 1.2 *(Continued)*

Extraction Method	Source	Conditions	Pretreatment	Yield	Characteristics	References
Acetic acid	Nile tilapia skins	0.7 M acetic acid for 5 h at 70 rpm.	10% (v/v) butanol for 16 h at 10°C; 0.7 M NaOH at 10°C and 400 rpm, sample size: 1 cm and 6 g sample/200 mL NaOH for 18 h.	Yield: 21.4% (purity: 94%)		[58]
Lactic acid	*Centrolophus niger* (black ruff) skin	0.5 M lactic acid (1:25 w/v) for 72 h	0.1 M NaOH (1:30 w/v) for 36 h at 4°C; 10% butyl alcohol (1:30 w/v) for 48 h	Yield: 45%		[61]
Citric acid	*C. niger* (black ruff) skin	0.5 M citric acid (1:25 w/v) for 72 h	0.1 M NaOH (1:30 w/v) for 36 h at 4°C; 10% butyl alcohol (1:30 w/v) for 48 h	Yield: 31%		[61]
Tartaric acid	*C. niger* (black ruff) skin	0.5 M tartaric acid (1:25 w/v) for 72 h	0.1 M NaOH (1:30 w/v) for 36 h at 4°C; 10% butyl alcohol (1:30 w/v) for 48 h	Yield: 31%		[61]
Formic acid	*C. niger* (black ruff) skin	0.5 M formic acid (1:25 w/v) for 72 h	0.1 M NaOH (1:30 w/v) for 36 h at 4°C; 10% butyl alcohol (1:30 w/v) for 48 h	Yield: 32%		[61]
Acetic acid and ultrafine bubbles	Acid-demineralized tilapia (*O. niloticus*) scales	125 g scales + 5 L of 0.1 M acetic acid + 3 L/min CO_2 for 5 h.		Yield: 1.58%		[62]
Acetic acid	Nile tilapia skins	0.35 M acetic acid (1:25 skin:solution) at 20°C for 65 h	0.1 M NaOH at 4°C for 48 h (1:20 skin:solution ratio); 20% ethanol at 4°C for 24 h (1:20 skin:solution).	Yield: 19%		[11]

TABLE 1.2 *(Continued)*

Extraction Method	Source	Conditions	Pretreatment	Yield	Characteristics	References
Acetic acid	Nile tilapia skins	0.5 M acetic acid for 24 h	0.1 M NaOH for 24 h (1:20 skin:solution ratio); 0.5 M EDTA at 5°C (1:10 skin:solution ratio) for 5 days; 10% butanol at 5°C for 24 h (1:20 skin:solution).	Protein concentration: 80%		[59]
Acetic acid	Mesoglea of jellyfish (*Rhopilema esculentum* Kishinouye)	0.6 M acetic acid at 4°C for 72 h.		0.12% (wet weight basis)		[60]
Acetic acid + Pepsin	Mesoglea of jellyfish (*R. esculentum* Kishinouye)	0.6 M acetic acid at 4°C for 72 h followed by a dispersion into 20 volumes of 0.5 M acetic acid containing 1% pepsin (w/w) at 4°C for 24 h.		0.28% (wet weight basis)		[60]
Acetic acid	Shabout (*Arabibarbus grypus* Heckel, 1843) skin	0.5 M acetic acid solution for 3 days	0.1 M NaOH for 3 days; 0.5 M EDTA for 5 days; 10% butyl alcohol (1:10 w/v) for 3 days.	6.79% (amino acid Pro + Hyp: 19.28%)		[9]
Acetic acid	Shabout (*A. grypus* Heckel, 1843) bones			2.57% (amino acid Pro + Hyp: 19.55%)		[9]

gel characteristics such as higher gel strength and viscosity [47, 64]. However, noncollagen proteins can also play an important role in some gel quality parameters [64].

On the other hand, fish gelatin usually presents a lower concentration of Pro and Hyp imino acids and lower levels of high MW proteins than mammalian gelatin, 22–25% for fish gelatin versus 30% for mammalian gelatin [28, 33, 36, 67, 68]. This results in fish gelatin with inferior gelation and rheological properties when compared with mammalian gelatin because the former has lower gelling and melting temperatures [54, 69], storage moduli, and gel strength [70]. It is worth mentioning that the gel stability is given by the superhelix structure, which is stabilized by steric restrictions imposed by pyrrolidine rings of the imino acids, in addition to the hydrogen bonds formed between amino acid residues [33].

Besides, it has been reported that collagen and gelatin derived from cold-water fish (such as cod, whiting, and halibut) have a lower content of Hyp and a lower thermal stability and gel strength and melting temperatures than the collagen and gelatin from warm-water fish (such as big eye tuna, black carp, and tilapia) [47, 48, 71–75].

The bloom strength is an important quality parameter to determine the gel strength of gelatin, and it depends on amino acid composition, especially Hyp content [64, 76]. The pH of the gelatin solution is also a significant value in establishing its subsequent use, and a value around neutral could allow it to be used in a broader range of products [64]. Besides, it has been reported that the pH of the gelatin solution could be affected by the chemical treatment used during extraction [64, 77].

Gelatin solution viscosity is a specific measurement as a function of time, strain, and frequency by the shear storage modulus (G'), loss modulus (G''), and loss factor ($\tan \delta$) data. According to Alias and Sarbon [78], when the shear storage modulus is greater than the loss modulus ($G' > G''$), the gelling material can exhibit quite a solid-like texture at a high frequency and fast timescale. However, when the loss modulus is greater than the storage modulus ($G'' > G'$), the behavior is much liquid-like at low frequency/long time scales [78].

The design of experiments is a valuable tool for optimizing gelatin extraction. Variables such as temperature and time of extraction; pH; and concentration of acid, alkali, or enzyme have been studied to determine their influence on physical and functional properties of gelatin [26, 49, 66, 77].

Using a screening experimental design, Alfaro et al. [77] found that the extraction temperature is the variable with the highest influence on the gelatin viscosity, followed by the sulfuric acid concentration and the extraction time, and based on these results, they performed a central composite rotational design for optimizing the gelatin extraction process. Liao et al. [26] reported a similar trend for gelatin extraction yield. They concluded that the high energy provided by high temperatures helps to break down the triple helix structure of collagen and the hydrogen bond between the α-chains, allowing collagen hydrolysis.

Cho et al. [54] showed that the extraction temperature is an important parameter in the processing of gelatins. They showed that temperatures between 40°C and 60°C caused an increase in rheological properties, and when the temperature increased from 60°C to 70°C, this property suffered a reduction. Higher extraction temperatures lead to protein degradation due to protein fragmentation, lowering gelling ability, which implies a decrease in gelatin content when the extraction process is performed at high temperatures, which could be the key to the process. However, the protein fragmentation and the relative contents of chains α, β, and γ are also relative to the MW property; fish and mammalian gelatins possess a polydispersed MW distribution associated with the gelatin source and the parameters of the extraction process [33].

Acidic pH is also reported to have an important influence on collagen hydrolysis and could result in protein loss in case of overhydrolysis [26]. Therefore, high temperature affects the gel strength and the viscosity of gelatin, and it takes a long time to form a reasonable gel network due to the presence of short protein peptides, which intensify in an acidic environment. Besides, high temperatures could result in nonenzymatic reactions causing higher yellowness of gelatin [26].

Zarubin et al. [47] extracted gelatin from the wastes of Alaska pollock and Pacific codfishes and found that the extracted product met the standardized requirements established for mammalian gelatin. The main differences were the color and the odor. Gelatin from fish presented a smoky white color, while the standard established has to be light yellow to yellow. Regarding the odor, the gelatin fish had a weak fishy odor, while the mammalian gelatin is odorless.

Even though higher temperatures and longer extraction time resulted in higher Hyp recovery, these temperatures caused more protein degradation by decreasing the band intensity of α-, β-, and γ-chains, affecting the gel strength. Therefore, the extraction conditions must be carefully evaluated [65].

Moosavi-Nasab et al. [79] reported that in the presence of protease extracted from Rainbow trout pyloric caeca, the Hyp content and the gelatin extraction yield obtained from fish skins and scales increased. Table 1.3 presents some gelatin extraction methods found in the literature.

1.2.2.3 OTHER ADDED-VALUE COMPOUNDS EXTRACTED FROM FISH BY-PRODUCTS

Different types of enzymes have been reported in the literature for obtaining hydrolysates [36, 48, 80]. Vázquez et al. [48] compared two enzymes (alcalase and papain) for two periods (2 and 4 h). Alcalase for 4 h was more effective than the other tested conditions, resulting in a bioactive, with large digestibility and high antihypertensive activity hydrolysate. Mudgil et al. [81] found similar results using protease, alcalase, or both. However, the authors reported that protease was more effective than alcalase for dipeptidyl peptidase-IV (DPP-IV) inhibitory capacity, which is important for the management of diabetes and reducing the risk of cardiovascular disease [80, 81]. Angiotensin-converting enzyme (ACE) is a dipeptidyl carboxypeptidase that promotes sodium retention and vasoconstriction, and its inhibition could reduce the risk associated with hypertension [81]. Hydrolysates obtained using a mixture of proteases (such as Purafect®), protease or alcalase alone, or a combination of both have shown inhibition activity to ACE and, consequently, antihypertensive properties [36, 81]. Regarding the enzyme concentration, Mudgil et al. [81] reported no difference in the degree of hydrolysis for a constant substrate concentration.

Cheung and Li-Chan [80] compared different enzymes for obtaining protein hydrolysates. The authors found that the whole hydrolysates were not as bioactive as the permeate obtained by ultrafiltration, containing the <3 kDa fraction. Interestingly, the ultrafiltrated hydrolysates with low hydrolysis degree presented higher inhibitory activity against ACE and DPP-IV. As an alternative to ultrafiltration, successive hydrolysis combining two enzymes allowed the preparation of bioactive hydrolysates without using the ultrafiltration process [80]. In this case, better results were obtained using 4% papain for 2 h followed by 1% CorolaseN for 2 h for obtaining hydrolysates with dual bioactivity.

TABLE 1.3 ethods for Gelatin Extraction

Source	Conditions	Pretreatment	Yield	References
Tuna (*Katsuwonus pelamis*) skin	Distilled water at 45°C for 12 h 1:10 (w/v) skin:water ratio	Warm tap water (40°C) for 10 min; 0.1 M NaOH room temperature 1:10 (w/v) skin:solution for 1 h (two times); 0.2 M acetic acid 4°C 1:10 (w/v) skin:solution for 12 h (two times)	Hyp: 6.3%	[64]
Alaska pollock (*Gadus chalcogrammus*) + Pacific cod (*G. macrocephalus*) scales + Tails and fins + Spinal bones + Viscera (without liver)	Mixture of "food collagenase" and "protepsin" at 40°C for 4 h; enzyme:sample ratio 1:1000 (w/w); enzyme inactivation at 70°C for 15 min	10% citric acid for 60 min; heat at 80–90°C for 20–25 min in water (1:2 sample:water); cooled to 35–40°C	18% Hyp: 8.51%	[47]
Cobia (*Rachycentron canadum*) skin	Distilled water (1:1 ratio, kg/L) at 52°C for 120 min at pH 4	3 M NaOH (1:1 kg/L) at pH 11 and room temperature for 15 min; 3 M NaOH (1:1 kg/L) at pH 11 and room temperature for 60 min; 3 M HCl at pH 2 for 15 min	79.5% Hyp: 962 mg/kg	[71]
Croaker (*Micropogonias furnieri*) skin	Distilled water (1:1 ratio, kg/L) at 52°C for 120 min at pH 4	3 M NaOH (1:1 kg/L) at pH 11 and room temperature for 15 min; 3 M NaOH (1:1 kg/L) at pH 11 and room temperature for 60 min; 3 M HCl at pH 2 for 15 min	75.6% Hyp: 877 mg/kg	[71]
Carp (*C. carpio*) skin	Warm distilled water at 45°C for 60 min (1:3 ratio w/v); filtered using a double cloth with a mesh diameter of 72 μm; filtered using a qualitative medium filter paper and dried by lyophilization	2.6% NaCl solution (1:6 ratio w/v) for 10 min at <16°C; filtered using a cloth with a mesh diameter of 72 μm; centrifuged for 5 min at 1000 g (twice); tap water (6:1 ratio v/w) for 10 min at <18°C; centrifuged for 5 min at 1000 g (repeated three times)	Hyp + Pro: 175.94 residues/1000 residues	[50]

TABLE 1.3 *(Continued)*

Source	Conditions	Pretreatment	Yield	References
	Warm distilled water at 45°C for 60 min (1:3 ratio w/v); filtered using a double cloth with a mesh diameter of 72 μm; filtered using a qualitative medium filter paper and dried by lyophilization	0.1 M NaOH for 6 h (1:3 ratio w/v); food grade ethanol (96.5%) (1:2 ratio w/v) overnight	Hyp + Pro: 185.21 residues/1000 residues	[50]
	Warm distilled water at 45°C for 60 min (1:3 ratio w/v); filtered using a double cloth with a mesh diameter of 72 μm; filtered using a qualitative medium filter paper and dried by lyophilization	0.2% NaOH for 2 h (1:6 ratio w/v); 0.2% H_2SO_4 for 2 h (1:6 ratio w/v); 1% aqueous citric acid for 2 h (1:6 ratio w/v)	Hyp + Pro: 144.86 residues/1000 residues	[50]
Salmon salar skin	Aqueous medium (1:2 ratio) at 45°C for 16 h. Purification: filtration (500 μm), active charcoal adsorption (1.5% w/v for 2 h), centrifugation, and oven-dried.	0.05 M NaOH for 30 min (1:4 ratio w/v) at 22°C; 0.02 M H_2SO_4 for 30 min (1:4 ratio w/v) at 22°C; 0.052 M citric acid for 30 min (1:4 ratio w/v) at 22°C	Hyp + Pro: 18%	[48]
Black-barred halfbeak (*Hemiramphus far*) skin	Pretreated skin was incubated at 50°C for 18 h and centrifugated at 8000 rpm for 30 min at 4°C; freeze-dried.	0.05 M NaOH for 2 h (1:10 ratio w/v) at 4°C; 0.1 M acetic acid for 48 h (1:10 ratio w/v) at 4°C; pH adjusted to 7.5	Hyp + Pro: 24.35%	[36]
Tilapia skin	Distilled water (1:8 w/v ratio) for 6 h at 75°C and pH 3	0.1 M NaOH for 2 h (1:10 ratio w/v) at room temperature; 0.05 M acetic acid for 1 h (1:10 ratio w/v) at room temperature; pH adjusted to neutral	Hyp + Pro: 199 residues/1000 residues	[26]

Several tests of antioxidant capacity have demonstrated the high antioxidant potential of hydrolysates obtained from fish by-products [36, 66]. In any case, the bioactive properties of hydrolysate peptides depend on the degree of hydrolysis, and a very high value can adversely affect these properties [48, 81].

Mirzapour-Kouhdasht et al. [66] employed a systematic methodology. They first extracted actinidin from kiwifruit at 37°C and pH 7.5. This actinidin was then combined with alcalase, leading to the production of hydrolysates with a degree of hydrolysis of 33.34%. These hydrolysates were subsequently subjected to ultrafiltration, resulting in different fractions. Each of these fractions exhibited high 1,1-diphenyl-2-picrylhydrazyl (DPPH) radical inhibitory percentages, indicating their significant antioxidant activity. To further purify the hydrolysates, a sequential procedure was implemented. This procedure involved ultrafiltration, ion exchange chromatography, and reversed-phase high-performance liquid chromatography, ultimately yielding a hydrolysate fraction with remarkably high antioxidant and anticancer activities [66].

HAp can be obtained from fish scales using alkaline treatments [39, 42]. Zainol et al. [39] used 5 M NaOH for extracting HAp from fish scales. The product was highly crystalline with a Ca/P ratio of 1.83, indicating the presence of a secondary phase, another calcium phosphate, namely β-tricalcium phosphate. Pon-On et al. [82] obtained flat-plate nanocrystals with a Ca/P ratio of 2.01 with higher bioactivity, forming more new apatite than synthesized HAp. Hence, fish scale is an important source to extract HAp for biomedical applications such as bone scaffolds or regenerative material [39, 82].

Natural bones are composed of HAp and collagen, and considerable research has been carried out to produce HAp /collagen composites that reproduce the characteristics of natural bones. Recently, Mudhafar et al. [38] evaluated the use of silver nanoparticles with antimicrobial characteristics in HAp /collagen composites obtained from fish scales for use as bone filler.

The investigation of HAp formation in the presence of functional organic molecules such as collagen has opened up new possibilities. This approach allows the preparation of the nano-sized HAp crystals from fish scales [83]. Chai and Tagaya [83] concluded that it is possible that the formation of HAp using calcium and phosphate ions, which were accumulated in the scale by the immersion process, allowing the nanostructured

HAp to grow. These nanoparticles were arranged in the internal fibrillary plate layer, showing that the formation occurred along with the collagen fibers inside the scales. This breakthrough could revolutionize the production of bone fillers. HAp extracted from fish scales has also been used as a natural adsorbent with high adsorption efficiency. Kongsri et al. [42] used it for selenium adsorption in aqueous solution, achieving higher adsorption capacities than commercial HAp or chitosan.

1.3 IMPORTANCE AND APPLICATIONS OF COLLAGEN AND GELATIN FOR THE FOOD INDUSTRY

Bio-based films have been widely tested for food packaging. Collagen and gelatin also have been used for this purpose, mainly by combining with other biopolymers or by incorporating functional agents to prepare active and intelligent packaging.

1.3.1 COLLAGEN APPLICATIONS

1.3.1.1 FOOD PACKAGING

The development of food packaging from renewable raw materials is continuously growing due to the environmental impact of fossil-based packaging [84]. In general, a food container must meet minimum requirements, among which are preserving the sensory qualities of the food by preventing fat oxidation and discoloration, acting as a barrier to oxygen and moisture migration, presenting adequate mechanical strength and elasticity, and reducing microbial activity [85]. In this context, collagen films from fish waste are a good alternative for manufacturing food containers due to their excellent biocompatibility with most foods and the human body and low antigenicity [86]. However, collagen films are limited by their low thermal stability and mechanical properties, and their hydroxyl groups allow water to pass through the film [87]. Besides, the collagen film-forming ability was reported to depend on the collagen source, which influences the primary structures and conformation of collagen molecules [88]. Regarding this, Sun et al. [89] evaluated the film-forming properties of acid-soluble collagens extracted from the skin, head, fins, and bone of the loach. Collagen from fins showed the strongest tensile strength (TS), best extensibility, and barrier property; however, a low extraction yield was achieved.

Nowadays, most research is focused on developing collagen films with other components that improve their properties. Different approaches could be performed to enhance the bio-based film characteristics for use in food packaging. One is adding fillers to the biopolymer matrix (physical method). Chemical modifications such as cross-linking reactions give rise to chemical methods for improving the film characteristics. Enzymatic methods allow biopolymer modification without producing toxic substances and using highly specific reactions [90].

Costa et al. [91] prepared collagen/starch films in different proportions. Films with a higher collagen content were thinner than the starch-rich ones. The thickness was attributed to the low viscosity of collagen solutions, which allows the molecular reorganization to be more homogeneous during drying. In contrast, starch films tend to be thicker due to the gelatinization process that occurs when starch is heated in water. The flexibility and the low water vapor permeability (WVP) of collagen films were attributed to the peptide chains of the amino acids present in the fish collagen that are organized parallel to an axis promoting elasticity and a great degree of organization of the protein molecules in the film network, resulting in a film with similar characteristics to low-density polyethylene [91]. Films containing 4.5%/1.5% collagen/starch showed the lowest water solubility value, which could be explained by the interaction between protein-free sulfhydryl groups and starch-free hydroxyl groups from collagen and starch, respectively. It has been reported that the thermal stability of collagen films is higher than that of collagen/starch ones, which is suitable for packed cooked products [91].

Qu et al. [86] prepared a film composed of collagen from tuna skin and chitosan by sweep frequency ultrasound technique, with good TS and elongation at break (EAB), where ultrasound improved the antioxidant and antibacterial properties of the film by promoting the cross-linking of biopolymers, allowing to obtain films with better characteristics for food packaging applications. On the other hand, Wang et al. [92] prepared pilot-scale sodium alginate films with collagen, which exhibited higher TS and EAB than collagen/chitosan and relatively good oxygen permeability (OP).

Fish gelatin nanoparticles of different sizes were prepared by adjusting the gelatin/glutaraldehyde ratio for fabricating collagen composite films, showing increased mechanical properties and decreased WVP [93]. The nanoparticles filled the gaps between collagen fibers, increasing the tortuous path of water vapor during vapor diffusion. This effect was more pronounced for small-sized nanoparticles than medium- or large-sized ones [93]. Other collagen composites and their improvements for food packaging can be seen in Table 1.4.

TABLE 1.4 Films for Food Packaging Based on Fish Collagen

Films	Improvements	References
Tuna skin collagen/chitosan (extraction by ultrasound)	Chitosan improved film-forming ability, antioxidant, and antimicrobial properties. Ultrasound improved film thermostability and mechanical properties.	[86]
Fish collagen/sodium alginate	The formation of hydrogen bonds between collagen and sodium alginate improved the film's thermal stability and physical and barrier properties.	[92]
Hydrolyzed collagen/sodium alginate/SiO_2	SiO_2 slightly improved the thermal stability and significantly improved the TS and EAB of the film.	[94]
Grass carp skin collagen/chitosan/lemon essential oil (LEO)	The films had lower OP, higher TS, and higher EAB, in addition to acting as an antioxidant and antimicrobial in pork preservation.	[95]
Red stingray skin collagen/pullulan (preparation using electrospinning)	Pullulan improved the spinnability of collagen for the electrospinning process. In addition, the electrospinning prepared film significantly increased the TS and elongation at the break of the film fibers.	[96]
Bocachico scales collagen/gelatin	The films were transparent, increased their mechanical strength and showed low water solubility	[97]
Tilapia skin, grass carp, and silver carp collagen	The film based on tilapia collagen had the highest TS, while the surface of the film based on carp collagen had the thickest microstructure.	[88]
Black ruff skin collagen/chitosan/pomegranate peel extract (PPE)	The addition of 5% PPE to the film decreased its water solubility and increased its antimicrobial activity.	[61]

1.3.1.2 BIOMEDICAL APPLICATIONS

Among the main properties that give relevance to fish collagen compared to that of terrestrial origin, in addition to its biocompatibility with the human body and its biodegradability, are the reduction of allergic reactions and transmission of prion diseases, such as large-scale BSE in bovines and Creutzfeldt–Jakob disease in humans, making them reliable alternative sources for biomedical applications [98, 99]. Collagen is known to account for approximately two-thirds of the dry weight of human skin and one-third of all proteins in the body, in addition to the fact that at the soft connective tissue level, the fibrils form a highly organized three-dimensional scaffold surrounding cells, thus playing a critical role during the four phases of wound repair in the skin: hemostasis, inflammation, proliferation, and remodeling [100, 101]. Based on that, many biomedical applications of collagen and its composites have been reported; among them, Zhang et al. [102] demonstrated the biocompatibility of type I collagen extracted from tilapia skin in mice. In that study, they did not observe any incidence of death or allergic reactions in mice injected with collagen; however, female mice showed a greater increase in body weight over time than male mice. On the other hand, Vigneswari et al. [103] studied the effect of a collagen dressing with polyhydroxyalkanoates, prepared with the electrospinning technique, for wound healing in rats. Their results showed that nanofibers of collagen/poly(3-hydroxybutyrate-*co*-4-hydroxybutyrate) peptides possess a significantly elevated healing rate, where, in only 14 days, a wound closed 98% and without the formation of inflamed scabs, compared to 63% of another wound closure with a conventional control gauze. A growing interest in wound dressing promotes the development of collagen–alginate films that can also be used as ink for bioprinting [57]. In another study, the authors prepared gellan gum/collagen hydrogel films containing gatifloxacin (antimicrobial agent) for wound dressing materials, promoting the formation of blood vessels and hair follicles [104]. Besides, lyophilized collagen sponges were cross-linked with 1-ethyl-3-(3-dimethylaminopropyl) carbodiimide (EDC) and *N*-hydroxysuccinimide (NHS) for use to improve hemostatic ability [60]. Menezes et al. [11] cross-linked collagen–hyaluronic acid hydrogel using EDC/NHS for application in tissue engineering. Menezes et al. [11] cross-linked collagen–hyaluronic acid hydrogel using EDC/NHS for application in tissue engineering.

Another collagen biomedical application includes the use in drug delivery systems, which is the transport of pharmaceuticals to specific areas of the body and at specific release rates to achieve therapeutic effects without posing any risk to the patient's health [105]. One way to achieve this is by encapsulating the drug in the network of a biomaterial that is biocompatible with the body, so the biodegradability, biocompatibility, and availability of fish collagen make it a tentative alternative for drug delivery in the body [106]. However, the material's mechanical resistance limits its application; therefore, collagen composites must be prepared to solve this limitation. That said, Cao et al. [107] fabricated chitosan/chondroitin sulfate/silver carp skin collagen scaffolds with poly(lactide-*co*-glycolide) microspheres loaded with basic fibroblast growth factor (bFGF) as a drug delivery system for tissue engineering. Their results showed that bFGF retained its structural integrity, bioactivity, and biocompatibility after release. In addition, they mention that the release rate can be adjusted by modifying the size of the microspheres or the collagen/chitosan ratio in the scaffold. On the other hand, Mandal et al. [108] studied the behavior of magnetite (Fe_3O_4) nanoparticles with cross-linked collagen extracted from *Lates calcarifer* scales. Their results indicated that the prepared composite exhibited the magnetic properties of magnetite and good biodegradability and biocompatibility due to the cross-linked collagen being a potential nanobiocomposite for targeted cancer therapies.

1.3.2 GELATIN APPLICATIONS

1.3.2.1 FOOD PACKAGING

Gelatin is one of the most studied biomaterials nowadays for food packaging because unlike collagen, it has a high capacity for the formation of highly extensible transparent films, with barrier properties to ultraviolet (UV) light and oxygen, in addition to presenting good antioxidant and antibacterial activity [109–111]. However, due to the low imino acid content of fish gelatin (Pro and Hyp), some of its properties are inferior to mammalian gelatin (such as porcine and bovine), including its lower melting and gelling temperature, and low rheological parameters such as gel strength and elastic modulus. Because of this, it is essential to modify the properties of films prepared from fish gelatin for application in food packaging.

The water solubility and the WVP of gelatin/chitosan blend films increased with the increase of gelatin proportion due to the hydrophobic nature of chitosan [112]. Tagrida et al. [112] added betel leaf ethanolic extract to gelatin/chitosan films to enhance elasticity, heat-seal ability, UV and visible light barrier, WVP, and antioxidant and antibacterial activities. The authors concluded that adding betel leaf extract caused a crosslinking between gelatin and chitosan, which led to decreased hydrophilic functional groups and reduced WVP. Riahi et al. [113] reported similar results by adding metal-organic frameworks ZIF-8 (Zn-MOFs) with dodecahedral morphology to gelatin-based films. These films were thicker and presented higher TS and WVP than those prepared by Tagrida et al. [112]. However, the presence of Zn-MOFS was helpful in retarding the texture deterioration of packed tomatoes, preventing microorganism growth [113]. Li et al. [114] extracted gelatin from fish scale waste and prepared composite films using gelatin/sodium alginate as a matrix with carvacrol-loaded ZIF-8 nanoparticles as active nanofiller to confer antibacterial properties to the composite. Films with and without nanofillers were prepared; those prepared with nanofillers resulted in excellent UV-light shielding, increased mechanical properties, and decreased water solubility and WVP.

Kilic et al. [115] performed a mechanical improvement of fish gelatin films using physical cross-linking with UV radiation and chemical cross-linking with carbon dots (CDs), also incorporating red cabbage anthocyanins into the film to observe the variation of food pH and determine the freshness of the food. Such films designed for continuous monitoring of food conditions by a colorimetric method are called "smart packaging" [116]. The results indicated that the UV-treated films prepared with a concentration of 100 mg/L of CDs showed the highest mechanical traction and sensitivity to the ammonia released during food decomposition. Another way to improve gelatin films is to increase their antibacterial activity during the initial phase of food preservation. Thus, Wu et al. [117] prepared gelatin films from silver carp skin, with the addition of cinnamaldehyde (the main component of cinnamon essential oil), using sulfobutyl-ether-β-cyclodextrin as an inclusion complex. In addition to significantly improving the antibacterial activity of gelatin against *Pseudomonas aeruginosa* up to 98% in the initial phase, the most interesting aspect of their results was the improvement in the EAB and the barrier properties to visible and UV radiation up to about 42% compared to pure gelatin.

Bioactive compounds are added to collagen and gelatin films to improve anti-bacterial and antioxidant activities and even to impart pH sensitivity [90, 114]. Salem et al. [118] used *Lepidium sativum* extract and glycerol as plasticizers to develop a fish film as active packaging for cheese preservation. The results showed high antioxidant potential, but no significant difference was seen in the mechanical parameters regarding thickness, TS, and EAB. Moreover, gelatin can form polyelectrolyte complexes with ionic polysaccharides, being part of intramolecular electrostatic interactions [70], and even when the production technique is used to compare the physicochemical properties [119]. Essential oil-based pickering emulsions with core-shell structures have also been proposed for producing collagen and gelatin films with enhanced antibacterial and antioxidant activities by the slow release of the essential oil. However, many practical problems have been observed using this approach, mainly related to the decrease of mechanical and water vapor barrier properties, and the appropriate film preparation methodology has to be considered [90, 120].

Besides the typical casting process for preparing bio-based films, recently, electrospinning technology has been tested to prepare gelatin nanofibers with appropriate characteristics for food packaging [121]. Heydarian and Shavisi [121] prepared gelatin-xanthan gum mats by electrospinning; chitin nanofibers and anthocyanins were added to monitor the quality of shrimps, extending the shelf life under cooled storage for 10 days. Further improvements in the films made from fish gelatin can be seen in Table 1.5.

1.3.2.2 FOOD INDUSTRY

Most of the gelatin extracted from different sources is destined for the food industry due to its great versatility in food, including the ability to form gels, water retention, emulsifying, thickening, stabilizing, and antioxidant capacity. Gelatin has applications in the production of candies, edible films, clarification of fruit juices, and processing of dairy products, among others [109, 131]. However, as mentioned in the previous section, fish gelatin may present some limitations in stability at high temperatures and low rheological properties compared to gelatin from bovine or porcine sources. Therefore, modifications are necessary to improve the properties of fish gelatin. Nevertheless, Duan et al. [132] incorporated fish gelatin into ice-cream samples and beer, which compared to calf bone gelatin showed a better mouth feel and clarification effect, respectively [132].

Among the most used improvements in the food industry are the addition of electrolytes (e.g., $CaCl_2$, $MgCl_2$, and NaH_2PO_4) and nonelectrolytes (e.g., sugar, polysaccharides, and glycerol), as well as physical treatments such as high pressure, ultrasound, UV irradiation, among others [74, 133]. Once that is done, there are several applications of fish gelatin, some of which are summarized in Table 1.6.

TABLE 1.5 Films for Food Packaging Based on Fish Gelatin

Films	Improvements	TS (MPa) (Better Results)	EAB (%)	References
Cold water fish skin gelatin/haskap berries (*Lonicera aerulea* L.) extract	It has increased film crystallinity, light, water vapor barrier, TS, and improved antioxidant capacity. In addition, the films were pH sensitive, with a color change that allowed them to determine the freshness of the preserved food.	51.5 ± 4.44	3.69 ± 0.44	[122]
Fish gelatin powder/*Aloe vera* gel	Increasing the concentration of *A. vera* decreased the water solubility and increased the antioxidant properties of the film, without causing a significant change in thickness, WVP, and color. However, as the *A. vera* concentration increased, the mechanical strength decreased.	11.82 ± 1.18	1.88 ± 0.18	[123]
Fish skin gelatin/chicken protein isolate/phenolic compounds (gallic acid or tannic acid)	Phenolic compounds increased the TS and EAB of the film and decreased the film solubility.	25.00 ± 0.39 for gallic acid	29.81 ± 0.91 for gallic acid	[124]
	Films with tannic acid had a more compact structure and lower vapor permeability. Gallic acid films had higher DPPH radical scavenging activity.	30.84 ± 0.42 for tannic acid	12.03 ± 1.04 for tannic acid	
Tilapia skin gelatin/ graphene oxide	Graphene oxide increased the TS, Young's modulus, and UV barrier. In addition, it reduced water vapor permeability and solubility.	18.60 ± 2.12	41.71 ± 1.44	[125]
Tilapia skin gelatin/ graphene oxide/cinnamon bark oil with ferulic acid	Cross-linking of the gelatin/graphene oxide film with ferulic acid increased the mechanical strength, antioxidant activity, water vapor barrier, and thermal stability of the film. In addition, it decreased WVP, UV light transmittance, and opacity.	23.0 ± 1.20	22.00±5.00	[126]
Fish gelatin/ chitosan loaded with nanoemulsified α-tocopherol (α-TP) and thyme essential oil (TEO)	The incorporation of α-TP drastically reduced the water solubility of the films. In addition, the inclusion of α-TP and TEO extended the shelf life of cold-preserved fish to 15 days.	–	–	[127]

TABLE 1.5 *(Continued)*

Films	Improvements	TS (MPa) (Better Results)	EAB (%)	References
Big eye tuna skin gelatin/spent coffee ground (SCGE)	The water solubility, moisture content, transmittance, and water vapor permeability of the films decreased with increasing SCGE, in addition to showing high antioxidant and antimicrobial activities. On the other hand, the surface hydrophobicity and transparency of the film increased with increasing SCGE.	43.14 ± 1.22	89.63 ± 4.18	[128]
Tilapia gelatin/polylactic acid/sodium alginate bilayer films	The bilayer improved oxygen and water vapor barrier capabilities and exhibited strong antioxidant and antibacterial activity.	34.15 ± 0.34	6.40 ± 0.94	[129]
Tilapia skin gelatin/palm wax	Films with 15% palm wax increased film thickness, opacity, flexibility, and TS. On the other hand, it reduced solubility, swelling, and WVP.	12.58 ± 0.76	72.86 ± 2.76	[130]

TABLE 1.6 Some Applications of Fish Gelatin in the Food Industry

Gelatin Source	Modification	Application	Results/Observations	References
Carp skin (species not specified)	Hydrolysis with alkalase	Inhibition of ice crystal growth in an ice-cream mixture.	Short gelatin polypeptides (1–2.5 kDa) have crystal growth inhibitory properties, regardless of whether they come from bovine or fish sources.	[134]
Tilapia skin (species not specified)	Addition of xanthan gum	Preparation of low-fat stirred yogurt	A 1:99 (w/w) ratio of xanthan gum:gelatin has better water holding capacity and consistency in yogurt than gelatin of porcine origin, with very similar viscosity, pseudoplasticity, and thixotropy.	[135]
Cod skin (*G. morhua*)	–	Bread making from wheat flour (*Triticum aestivum*)	The addition of gelatin to the wheat dough formed a gelatin–glutelin complex with high water-holding capacity, increasing the maximum dough strength, rupture time, and final gas holding volume.	[136]
Fish skin and fish bone (unspecified fish)	Addition of high-acyl gellan gum	3D-printing edible ink	The composite hydrogel exhibited shear-thinning behavior, facilitating extrusion. The gellan gum provided the necessary structure and stability for printing.	[137]
Cod, haddock and haddock skins	Addition of sodium alginate	Formation and stability of foams in foods	The addition of sodium alginate to gelatin suppressed foam formation but increased its stability. Under acidic conditions (below the EIP of gelatin) and at an 80:20 gelatin:alginate ratio, the mixture showed the lowest foam suppressing effect and the highest stability.	[138]
Red tilapia skin (*Oreochromis* sp.)	Extraction by the high-pressure processing method (HPP)	Gelatin desserts elaboration	The gelatin dessert prepared with HPP gelatin presented greater shelf life, hardness, adhesiveness, gumminess, chewiness, stiffness, and firmness than the gelatin dessert prepared with commercial fish gelatin.	[139]
Red tilapia scale (*Oreochromis* sp.)	Extrusion pretreatment and subsequent hydrolysis with proteases	Gelatin with antihypertensive and antioxidant properties	The results indicated that extrusion treatment improved the yield and antioxidant activity. In addition, hydrolysis with pepsin + pancreatin had the highest antioxidant activity and the highest yield.	[140]

1.4 CRITICAL ANALYSIS OF THE CHARACTERISTICS OF THE COMPOUNDS

1.4.1 MECHANICAL PROPERTIES

The mechanical properties of food films are of the utmost importance as they determine whether the food product can be distributed without losses during transport and storage while providing an adequate level of protection. TS and EAB, defined as the maximum stress the film can withstand before breaking and the ability of the film to withstand shape changes without breaking, respectively, are two of the critical parameters used to evaluate the mechanical properties of films [141].

The TS and EAB of films composed of collagen will depend on several factors, such as the species origin, the extraction method, the reticulation degree, and the presence of other compounds that improve their qualities [142]. In addition, it is well known that the MW distribution and amino acid content of collagen play a key role in the mechanical properties of the resulting films [88]. In fact, it has been observed that a lower amount of Pro + Hyp residues increases the deformation at the break of the films because the pyrrolidine rings can impose conformational constraints, imparting a certain degree of molecular stiffness that can affect film deformability [69].

However, due to the hydrophilic properties of collagen, the degree of hydration of the film can be very high, so that the EAB value will be high and the TS very low [142]. Because of this, it is necessary to combine fish collagen with different compounds or to modify the degree of reticulation of fish collagen to improve the TS. For example, Sripriya and Kumar [143] performed reticulation of collagen film extracted from the swim bladders of the freshwater fish Rohu (*Labeo rohita*) by UV irradiation. Their results indicated that the TS of the reticulated film was approximately seven times higher than the unreticulated film. On the other hand, Ben Slimane and Sadok [142] fabricated films composed of smooth hound (*Mustelus mustelus*) skin collagen with chitosan in different proportions. Their results showed that the films with 50% and 75% chitosan exhibited similar or even higher TS than commercial films, but with very low EAB values for both cases.

Furthermore, Niu et al. [144] observed that the TS of gelatin films from Nile tilapia (*O. niloticus*) skin can become considerably higher than that of pure collagen films due to the higher film-forming capacity of gelatin;

however, with EAB values that can become very low, probably due to the lower β-chain content [145]. Kchaou et al. [146], in contrast, observed that the mechanical properties of type A gelatin films can be significantly increased by Maillard reactions. Their results indicated that treating the films with glucose:lysine at a 0.5:1 ratio can increase the TS up to 23% and the EAB up to 54% with respect to that of pure gelatin film due to cross-linking and plasticizing action of glucose, making it possible to obtain films with mechanical properties similar to mammalian gelatin films.

1.4.2 BARRIER PROPERTIES

Food packaging films must have good mechanical strength and be able to prevent the penetration of gases, flavors/odors, water vapor, water, and other substances that could compromise the sensory and hygienic integrity of the food product [147]. Among the different barrier properties of collagen films, the most common are WVP, OP, UV light barrier properties, and optical transparency.

WVP of films is a critical parameter that affects the shelf life of food products by preventing water vapor penetration through film boundaries, microbial growth, and modification of food texture and functionality [148]. Because the primary function of films is to prevent moisture transfer between the food and the environment, WVP values should be as low as possible. In this regard, Zhuang et al. [149] prepared films of tilapia skin collagen (TSC)/Pachyrhizus starch (PS)/Rambutan peel phenolic (RPP) compounds. The authors observed that increasing the PS concentration increased the degree of cross-linking between the hydrogen bonds of TSC and PS, increasing the order and crystallinity of the films and hindering the free diffusion of water molecules. The starch's high crystallinity also assisted the film's overall crystallinity. In addition, the hydroxyl groups of RPP increased the density of hydrogen bonds between TSC and PS, making the molecular structure even denser and the WVP value lower.

Meanwhile, the films elaborated from fish gelatin present a lower content of Pro + Hyp than collagen films due to the partial hydrolysis of the triple helix. The reduction in the content of hydroxyl groups of Hyp, responsible for the formation of hydrogen bonds with water, causes a decrease in the hydrophilic behavior of the film, resulting in a lower value of the WVP [73, 145]. In fact, Gumus et al. [150] observed that films prepared from *Sparus aurata* skin gelatin with wine lees (WL) and carrageenan decreased WVP value to values similar to bovine origin films, which could be because the presence of WL induced the formation of a film matrix by cross-linking interactions between the phenolic compounds in WL and gelatin, obtaining a microstructure similar to bovine gelatin films.

On the other hand, the presence of oxygen in food packaging generally harms the quality and shelf life of food products due to the rapid oxidation of fats and/or vitamins present in them or due to the growth of microorganisms such as aerobic bacteria, yeasts and molds, which can cause off-flavors, loss of nutritional value and color changes [148, 151, 152]. Therefore, oxygen-sensitive foods should be wrapped with films with a low OP. Jiang et al. [95] observed that a film prepared from grass carp scale collagen in conjunction with chitosan nanoparticles and LEO allowed the reduction of the OP by increasing the LEO concentration to 40% LEO. Above this point, the OP value increased significantly because the film probably presented a loose texture and pores distributed throughout the matrix, causing an increase in oxygen solubility in the lipid phase, increasing permeability. Therefore, it is extremely important that the film matrix be sufficiently compact and crystalline to prevent oxygen penetration and that the oxygen solubility in the matrix be as low as possible. Furthermore, Jafari et al. [28] observed that by incorporating *Cirish fructans* (CF) into an ultrasonically extracted cold water fish gelatin matrix, the crystallinity of the film increases with increasing CF concentration due to its semi-crystalline nature, reducing oxygen transfer and decreasing the OP value.

An important property of protein-based films is that the peptide bonding and aromatic amino acid residues in the film matrix (such as tyrosine, phenylalanine, and tryptophan) allow them to strongly absorb a particular wavelength range within the UV spectrum, thus reducing lipid peroxidation and UV damage, such as nutrient losses, discoloration, and off-flavors [95, 153, 154]. In addition, fish protein-based films generally also have high optical transparency, one of the most valuable characteristics for consumers, as it allows them to assess the freshness and appearance of the food by visual inspection before purchase [155]. Slimane et al. [142] observed that collagen extracted from smooth hound (*M. mustelus*) skins exhibited maximum UV light absorption at 235 and 240 nm for extractions performed with acetic acid and pepsin, respectively. However, when preparing the films with such collagen and using chitosan as an adjuvant, the UV light barrier property was further improved by increasing the chitosan concentration, maintaining a transmittance higher than 70% in the range of 400–800 nm for films with a chitosan concentration lower than 75%. Thus, a film with a good UV light barrier and a relatively high optical transparency was obtained. In turn, Vejdan et al. [156] prepared a bilayer film of cold-water fish gelatin with agar, which presented a transmittance close to 90% in the visible light region but with a poor barrier to UV light. However, by including TiO_2 nanoparticles in the structure, the transmittance of the film in the UV region was reduced below 10% because the presence of TiO_2 can absorb a wide range of wavelengths in the UV region, thus improving the UV light barrier properties of the film.

1.4.3 ANTIMICROBIAL PROPERTIES

When it comes to ensuring food safety, it is crucial to consider the potential presence of pathogenic microorganisms in the food. These microorganisms cause food spoilage and can lead to food intoxication in consumers. Therefore, it is essential to have packaging that can act as an active barrier, reducing microbial damage through an interaction between the film and the food [157, 158]. In this context, Qu et al. [86] developed a film of tuna skin collagen and chitosan enhanced with ultrasound. Their results demonstrated a remarkable inhibition capacity against *Escherichia coli*, *Staphylococcus aureus*, *Bacillus subtilis*, and *Listeria monocytogenes*. This was due to the film's positively charged amino groups (from the collagen), which, when directed toward the negatively charged microbial cell membrane, could damage it during the binding and neutralization process, leading to endoplasmic leakage and cell death. However, Kulkarni et al. [159] observed that a film prepared using collagen from *Cirrhinus mrigala* and chitosan showed no significant antimicrobial activity against *E. coli*, *S. aureus*, and *B. subtilis*. Conversely, when a 5% concentration of neem (*Azadirachta indica*) powder related to the total weight of the film was added to the film, the antimicrobial activity was significant. This underscores the fact that the antimicrobial activity of a film is influenced by various factors, including the fish species and source used, the extraction method, and the presence of other antimicrobial components, among other factors.

Similarly to collagen, films prepared from fish gelatin generally do not exhibit significant antimicrobial properties. Pouralkhas et al. [160] prepared films from the gelatin of cold-water fish skins, in which they did not observe any significant antimicrobial activity. However, when fucoidan extracted from brown algae (*Sargassum tenerrimum*) was added to the films, the antimicrobial activity significantly increased against *S. aureus* and *E. coli*, in addition to showing a bacteriostatic effect against Gram-positive and Gram-negative bacteria. The antimicrobial activity occurs since sulfated polysaccharides, such as fucoidan, adhere to the surface of the bacteria, causing damage and nutrient leakage.

1.5 FEASIBILITY OF RECOVERING FISH WASTE FOR APPLICATION IN FOOD PACKAGING

Globally, the collagen market was valued at USD 5.1 billion in 2023 and is expected to grow, reaching a valuation of USD 7.4 billion by 2030. Collagen is mainly destined for cosmetics, healthcare products, and the food industry. Regarding the type of collagen, fish collagen occupies the third position in the collagen market [161]. On the other hand, the gelatin market was valued at USD 3.7 billion in 2023 with a compound annual growth rate (CAGR) of 6.6% from 2023 to 2028. This product is destined for the food and beverage industries, followed by pharmaceuticals, health and nutrition, and cosmetic and personal care products, due to its function as a thickener, stabilizer, and gelling agent. Besides, there has been a growing interest in functional and convenience food [161]. The Markets and Markets report reveals that the gelatin industry is concerned about raw material availability. Also, it is worth mentioning that the raw material costs constitute 50–55% of the total cost in the gelatin market [161]. Collagen and gelatin from fish by-products meet the Halal criteria, further enhancing their market potential [131].

The fish by-products are composed of heads (9–12% of total fish weight), viscera (12–18%), skin (1–3%), bones (9–15%), and scales (5%) [1]. As mentioned, collagen and gelatin are obtained from skin, bones, or scales.

The total projected fisheries production to 2030 is expected to reach 202,112,000 tons of live weight, of which 106,445,000 tons are from aquaculture. The expected growth is 13.7% from 2020 to 2030 [1].

Considering that around 15–23% of fish total weight (skin, bones, and scales) could be processed to obtain collagen or gelatin, it is estimated that at least 30,000,000 tons of by-products are available as raw materials for these products.

Besides, the price of gelatin is around a few dollars per kilogram, while the price of collagen can reach USD 40 per kilogram [162]. Regarding gelatin production, Zarubin et al. [47] reported that in Russia the production cost of gelatin from porcine and bovine bone is about USD 2350–2650 per ton of product, including raw materials costs, equipment, and labor. In contrast, the production cost of fish gelatin is significantly lower, ranging from USD 1200 to 1300 per ton of product. This cost advantage is attributed to the lower cost of raw materials and to the simpler production line, underscoring the competitiveness of fish gelatin in the market.

The utilization of fish by-products to produce added-value products contributes to the industry's profitability and reinforces the principles of a circular bioeconomy, a sustainable practice that industries should strive for. This approach not only presents a lucrative opportunity but also underscores the sustainability of the fish industry, as these by-products that would otherwise go to waste are effectively utilized. This waste management also reduces pressure on landfills.

It is worth mentioning that despite the promising advancements in fish waste management for packaging applications, several significant challenges remain to be addressed. These include scalability of production, technological feasibility, and regulatory considerations. Salwa et al. [163] concluded that the whole life cycle of bio-based packaging material has a lower environmental impact and is more environmentally friendly. However, the highest impact of bio-based packaging is in the manufacturing stage because industrial processing has yet to be optimized.

Collaboration between industry stakeholders, academia, and regulatory bodies is crucial to overcoming the industrialization of biopolymers from fish by-products and realizing their full potential as valuable resources for sustainable food packaging.

ACKNOWLEDGMENTS

The authors are grateful to CONACYT/PROCIENCIA II/FEEI Paraguay for the financial support throughout the PINV01-89 project.

KEYWORDS

- **fish collagen**
- **fish gelatin**
- **fish waste valorization**
- **film packaging**

REFERENCES

1. FAO. (2022). *The State of World Fisheries and Aquaculture 2022* FAO: Rome. https://doi.org/10.4060/cc0461en.
2. Willett, W., Rockström, J., Loken, B., Springmann, M., Lang, T., Vermeulen, S., Garnett, T., Tilman, D., DeClerck, F., Wood, A., et al. (2019). Food in the Anthropocene: The EAT–lancet commission on healthy diets from sustainable food systems. *The Lancet*, 393, 447–492. doi: 10.1016/S0140-6736(18)31788-4.
3. Coppola, D., Lauritano, C., Palma Esposito, F., Riccio, G., Rizzo, C., de Pascale, D. (2021). Fish waste: From problem to valuable resource. *Marine Drugs*, 19, 116. doi: 10.3390/md19020116.
4. Venkata Mohan, S., Varjani, S., Pant, D., Sauer, M., Chang, J.-S. (2020). Circular bioeconomy approaches for sustainability. *Bioresource Technology*, *318*, 124084. doi: 10.1016/j.biortech.2020.124084.
5. Oslan, S.N.H., Shapawi, R., Mokhtar, R.A.M., Noordin, W.N.Md., Huda, N. (2022). Characterization of acid- and pepsin-soluble collagen extracted from the skin of purple-spotted bigeye snapper. *Gels*, 8, 665. doi: 10.3390/gels8100665.
6. Ideia, P., Pinto, J., Ferreira, R., Figueiredo, L., Spínola, V., Castilho, P.C. (2020). Fish processing industry residues: A review of valuable products extraction and characterization methods. *Waste Biomass Valorization*, 11, 3223–3246. doi: 10.1007/s12649-019-00739-1.
7. El-Rashidy, A.A., Gad, A., Abu-Hussein, A.E.H.G., Habib, S.I., Badr, N.A., Hashem, A.A. (2015). Chemical and biological evaluation of Egyptian Nile Tilapia (*Oreochromis niloticas*) fish scale collagen. *International Journal of Biological Macromolecules*, 79, 618–626. doi: 10.1016/j.ijbiomac.2015.05.019.
8. González-González, D.C., Rodríguez-Félix, D.E., García-Sifuentes, C.O., Castillo-Ortega, M.M., Encinas-Encinas, J.C., Santacruz Ortega, H.D.C., Romero-García, J. (2022). Collagen scaffold derived from Tilapia (*Oreochromis niloticus*) skin: Obtention, structural and physico-chemical properties. *Journal of Aquatic Food Product Technology*, 31, 374–387. doi: 10.1080/10498850.2022.2048332.
9. Göçer, M. (2022). Extraction and characterization of collagen from the skin and bone of shabout (*Arabibarbus grypus* Heckel, 1843). *Iranian Journal of Fisheries Sciences*, 21, 671–687. doi: 10.22092/ijfs.2022.126898.
10. Solari, A., Córdova, J.S. (2015). Extracción de colágeno proveniente del procesamiento de engraulis ringens "ANCHOVETA." *Ciencia e Investgacion*, 18, 65–68.

11. Menezes, M.do L.L.R., Ribeiro, H.L., Abreu, F.de O.M.da S., Feitosa, J.P.de A., Filho, M.de S.M.de S. (2020). Optimization of the collagen extraction from Nile Tilapia skin (*Oreochromis niloticas*) and its hydrogel with hyaluronic acid. *Colloids Surfaces B Biointerfaces*, 189, 110852. doi: 10.1016/j.colsurfb.2020.110852.
12. Zhang, F., Wang, A., Li, Z., He, S., Shao, L. (2011). Preparation and characterisation of collagen from freshwater fish scales. *Food and Nutrition Sciences*, 2, 818–823. doi: 10.4236/fns.2011.28112.
13. Liu, D., Nikoo, M., Boran, G., Zhou, P., Regenstein, J.M. (2015). Collagen and gelatin. *Annual Review of Food Science and Technology*, 6, 527–557. doi: 10.1146/annurev-food-031414-111800.
14. Suki, B. (2022). *The Collagen Molecule. In Structure and Function of the Extracellular Matrix* (pp. 29–54). Elsevier.
15. Hulmes, D.J.S. (2002). Building collagen molecules, fibrils, and suprafibrillar structures. *Jornal of Structural Biology*, 137, 2–10. doi: 10.1006/jsbi.2002.4450.
16. Morimoto, K., Kunii, S., Tonomura, B. (2021). Defective chicken skin collagen molecules, hydrolyzed by actinidain protease, assemble to form loosely packed fibrils that promote cell spheroid formation. *International Journal of Biological Macromolecules*, 167, 1066–1075. doi: 10.1016/j.ijbiomac.2020.11.061.
17. Espinales, C., Romero-Peña, M., Calderón, G., Vergara, K., Cáceres, P.J., Castillo, P. (2023). Collagen, protein hydrolysates and chitin from by-products of fish and shellfish: An overview. *Heliyon*, 9, e14937. doi: 10.1016/j.heliyon.2023.e14937.
18. Gahlawat, S., Nanda, V., Shreiber, D.I. (2024). Designing collagens to shed light on the multi-scale structure–function mapping of matrix disorders. *Matrix Biology Plus*, 21, 100139. doi: 10.1016/j.mbplus.2023.100139.
19. Revell, C.K., Jensen, O.E., Shearer, T., Lu, Y., Holmes, D.F., Kadler, K.E. (2021). Collagen fibril assembly: New approaches to unanswered questions. *Matrix Biology Plus*, 12, 100079. doi: 10.1016/j.mbplus.2021.100079.
20. Ricard-Blum, S. (2011). The collagen family. *Cold Spring Harbor Perspectives in Biology*, 3, a004978. doi: 10.1101/cshperspect.a004978.
21. Alam, M.R., Shahid, M.A., Alimuzzaman, S., Khan, A.N. (2022). Sources, extractions and applications of bio-maker collagen—A review. *Biomedical Engineering Advances*, 4, 100064. doi: 10.1016/j.bea.2022.100064.
22. Calvo, A.C., Moreno, L., Moreno, L., Toivonen, J.M., Manzano, R., Molina, N., de la Torre, M., López, T., Miana-Mena, F.J., Muñoz, M.J., et al. (2019). Type XIX collagen: A promising biomarker from the basement membranes. *Neural Regeneration Research*, 15, 988–995. doi: 10.4103/1673-5374.270299.
23. Myllyharju, J., Kivirikko, K.I. (2001). Collagens and collagen-related diseases. *Annuals of Medicine*, 33, 7–21. doi: 10.3109/07853890109002055.
24. Owczarzy, A., Kurasiński, R., Kulig, K., Rogóż, W., Szkudlarek, A., Maciążek-Jurczyk, M. (2020). Collagen-structure, properties and application. *Engineering of Biomaterials*, 23 (156). doi: 10.34821/eng.biomat.156.2020.17-23.
25. Shoulders, M.D., Raines, R.T. (2009). Collagen structure and stability. *Annual Review of Biochemistry*, 78, 929–958. doi: 10.1146/annurev.biochem.77.032207.120833.
26. Liao, W., Zhu, Y., Lu, Y., Wang, Y., Dong, X., Xia, G., Shen, X. (2021). Effect of extraction variables on the physical and functional properties of Tilapia gelatin. *LWT*, 146, 111514. doi: 10.1016/j.lwt.2021.111514.

27. Tan, C.-C., Karim, A.A., Uthumporn, U., Ghazali, F.C. (2020). Effect extraction temperature on the emulsifying properties of gelatin from black Tilapia (*Oreochromis mossambicus*) skin. *Food Hydrocolloids*, 108, 106024. doi: 10.1016/j.foodhyd.2020.106024.
28. Jafari, M., Afkhami, R., Sedaghat, N. (2023). Preparation and characterization of active Cirish fructans–fish gelatin film: Physicochemical, antioxidant, and antimicrobial properties. *Food Science and Nutrition*, 11, 709–721. doi: 10.1002/fsn3.3106.
29. Fonkwe, L.G., Narsimhan, G., Cha, A.S. (2003). Characterization of gelation time and texture of gelatin and gelatin–polysaccharide mixed gels. *Food Hydrocolloids*, 17, 871–883. doi: 10.1016/S0268-005X(03)00108-5.
30. León-López, A., Morales-Peñaloza, A., Martínez-Juárez, V.M., Vargas-Torres, A., Zeugolis, D.I., Aguirre-Álvarez, G. (2019). Hydrolyzed collagen—Sources and applications. *Molecules*, 24, 4031. doi: 10.3390/molecules24224031.
31. Nhari, R.M.H.R., Ismail, A., Che Man, Y.B. (2012). Analytical methods for gelatin differentiation from bovine and porcine origins and food products. *Journal of Food Science*, 77, R42–R46. doi: 10.1111/j.1750-3841.2011.02514.x.
32. Baydin, T., Aarstad, O.A., Dille, M.J., Hattrem, M.N., Draget, K.I. (2022). Long-term storage stability of type a and type b gelatin gels: The effect of bloom strength and co-solutes. *Food Hydrocolloids*, 127, 107535. doi: 10.1016/j.foodhyd.2022.107535.
33. Karim, A.A., Bhat, R. (2009). Fish gelatin: Properties, challenges, and prospects as an alternative to mammalian gelatins. *Food Hydrocolloids*, 23, 563–576. doi: 10.1016/j.foodhyd.2008.07.002.
34. Peng, J., Zi, Y., Xu, J., Zheng, Y., Huang, S., Hu, Y., Liu, B., Wang, X., Zhong, J. (2022). Effect of extraction methods on the properties of Tilapia scale gelatins. *International Journal of Biological Macromolecules*, 221, 1150–1160. doi: 10.1016/j.ijbiomac.2022.09.094.
35. Liu, Y., Xia, L., Jia, H., Li, Q., Jin, W., Dong, X., Pan, J. (2017). Physiochemical and functional properties of chum salmon (*Oncorhynchus keta*) skin gelatin extracted at different temperatures. *Journal of the Science of Food and Agriculture*, 97, 5406–5413. doi: 10.1002/jsfa.8431.
36. Abdelhedi, O., Nasri, R., Mora, L., Toldrá, F., Nasri, M., Jridi, M. (2017). Collagenous proteins from black-barred halfbeak skin as a source of gelatin and bioactive peptides. *Food Hydrocolloids*, 70, 123–133. doi: 10.1016/j.foodhyd.2017.03.030.
37. Neves Granito, R., Muniz Renno, A.C., Yamamura, H., Cruz de Almeida, M., Menin Ruiz, P.L., Araki Ribeiro, D. (2018). Hydroxyapatite from fish for bone tissue engineering: A promising approach. *International Journal of Molecular and Cellular Medicine*, 7, 80–90. doi: doi: 10.22088/ijmcm.bums.7.2.80.
38. Mudhafar, M., Zainol, I., Alsailawi, H.A., Aiza Jaafar, C.N. (2022). Synthesis and characterization of fish scales of hydroxyapatite/collagen–silver nanoparticles composites for the applications of bone filler. *Journal of the Korean Ceramic Society*, 59, 229–239. doi: 10.1007/s43207-021-00154-0.
39. Zainol, I., Adenan, N.H., Rahim, N.A., Jaafar, C.N.A. (2019). Extraction of natural hydroxyapatite from tilapia fish scales using alkaline treatment. *Materials Today Proceedings*, 16, 1942–1948. doi: 10.1016/j.matpr.2019.06.072.

40. Verma, G., Barick, K.C., Shetake, N.G., Pandey, B.N., Hassan, P.A. (2016). Citrate-functionalized hydroxyapatite nanoparticles for pH-responsive drug delivery. *RSC Advances*, 6, 77968–77976. doi: 10.1039/C6RA10659E.
41. De Lama-Odría, M. del C., del Valle, L.J., Puiggalí, J. (2022). Hydroxyapatite biobased materials for treatment and diagnosis of cancer. *International Journal of Molecular Sciences*, 23, 11352. doi: 10.3390/ijms231911352.
42. Kongsri, S., Janpradit, K., Buapa, K., Techawongstien, S., Chanthai, S. (2013). Nanocrystalline hydroxyapatite from fish scale waste: Preparation, characterization and application for selenium adsorption in aqueous solution. *Chemical Engineering Journal*, 215–216, 522–532. doi: 10.1016/j.cej.2012.11.054.
43. Monteiro, M., Medina, L., Casanova, P., Espinola, M., Machado, A., Ribeiro, A.A., Riquet, A., Azevedo Rangel Miguel, T.B., de Castro Miguel, E., Quevedo Nogueira, R. E.F. (2023). Evaluation of the acute toxicity by Artemia salina of hydroxyapatite nanoparticles obtained via sol-gel in an aqueous medium without using additives. *Ceramics International*, 49, 38509–38518. doi: 10.1016/j.ceramint.2023.09.181.
44. Song, Z., Liu, H., Chen, L., Chen, L., Zhou, C., Hong, P., Deng, C. (2021). Characterization and comparison of collagen extracted from the skin of the Nile tilapia by fermentation and chemical pretreatment. *Food Chemistry*, 340, 128139. doi: 10.1016/j.foodchem.2020.128139.
45. Quintero, J., Zapata, J.E. (2017). Optimización de la extracción del colágeno soluble en ácido de subproductos de tilapia roja (Oreochromis Spp) mediante un diseño de superficie de respuesta. *Informacion Tecnologica*, 28, 109–120. doi: 10.4067/S0718-07642017000100011.
46. Chinh, N.T., Manh, V.Q., Trung, V.Q., Lam, T.D., Huynh, M.D., Tung, N.Q., Trinh, N. D., Hoang, T. (2019). Characterization of collagen derived from tropical freshwater carp fish scale wastes and its amino acid sequence. *Natural Product Communications*, 14, 1934578X1986628. doi: 10.1177/1934578X19866288.
47. Zarubin, N.Yu., Kharenko, E.N., Bredikhina, O.V., Arkhipov, L.O., Zolotarev, K.V., Mikhailov, A.N., Nakhod, V.I., Mikhailova, M.V. (2021). Application of the Gadidae fish processing waste for food grade gelatin production. *Marine Drugs*, 19, 455. doi: 10.3390/md19080455.
48. Vázquez, J.A., Hermida-Merino, C., Hermida-Merino, D., Piñeiro, M.M., Johansen, J., Sotelo, C.G., Pérez-Martín, R.I., Valcarcel, J. (2021). Characterization of gelatin and hydrolysates from valorization of farmed salmon skin by-products. *Polymers (Basel)*, 13, 2828. doi: 10.3390/polym13162828.
49. Gál, R., Mokrejš, P., Pavlačková, J., Janáčová, D. (2022). Cyprinus carpio skeleton byproduct as a source of collagen for gelatin preparation. *International Journal of Molecular Sciences*, 23, 3164. doi: 10.3390/ijms23063164.
50. Tkaczewska, J., Morawska, M., Kulawik, P., Zając, M. (2018). Characterization of carp (*Cyprinus carpio*) skin gelatin extracted using different pretreatments method. *Food Hydrocolloids*, 81, 169–179. doi: 10.1016/j.foodhyd.2018.02.048.
51. Jamilah, B., Harvinder, K.G. (2002). Properties of gelatins from skins of fish—black tilapia (*Oreochromis mossambicus*) and red tilapia (*Oreochromis nilotica*). *Food Chemistry*, 77, 81–84. doi: 10.1016/S0308-8146(01)00328-4.
52. Gudmundsson, M., Hafsteinsson, H. (1997). Gelatin from cod skins as affected by chemical treatments. *Journal of Food Science*, 62, 37–39. doi: 10.1111/j.1365-2621.1997.tb04363.x.

53. Arnesen, J.A., Gildberg, A. (2007). Extraction and characterisation of gelatine from Atlantic salmon (*Salmo salar*) skin. *Bioresource Technology*, 98, 53–57. doi: 10.1016/j.biortech.2005.11.021.
54. Cho, S.M., Gu, Y.S., Kim, S.B. (2005). Extracting optimization and physical properties of yellowfin tuna (*Thunnus albacares*) skin gelatin compared to mammalian gelatins. *Food Hydrocolloids*, 19, 221–229. doi: 10.1016/j.foodhyd.2004.05.005.
55. Rahman, M.S., Al-Saidi, G.S., Guizani, N. (2008). Thermal characterisation of gelatin extracted from yellowfin tuna skin and commercial mammalian gelatin. *Food Chemistry*, 108, 472–481. doi: 10.1016/j.foodchem.2007.10.079.
56. Bi, C., Li, X., Xin, Q., Han, W., Shi, C., Guo, R., Shi, W., Qiao, R., Wang, X., Zhong, J. (2019). Effect of extraction methods on the preparation of electrospun/electrosprayed microstructures of tilapia skin collagen. *Journal of Bioscience Bioengineering*, 128, 234–240. doi: 10.1016/j.jbiosc.2019.02.004.
57. Heng, T.T., Tey, J.Y., Soon, K.S., Woo, K.K. (2022). Utilizing fish skin of ikan belida (*Notopterus lopis*) as a source of collagen: Production and rheology properties. *Marine Drugs*, 20, 525. doi: 10.3390/md20080525.
58. Giraldo-Rios, D.E., Rios, L.A., Zapata-Montoya, J.E. (2020). Kinetic modeling of the alkaline deproteinization of Nile-tilapia skin for the production of collagen. *Heliyon*, 6, e03854. doi: 10.1016/j.heliyon.2020.e03854.
59. Costa, F.T., Oliveira, T.P., Droval, A.A., Marques, L.L.M., Fuchs, R.H.B., Cardoso, F.A.R. (2024). Evaluation of physicochemical properties of Nile tilapia skin collagen extracted in acid médium. *Brazilian Journal of Biology*, 84. doi: 10.1590/1519-6984.255440.
60. Cheng, X., Shao, Z., Li, C., Yu, L., Raja, M.A., Liu, C. (2017). Isolation, characterization and evaluation of collagen from jellyfish *Rhopilema esculentum* Kishinouye for use in hemostatic applications. *PLoS One*, 12, e0169731. doi: 10.1371/journal.pone.0169731.
61. Bhuimbar, M.V., Bhagwat, P.K., Dandge, P.B. (2019). Extraction and characterization of acid soluble collagen from fish waste: development of collagen-chitosan blend as food packaging film. *Journal of Environmental Chemistry Engineering*, 7, 102983. doi: 10.1016/j.jece.2019.102983.
62. Kuwahara, J. (2021). Extraction of type i collagen from tilapia scales using acetic acid and ultrafine bubbles. *Processes*, 9, 288. doi: 10.3390/pr9020288.
63. Lin, L., Regenstein, J.M., Lv, S., Lu, J., Jiang, S. (2017). An overview of gelatin derived from aquatic animals: Properties and modification. *Trends Food Science and Technology*, 68, 102–112. doi: 10.1016/j.tifs.2017.08.012.
64. Aksun Tümerkan, E.T., Cansu, Ü., Boran, G., Regenstein, J. Mac, Özoğul, F. (2019). Physiochemical and functional properties of gelatin obtained from tuna, frog and chicken skins. *Food Chemistry*, 287, 273–279. doi: 10.1016/j.foodchem.2019.02.088.
65. Kittiphattanabawon, P., Benjakul, S., Sinthusamran, S., Kishimura, H. (2016). Gelatin from clown featherback skin: Extraction conditions. *LWT*, 66, 186–192. doi: 10.1016/j.lwt.2015.10.029.
66. Mirzapour-Kouhdasht, A., Moosavi-Nasab, M., Krishnaswamy, K., Khalesi, M. (2020). Optimization of gelatin production from barred mackerel by-products: characterization and hydrolysis using native and commercial proteases. *Food Hydrocolloids*, 108, 105970. doi: 10.1016/j.foodhyd.2020.105970.

67. Al-Saidi, G.S., Al-Alawi, A., Rahman, M.S., Guizani, N. (2012). Fourier transform infrared (FTIR) spectroscopic study of extracted gelatin from shaari (*Lithrinus microdon*) skin: Effects of extraction conditions. *International Food Research Journal*, 19, 1167–1173.
68. Derkach, S.R., Kuchina, Y.A., Baryshnikov, A.V., Kolotova, D.S., Voron'ko, N.G. (2019). Tailoring cod gelatin structure and physical properties with acid and alkaline extraction. *Polymers (Basel)*, 11, 1724. doi: 10.3390/polym11101724.
69. Gómez-Guillén, M.C., Giménez, B., López-Caballero, M.E., Montero, M.P. (2011). Functional and bioactive properties of collagen and gelatin from alternative sources: A review. *Food Hydrocolloids*, 25, 1813–1827. doi: 10.1016/j.foodhyd.2011.02.007.
70. Derkach, S.R., Kolotova, D.S., Voron'ko, N.G., Obluchinskaya, E.D., Malkin, A.Y. (2021). Rheological properties of fish gelatin modified with sodium alginate. *Polymers (Basel)*, 13, 1–18. doi: 10.3390/polym13050743.
71. Silva, R.S.G., Bandeira, S.F., Pinto, L.A.A. (2014). Characteristics and chemical composition of skins gelatin from cobia (*Rachycentron canadum*). *LWT*, 57, 580–585. doi: 10.1016/j.lwt.2014.02.026.
72. Ninan, G., Joseph, J., Aliyamveettil, Z.A. (2014). A comparative study on the physical, chemical and functional properties of carp skin and mammalian gelatins. *Journal of Food Science Technology*, 51, 2085–2091. doi: 10.1007/s13197-012-0681-4.
73. Said, N.S., Sarbon, N.M. (2022). Physical and mechanical characteristics of gelatin-based films as a potential food packaging material: A review. *Membranes (Basel)*, 12, 442. doi: 10.3390/membranes12050442.
74. Derkach, S.R., Voron'ko, N.G., Kuchina, Y.A., Kolotova, D.S. (2020). Modified fish gelatin as an alternative to mammalian gelatin in modern food technologies. *Polymers (Basel)*, 12, 3051. doi: 10.3390/polym12123051.
75. Ashrafi, A., Babapour, H., Johari, S., Alimohammadi, F., Teymori, F., Nafchi, A.M., Shahrai, N.N., Huda, N., Abedinia, A. (2023). Application of poultry gelatin to enhance the physicochemical, mechanical, and rheological properties of fish gelatin as alternative mammalian gelatin films for food packaging. *Foods*, 12, 670. doi: 10.3390/foods12030670.
76. Chandra, M.V., Shamasundar, B.A. (2015). Rheological properties of gelatin prepared from the swim bladders of freshwater fish Catla. *Food Hydrocolloids*, 48, 47–54. doi: 10.1016/j.foodhyd.2015.01.022.
77. Alfaro, A.T., Biluca, F.C., Marquetti, C., Tonial, I.B., de Souza, N.E. (2014). African catfish (*Clarias gariepinus*) skin gelatin: Extraction optimization and physical-chemical properties. *Food Research International*, 65, 416–422. doi: 10.1016/j.foodres.2014.05.070.
78. Alias, S.A., Mhd Sarbon, N. (2019). Rheological, physical, and mechanical properties of chicken skin gelatin films incorporated with potato starch. *NPJ Science of Food*, 3, 26. doi: 10.1038/s41538-019-0059-3.
79. Moosavi-Nasab, M., Yazdani-Dehnavi, M., Mirzapour-Kouhdasht, A. (2020). The effects of enzymatically aided acid-swelling process on gelatin extracted from fish by-products. *Food Science and Nutrition*, 8, 5017–5025. doi: 10.1002/fsn3.1799.
80. Cheung, I.W.Y., Li-Chan, E.C.Y. (2017). Enzymatic production of protein hydrolysates from steelhead (*Oncorhynchus mykiss*) skin gelatin as inhibitors of

dipeptidyl-peptidase IV and angiotensin-I converting enzyme. *Journal of Functional Foods*, 28, 254–264. doi: 10.1016/j.jff.2016.10.030.

81. Mudgil, P., Jobe, B., Kamal, H., Alameri, M., Al Ahbabi, N., Maqsood, S. (2019). Dipeptidyl peptidase-IV, α-amylase, and angiotensin I converting enzyme inhibitory properties of novel camel skin gelatin hydrolysates. *LWT*, 101, 251–258. doi: 10.1016/j.lwt.2018.11.014.
82. Pon-On, W., Suntornsaratoon, P., Charoenphandhu, N., Thongbunchoo, J., Krishnamra, N., Tang, I.M. (2016). Hydroxyapatite from fish scale for potential use as bone scaffold or regenerative material. *Materials Science and Engineering C Materials for Biological Applications*, 62, 183–189. doi: 10.1016/j.msec.2016.01.051.
83. Chai, Y., Tagaya, M. Simple preparation of hydroxyapatite nanostructures derived from fish scales. *Materials Letters*, 222, 156–159. doi: 10.1016/j.matlet.2018.04.009.
84. Zhao, X., Wang, Y., Chen, X., Yu, X., Li, W., Zhang, S., Meng, X., Zhao, Z.-M., Dong, T., Anderson, A., et al. (2023). Sustainable bioplastics derived from renewable natural resources for food packaging. *Matter*, 6, 97–127. doi: 10.1016/j.matt.2022.11.006.
85. Lionetto, F., Esposito Corcione, C. (2021). Recent applications of biopolymers derived from fish industry waste in food packaging. *Polymers (Basel)*, 13, 2337. doi: 10.3390/polym13142337.
86. Qu, W., Guo, T., Zhang, X., Jin, Y., Wang, B., Wahia, H., Ma, H. (2022). Preparation of tuna skin collagen-chitosan composite film improved by sweep frequency pulsed ultrasound technology. *Ultrasonics Sonochemistry*, 82, 105880. doi: 10.1016/j.ultsonch.2021.105880.
87. Sommer, A., Dederko-Kantowicz, P., Staroszczyk, H., Sommer, S., Michalec, M. (2021). Enzymatic and chemical cross-linking of bacterial cellulose/fish collagen composites—A comparative study. *International Journal of Molecular Sciences*, 22, 3346. doi: 10.3390/ijms22073346.
88. Tang, L., Chen, S., Su, W., Weng, W., Osako, K., Tanaka, M. (2015). Physicochemical properties and film-forming ability of fish skin collagen extracted from different freshwater species. *Process Biochemistry*, 50, 148–155. doi: 10.1016/j.procbio.2014.10.015.
89. Sun, P., Ren, Y., Wang, S., Zhu, H., Zhou, J. (2021). Characterization and film-forming properties of acid soluble collagens from different by-products of loach (Misgurnus anguillicaudatus). *LWT*, 149, 111844. doi: 10.1016/j.lwt.2021.111844.
90. Ran, R., Xiong, Y., Zheng, T., Tang, P., Zhang, Y., Yang, C., Li, G. (2024). Active and intelligent collagen films containing laccase-catalyzed mulberry extract and pickering emulsion for fish preservation and freshness indicator. *Food Hydrocolloids*, 147, 109326. doi: 10.1016/j.foodhyd.2023.109326.
91. da Costa, G.F., Grisi, C.V.B., de Albuquerque Meireles, B.R.L., de Sousa, S., de Magalhães Cordeiro, A.M.T. (2022). Collagen films, cassava starch and their blends: physical–chemical, thermal and microstructure properties. *Packaging Technology and Science*, 35, 229–240. doi: 10.1002/pts.2621.
92. Wang, Z., Hu, S., Wang, H. (2017). Scale-up preparation and characterization of collagen/sodium alginate blend films. *Journal of Food Quality*, 2017, e4954259. doi: 10.1155/2017/4954259.
93. Nuerjiang, M., Bai, X., Sun, L., Wang, Q., Xia, X., Li, F. (2023). Size effect of fish gelatin nanoparticles on the mechanical properties of collagen film based

on its hierarchical structure. *Food Hydrocolloids*, 144, 108931. doi: 10.1016/j.foodhyd.2023.108931.

94. Marangoni Júnior, L., Rodrigues, P.R., Silva, R.G. da, Vieira, R.P., Alves, R.M.V. (2022). Improving the mechanical properties and thermal stability of sodium alginate/hydrolyzed collagen films through the incorporation of SiO_2. *Current Research Food Science*, 5, 96–101. doi: 10.1016/j.crfs.2021.12.012.
95. Jiang, Y., Lan, W., Sameen, D. E., Ahmed, S., Qin, W., Zhang, Q., Chen, H., Dai, J., He, L., Liu, Y. (2020). Preparation and characterization of grass carp collagen-chitosan-lemon essential oil composite films for application as food packaging. *International Journal of Biological Macromolecules*, 160, 340–351. doi: 10.1016/j.ijbiomac.2020.05.202.
96. Chen, J., Li, J., Li, Y., Wu, S. (2024). Fabrication and characterisation of collagen/pullulan ultra-thin fibers by electrospinning. *Food Chemistry X*, 21, 101138. doi: 10.1016/j.fochx.2024.101138.
97. Moreno-Ricardo, M.A., Gómez-Contreras, P., González-Delgado, Á.D., Hernández-Fernández, J., Ortega-Toro, R. (2024). Development of films based on chitosan, gelatin and collagen extracted from bocachico scales (*Prochilodus magdalenae*). *Heliyon*, 10, e25194. doi: 10.1016/j.heliyon.2024.e25194.
98. Silva, T.H., Moreira-Silva, J., Marques, A.L.P., Domingues, A., Bayon, Y., Reis, R.L. (2014). Marine origin collagens and its potential applications. *Marine Drugs*, 12, 5881–5901. doi: 10.3390/md12125881.
99. Wenz, B., Oesch, B., Horst, M. (2001). Analysis of the risk of transmitting bovine spongiform encephalopathy through bone grafts derived from bovine bone. *Biomaterials*, 22, 1599–1606. doi: 10.1016/S0142-9612(00)00312-4.
100. Chattopadhyay, S., Raines, R.T. (2014). Collagen-based biomaterials for wound healing. *Biopolymers*, 101, 821–833. doi: 10.1002/bip.22486.
101. Mathew-Steiner, S.S., Roy, S., Sen, C.K. (2021). Collagen in wound healing. *Bioengineering*, 8, 63. doi: 10.3390/bioengineering8050063.
102. Zhang, J., Jeevithan, E., Bao, B., Wang, S., Gao, K., Zhang, C., Wu, W. (2016). Structural characterization, in-vivo acute systemic toxicity assessment and in-vitro intestinal absorption properties of tilapia (*Oreochromis niloticus*) skin acid and pepsin solublilized type I collagen. *Process Biochemistry*, 51, 2017–2025. doi: 10.1016/J.PROCBIO.2016.08.009.
103. Vigneswari, S., Murugaiyah, V., Kaur, G., Abdul Khalil, H.P.S., Amirul, A.A. (2016). Simultaneous dual syringe electrospinning system using benign solvent to fabricate nanofibrous P(3HB-Co-4HB)/collagen peptides construct as potential leave-on wound dressing. *Materials Science and Engineering C Materials for Biological Applications*, 66, 147–155. doi: 10.1016/j.msec.2016.03.102.
104. Bakar, A.J.A., Azam, N.S.M., Sevakumaran, V., Ismail, W.I.W., Razali, M.H., Razak, S. I.A., Amin, K.A.M. (2023). Effectiveness of collagen and gatifloxacin in improving the healing and antibacterial activities of gellan gum hydrogel films as dressing materials. *International Journal of Biological Macromolecules*, 245, 125494. doi: 10.1016/j.ijbiomac.2023.125494.
105. Furtado, M., Chen, L., Chen, Z., Chen, A., Cui, W. (2022). Development of fish collagen in tissue regeneration and drug delivery. *Engineered Regeneration*, 3, 217–231. doi: 10.1016/j.engreg.2022.05.002.

106. Patra, J.K., Das, G., Fraceto, L.F., Campos, E.V.R., Rodriguez-Torres, M. del P., Acosta-Torres, L.S., Diaz-Torres, L.A., Grillo, R., Swamy, M.K., Sharma, S., et al. (2018). Nano based drug delivery systems: Recent developments and future prospects. *Journal of Nanobiotechnology*, 16, 71. doi: 10.1186/s12951-018-0392-8.
107. Cao, H., Chen, M.-M., Liu, Y., Liu, Y.-Y., Huang, Y.-Q., Wang, J.-H., Chen, J.-D., Zhang, Q.-Q. (2015). Fish collagen-based scaffold containing PLGA microspheres for controlled growth factor delivery in skin tissue engineering. *Colloids Surfaces B Biointerfaces*, 136, 1098–1106. doi: 10.1016/j.colsurfb.2015.10.022.
108. Mandal, A., Dhineshkumar, E., Murugan, E. (2023). Collagen biocomposites derived from fish waste: Doped and cross-linked with functionalized Fe_3O_4 nanoparticles and their comparative studies with a green approach. *ACS Omega*, 8, 24256–24267. doi: 10.1021/acsomega.3c01106.
109. Alfaro, A. da T., Balbinot, E., Weber, C.I., Tonial, I.B., Machado-Lunkes, A. (2015). Fish gelatin: Characteristics, functional properties, applications and future potentials. *Food Engineering Reviews*, 7, 33–44. doi: 10.1007/s12393-014-9096-5.
110. Etxabide, A., Mojío, D., Guerrero, P., de la Caba, K., Gómez-Estaca, J. (2024). Chitin nanowhisker-containing photo-crosslinked antimicrobial gelatin films. *Food Hydrocolloids*, 147, 109371. doi: 10.1016/j.foodhyd.2023.109371.
111. Zhao, Z., Li, Y., Du, Z. (2022). Seafood waste-based materials for sustainable food packing: From waste to wealth. *Sustainability*, 14, 16579. doi: 10.3390/su142416579.
112. Tagrida, M., Nilsuwan, K., Gulzar, S., Prodpran, T., Benjakul, S. (2023). Fish gelatin/chitosan blend films incorporated with betel (*Piper betle* L.) leaf ethanolic extracts: Characteristics, antioxidant and antimicrobial properties. *Food Hydrocolloids*, 137, 108316. doi: 10.1016/j.foodhyd.2022.108316.
113. Riahi, Z., Hong, S.J., Rhim, J.-W., Shin, G.H., Kim, J.T. (2023). High-performance multifunctional gelatin-based films engineered with metal-organic frameworks for active food packaging applications. *Food Hydrocolloids*, 144, 108984. doi: 10.1016/j.foodhyd.2023.108984.
114. Li, Y., Shan, P., Yu, F., Li, H., Peng, L. (2023). Fabrication and characterization of waste fish scale-derived gelatin/sodium alginate/carvacrol loaded ZIF-8 nanoparticles composite films with sustained antibacterial activity for active food packaging. *International Journal of Biological Macromolecules*, 230, 123192. doi: 10.1016/j.ijbiomac.2023.123192.
115. Kilic, B., Dogan, V., Kilic, V., Kahyaoglu, L.N. (2022). Colorimetric food spoilage monitoring with carbon dot and UV light reinforced fish gelatin films using a smartphone application. *International Journal of Biological Macromolecules*, 209, 1562–1572. doi: 10.1016/j.ijbiomac.2022.04.119.
116. Jafarzadeh, S., Yildiz, Z., Yildiz, P., Strachowski, P., Forough, M., Esmaeili, Y., Naebe, M., Abdollahi, M. (2024). Advanced technologies in biodegradable packaging using intelligent sensing to fight food waste. *International Journal of Biological Macromolecules*, 261, 129647. doi: 10.1016/j.ijbiomac.2024.129647.
117. Wu, J., Song, G., Huang, R., Yan, Y., Li, Q., Guo, X., Shi, X., Tian, Y., Wang, J., Wang, S. (2023). Fish gelatin films incorporated with cinnamaldehyde and its sulfobutyl ether-β-cyclodextrin inclusion complex and their application in fish preservation. *Food Chemistry*, 418, 135871. doi: 10.1016/j.foodchem.2023.135871.

118. Salem, A., Jridi, M., Abdelhedi, O., Fakhfakh, N., Nasri, M., Debeaufort, F., Zouari, N. (2021). Development and characterization of fish gelatin-based biodegradable film enriched with Lepidium sativum extract as active packaging for cheese preservation. *Heliyon*, 7, e08099. doi: 10.1016/j.heliyon.2021.e08099.
119. Chuaynukul, K., Nagarajan, M., Prodpran, T., Benjakul, S., Songtipya, P., Songtipya, L. (2018). Comparative characterization of bovine and fish gelatin films fabricated by compression molding and solution casting methods. *Journal of Polymers and the Environment*, 26, 1239–1252. doi: 10.1007/s10924-017-1030-5.
120. Ran, R., Zheng, T., Tang, P., Xiong, Y., Yang, C., Gu, M., Li, G. (2023). Antioxidant and antimicrobial collagen films incorporating pickering emulsions of cinnamon essential oil for pork preservation. *Food Chemistry*, 420, 136108. doi: 10.1016/j.foodchem.2023.136108.
121. Heydarian, A., Shavisi, N. (2023). Multifunctional food packaging materials: electrospun mats based on gelatin-xanthan gum containing chitin nanofibers and black barberry anthocyanins for freshness monitoring and enhancing the shelf life quality of pacific white shrimps. *Food Packaging and Shelf life*, 40, 101219. doi: 10.1016/j.fpsl.2023.101219.
122. Liu, J., Yong, H., Liu, Y., Qin, Y., Kan, J., Liu, J. (2019). Preparation and characterization of active and intelligent films based on fish gelatin and haskap berries (*Lonicera caerulea* L.) Extract. *Food Packaging and Shelf life*, 22, 100417. doi: 10.1016/j.fpsl.2019.100417.
123. Sui Chin, S., Han Lyn, F., Nur Hanani, Z.A. (2017). Effect of aloe vera (*Aloe barbadensis* Miller) Gel on the physical and functional properties of fish gelatin films as active packaging. *Food Packaging and Shelf life*, 12, 128–134. doi: 10.1016/j.fpsl.2017.04.008.
124. Nilsuwan, K., Arnold, M., Benjakul, S., Prodpran, T., de la Caba, K. (2021). Properties of chicken protein isolate/fish gelatin blend film incorporated with phenolic compounds and its application as pouch for packing chicken skin oil. *Food Packaging and Shelf life*, 30, 100761. doi: 10.1016/j.fpsl.2021.100761.
125. Nor Adilah, A., Gun Hean, C., Nur Hanani, Z.A. (2021). Incorporation of graphene oxide to enhance fish gelatin as bio-packaging material. *Food Packaging and Shelf life*, 28, 100679. doi: 10.1016/j.fpsl.2021.100679.
126. Maryam Adilah, Z.A., Han Lyn, F., Nabilah, B., Jamilah, B., Gun Hean, C., Nur Hanani, Z.A. (2022). Enhancing the physicochemical and functional properties of gelatin/graphene oxide/cinnamon bark oil nanocomposite packaging films using ferulic acid. *Food Packaging and Shelf life*, 34, 100960. doi: 10.1016/j.fpsl.2022.100960.
127. Liu, Y., Kai, Y., Yang, H. (2023). Biodegradable fish gelatin/chitosan-based active films alter chill-stored golden pomfret (*Trachinotus blochii*) metabolites mainly through modulating four metabolic pathways. *Food Packaging and Shelf life*, 36, 101046. doi: 10.1016/j.fpsl.2023.101046.
128. Getachew, A. T., Ahmad, R., Park, J.-S., Chun, B.-S. (2021). Fish skin gelatin based packaging films functionalized by subcritical water extract from spent coffee ground. *Food Packaging and Shelf life*, 29, 100735. doi: 10.1016/j.fpsl.2021.100735.
129. Chen, J., Li, Y., Wang, Y., Yakubu, S., Tang, H., Li, L. (2022). Active polylactic acid/tilapia fish gelatin-sodium alginate bilayer films: Application in preservation

of japanese sea bass (*Lateolabrax japonicus*). *Food Packaging and Shelf life*, 33, 100915. doi: 10.1016/j.fpsl.2022.100915.

130. Nurul Syahida, S., Ismail-Fitry, M.R., Ainun, Z.M.A., Nur Hanani, Z.A. (2020). Effects of palm wax on the physical, mechanical and water barrier properties of fish gelatin films for food packaging application. *Food Packaging and Shelf life*, 23, 100437. doi: 10.1016/j.fpsl.2019.100437.
131. Nurilmala, M., Suryamarevita, H., Husein Hizbullah, H., Jacoeb, A.M., Ochiai, Y. (2022). Fish skin as a biomaterial for halal collagen and gelatin. *Saudi Journal of Biological Sciences*, 29, 1100–1110. doi: 10.1016/j.sjbs.2021.09.056.
132. Duan, R., Zhang, J., Liu, L., Cui, W., Regenstein, J.M. (2018). The functional properties and application of gelatin derived from the skin of channel catfish (*Ictalurus punctatus*). *Food Chemistry*, 239, 464–469. doi: 10.1016/j.foodchem.2017.06.145.
133. Huang, T., Tu, Z., Shangguan, X., Sha, X., Wang, H., Zhang, L., Bansal, N. (2019). Fish gelatin modifications: A comprehensive review. *Trends in Food Science and Technology*, 86, 260–269. doi: 10.1016/j.tifs.2019.02.048.
134. Damodaran, S., Wang, S. (2017). Ice crystal growth inhibition by peptides from fish gelatin hydrolysate. *Food Hydrocolloids*, 70, 46–56. doi: 10.1016/j.foodhyd.2017.03.029.
135. Yin, M., Yang, D., Lai, S., Yang, H. (2021). Rheological properties of xanthan-modified fish gelatin and its potential to replace mammalian gelatin in low-fat stirred yogurt. *LWT*, 147, 111643. doi: 10.1016/j.lwt.2021.111643.
136. Sang, S., Ou, C., Fu, Y., Su, X., Jin, Y., Xu, X. (2022). Complexation of fish skin gelatin with glutentin and its effect on the properties of wheat dough and bread. *Food Chemistry X*, 14, 100319. doi: 10.1016/j.fochx.2022.100319.
137. Bian, M., Jiang, S., Liu, S., Zhang, L., Miao, S., Zhou, F., Zheng, B. (2024). Fish gelatin and gellan gum mixture as edible ink for 3D printing. *Journal of Food Engineering*, 362, 111762. doi: 10.1016/j.jfoodeng.2023.111762.
138. Phawaphuthanon, N., Yu, D., Ngamnikom, P., Shin, I.-S., Chung, D. (2019). Effect of fish gelatine-sodium alginate interactions on foam formation and stability. *Food Hydrocolloids*, 88, 119–126. doi: 10.1016/j.foodhyd.2018.09.041.
139. Yusof, N., Jaswir, I., Jamal, P., Jami, M.S. (2019). Texture Profile Analysis (TPA) of the jelly dessert prepared from halal gelatin extracted using high pressure processing (HPP). *Malaysian Journal of Fundamental and Applied Sciences*, 15, 604–608. doi: 10.11113/mjfas.v15n4.1583.
140. Shiao, W.-C., Wu, T.-C., Kuo, C.-H., Tsai, Y.-H., Tsai, M.-L., Hong, Y.-H., Huang, C.-Y. (2021). Physicochemical and antioxidant properties of gelatin and gelatin hydrolysates obtained from extrusion-pretreated fish (*Oreochromis* sp.) scales. *Marine Drugs*, 19, 275. doi: 10.3390/md19050275.
141. Shah, Y.A., Bhatia, S., Al-Harrasi, A., Afzaal, M., Saeed, F., Anwer, M.K., Khan, M.R., Jawad, M., Akram, N., Faisal, Z. (2023). Mechanical properties of protein-based food packaging materials. *Polymers (Basel)*, 15, 1724. doi: 10.3390/polym15071724.
142. Ben Slimane, E., Sadok, S. (2018). Collagen from cartilaginous fish by-products for a potential application in bioactive film composite. *Marine Drugs*, 16, 211. doi: 10.3390/md16060211.
143. Sripriya, R., Kumar, R. (2015). A novel enzymatic method for preparation and characterization of collagen film from swim bladder of fish rohu (*Labeo rohita*). *Food and Nutrition Sciences*, 6, 1468–1478. doi: 10.4236/fns.2015.615151.

144. Niu, L., Zhou, X., Yuan, C., Bai, Y., Lai, K., Yang, F., Huang, Y. (2013). Characterization of tilapia (*Oreochromis niloticus*) skin gelatin extracted with alkaline and different acid pretreatments. *Food Hydrocolloids*, 33, 336–341. doi: 10.1016/j.foodhyd.2013.04.014.
145. Zhang, Q., Wang, Q., Lv, S., Lu, J., Jiang, S., Regenstein, J.M., Lin, L. (2016). Comparison of collagen and gelatin extracted from the skins of nile tilapia (*Oreochromis niloticus*) and channel catfish (*Ictalurus punctatus*). *Food Bioscience*, 13, 41–48. doi: 10.1016/j.fbio.2015.12.005.
146. Kchaou, H., Benbettaïeb, N., Jridi, M., Abdelhedi, O., Karbowiak, T., Brachais, C.-H., Léonard, M.-L., Debeaufort, F., Nasri, M. (2018). Enhancement of structural, functional and antioxidant properties of fish gelatin films using maillard reactions. *Food Hydrocolloids*, 83, 326–339. doi: 10.1016/j.foodhyd.2018.05.011.
147. Tyagi, P., Salem, K.S., Hubbe, M.A., Pal, L. (2021). Advances in barrier coatings and film technologies for achieving sustainable packaging of food products–A review. *Trends in Food Science and Technology*, 115, 461–485. doi: 10.1016/j.tifs.2021.06.036.
148. Tyuftin, A.A., Kerry, J.P. (2021). Gelatin films: Study review of barrier properties and implications for future studies employing biopolymer films. *Food Packaging Shelf life*, 29, 100688. doi: 10.1016/j.fpsl.2021.100688.
149. Zhuang, Y., Ruan, S., Yao, H., Sun, Y. (2019). Physical properties of composite films from tilapia skin collagen with Pachyrhizus starch and rambutan peel phenolics. *Marine Drugs*, 17, 662. doi: 10.3390/md17120662.
150. Gumus, T., Kaynarca, G.B., Kamer, D.D.A. (2024). Optimization of an edible film formulation by incorporating carrageenan and red wine lees into fish gelatin film matrix. *International Journal of Biological Macromolecules*, 258, 128854. doi: 10.1016/j.ijbiomac.2023.128854.
151. Toldra, F. (2022). *Lawrie's Meat Science*. Woodhead Publishing.
152. Dey, A., Neogi, S. (2019). Oxygen scavengers for food packaging applications: A review. *Trends in Food Science and Technology*, 90, 26–34. doi: 10.1016/j.tifs.2019.05.013.
153. Martins, J.T., Cerqueira, M.A., Vicente, A.A. (2012). Influence of α-tocopherol on physicochemical properties of chitosan-based films. *Food Hydrocolloids*, 27, 220–227. doi: 10.1016/j.foodhyd.2011.06.011.
154. Roy, S., Ramakrishnan, R., Goksen, G., Singh, S., Łopusiewicz, Ł. (2024). Recent progress on UV-light barrier food packaging films—A systematic review. *Innovative Food Science & Emerging Technologies*, 91, 103550. doi: 10.1016/j.ifset.2023.103550.
155. Guzman-Puyol, S., Benítez, J.J., Heredia-Guerrero, J.A. (2022). Transparency of polymeric food packaging materials. *Food Research International*, 161, 111792. doi: 10.1016/j.foodres.2022.111792.
156. Vejdan, A., Ojagh, S. M., Adeli, A., Abdollahi, M. (2016). Effect of TiO_2 nanoparticles on the physico-mechanical and ultraviolet light barrier properties of fish gelatin/agar bilayer film. *LWT—Food Science and Technology*, 71, 88–95. doi: 10.1016/j.lwt.2016.03.011.
157. Bintsis, T. (2017). Foodborne pathogens. *AIMS Microbiology*, 3, 529–563. doi: 10.3934/microbiol.2017.3.529.

158. Anvar, A.A., Ahari, H., Ataee, M. (2021). Antimicrobial properties of food nanopackaging: A new focus on foodborne pathogens. *Frontiers in Microbiology*, 12, 690706. doi: 10.3389/fmicb.2021.690706.
159. Kulkarni, P., Maniyar, M., Nalawade, M., Bhagwat, P., Pillai, S. (2022). Isolation, biochemical characterization, and development of a biodegradable antimicrobial film from cirrhinus mrigala scale collagen. *Environmental Science and Pollution Research*, 29, 18840–18850. doi: 10.1007/S11356-021-17108-Y/TABLES/4.
160. Pouralkhas, M., Kordjazi, M., Ojagh, S.M., Farsani, O.A. (2023). Physicochemical and functional characterization of gelatin edible film incorporated with fucoidan isolated from Sargassum tenerrimum. *Food Science & Nutrition*, 11, 4124–4135. doi: 10.1002/fsn3.3402.
161. Markets and Markets. Gelatin Market Report: Size, Share & Forecast [Latest]. Retrieved April 21, 2024, from https://www.marketsandmarkets.com/Market-Reports/gelatin-market-850.html?_gl=1*kn8wns*_up*MQ..&gclid=Cj0KCQjwiYOxBhC5ARIsAIvdH538a-nFCixwQ7CZ-lTXBv1HLMPHd8LFLBAZT_13yugIOVFeplpy7K0aAk-7EALw_wcB.
162. Rajabimashhadi, Z., Gallo, N., Salvatore, L., Lionetto, F. (2023). Collagen derived from fish industry waste: Progresses and challenges. *Polymers (Basel)*, 15, 544. doi: 10.3390/polym15030544.
163. Salwa, H.N., Sapuan, S.M., Mastura, M.T., Zuhri, M.Y.M., Ilyas, R.A. (2021). *Life Cycle Assessment of Bio-Based Packaging Products. In Bio-based Packaging* (pp. 381–411). Wiley. doi: 10.1002/9781119381228.ch22.

CHAPTER 2

Bio-Based Materials from Nonconventional Sources of Mexican Semidesert Origin: An Alternative to Synthetic-Based Food Packaging

MARÍA HERNÁNDEZ-GONZÁLEZ,[1*]
FRANCISCO HERNÁNDEZ-CENTENO,[1]
HAYDEÉ YAJAIRA LÓPEZ-DE LA PEÑA,[1]
MÓNICA DANIELA ALVARADO-MACÍAS,[1]
ALEXIA DOMÍNGUEZ-VILLARREAL,[1]
CINTHYA GABRIELA ROCHA-CORTÉS,[1]
PAUL BARUK ZAMUDIO-FLORES,[2] and
ANA MARGARITA RODRÍGUEZ-HERNÁNDEZ[3]

[1]*Departamento de Ciencia y Tecnología de Alimentos, Universidad Autónoma Agraria Antonio Narro, Buenavista, Saltillo, Coahuila, Mexico*

[2]*Centro de Investigación en Alimentación y Desarrollo, A.C., Unidad Cuauhtémoc, Laboratorio de Carbohidratos, Empaques y Alimentos Funcionales, Cd. Cuauhtémoc, Cuauhtémoc, Chihuahua, Mexico*

[3]*Centro de Investigación en Química Aplicada, San José de los Cerritos, Saltillo, Coahuila, Mexico*

*Corresponding author

Sustainable Food Packaging Materials and Technology. Juan Manuel Tirado-Gallegos, Juan Daniel Rivaldi Chávez, and José Juan Buenrostro-Figueroa (Eds)

DOI: 10.1201/9781779641434-2

ABSTRACT

Bio-based materials derived from semidesert plants could offer a sustainable alternative in food packaging manufacturing. These materials include bioplastics, plant fibers, natural coatings, and functionalized materials for potential use in food packaging. Bio-based plastic materials, obtained from plants such as the Devil's claw root, could be molded into different shapes and sizes to suit various packaging needs. Plant fibers, such as those obtained from the *lechuguilla* plant, can be used in the production of paper or cardboard packaging, which are biodegradable and compostable. Additionally, some natural resins derived from semidesert plants can be used as internal coatings for these packages or their essential oils can be used to functionalize such materials, providing a protective barrier against moisture and acting as antimicrobial and/or antioxidant agents. These bio-based materials are not only environmentally friendly due to their renewable and biodegradable nature but also can contribute to reducing the environmental impact associated with conventional packaging.

2.1 INTRODUCTION

Packaging, such as bottles and food wrappers, constitutes a large portion of urban solid waste. Their widespread use and ease of disposal after a single use mean they accumulate rapidly in landfills and the environment, often ending up in rivers, seas, and oceans. Additionally, microplastics contain and absorb toxic pollutants that can be transferred to living organisms that ingest them, posing a serious threat to terrestrial and marine wildlife. Many of these packages are made of plastics that are durable and take hundreds of years to degrade. This means that the plastic discarded today will remain in the environment, breaking down into microplastics that can be harmful to wildlife and ecosystems. Furthermore, plastic packaging can release toxic chemicals during its manufacturing, use, and decomposition. These substances can contaminate soil and water, and some of them can be hazardous to human health.

The massive production of single-use packaging requires large amounts of natural resources, such as petroleum and water, and this contributes to the overexploitation of resources and their depletion. The problem of pollution caused by single-use packaging for processed foods is multifaceted and requires individual, business, and governmental actions to address it effectively and mitigate its impact on the environment and public health.

Not all single-use packages are easy to recycle. Some contain material mixtures or additives that complicate their processing in recycling facilities. In many places, recycling and waste management systems are not sufficient to handle the amount of discarded plastic packaging. This results in their accumulation in landfills, illegal dumps, or their dumping into bodies of water, creating a significant environmental footprint.

There are more sustainable alternatives to single-use packaging, such as compostable or biodegradable packaging, as well as promoting the use of reusable packaging. However, the adoption of these alternatives faces challenges, such as additional costs and a lack of adequate infrastructure for their recycling or composting.

The growing concern about environmental problems associated with synthetic materials used in food packaging has led to the search for more sustainable alternatives. One of these alternatives is bio-based materials, which are obtained from renewable sources such as starch, cellulose, and other materials from unconventional sources [1], some of which could be obtained from various botanical species that thrive in the semidesert. These bio-based materials offer several advantages compared to synthetic materials. They are biodegradable, meaning they do not cause environmental waste problems after use. Additionally, their production uses fewer fossil resources and generates fewer greenhouse gas emissions compared to synthetic materials. In this context, bio-based materials obtained from unconventional sources in the semidesert are presented as a promising alternative to synthetic-based food packaging.

This chapter summarizes advances in the development and proposals for the use of bio-based materials obtained from unconventional sources in the semidesert as an alternative to single-use synthetic-based food packaging.

2.2 IMPORTANCE OF FOOD PACKAGING

Packaging is a container or structure designed and prepared to contain, protect, handle, distribute, and present goods, including food products (Figure 2.1). Packaging materials play a crucial role in the preservation and marketing of processed foods [2], as they are essential not only for maintaining the quality and safety of processed foods but also for their efficient distribution and success in the market. In the context of processed foods, packaging is fundamental mainly for the following reasons:

- Packaging provides a protective barrier between the food and the environment, preventing contamination and microbial growth that can lead to food decomposition and health risks [3].
- Packaging designed to preserve food can extend the shelf life of products by reducing exposure to destructive agents such as light, moisture, and oxygen and by maintaining the qualities of the food by protecting it from mechanical forces and external odors [3, 4].
- Some packages are designed to withstand sterilization processes, such as canning, where high temperatures are used to eliminate pathogenic and spoilage microorganisms, thus ensuring food safety [5].
- Packaging provides an efficient way to handle, transport, and store food products, protecting them from physical and chemical damage during these processes [2].
- Packaging is a key tool in marketing, aiding in product identification and promotion. It also communicates important information to the consumer, such as ingredients, expiration dates, nutritional information, and product handling [6].

FIGURE 2.1 A container meets a set of functions that lead to food preservation.

Source: Image generated by authors with artificial intelligence.

2.3 POTENTIAL OF BIO-BASED MATERIALS FROM SEMIDESERT PLANTS

Bio-based materials have been presented as a possible sustainable alternative to reduce the environmental impact of food packaging. Bio-based materials can be defined as those materials obtained from living organisms, usually plants, from which their biopolymers are exploited, such as starch and cellulose, for the formulation of useful materials in various industrial applications, including the manufacture of food packaging [7].

Some bio-based materials can come from unconventional sources in the semidesert, such as plants adapted to extreme dryness and high temperatures. Studies have shown that these materials have interesting properties, such as a barrier to oxygen and moisture, biodegradability, and recyclability [8, 9]. Additionally, these bio-based materials can be produced sustainably, as they come from renewable and more sustainable sources, compared to other commercial crops, such as sugarcane, which requires large amounts of water for cultivation, in addition to generating a higher carbon footprint due to the burning process during its harvest. Additionally, the use of bio-based materials of semidesert origin can benefit the inhabitants of communities in semiarid regions by promoting the development of a sustainable economy and the conservation of natural resources, while being responsibly exploited [10–12].

Various plants, such as *lechuguilla*, *Agave*, jojoba, candelilla, and prickly pear, among others, contain biopolymeric compounds that can be extracted and used to produce bio-based materials with various industrial applications, not necessarily for food use [13].

There are also other plants in the semidesert from which useful natural compounds can be obtained for the development of bio-based materials; examples of these are aromatic plants that possess essential oils, whose chemical composition gives them antimicrobial and antioxidant properties, making them suitable for the production of food packaging with additional preservation and protection capabilities. An example of this is the essential oil of Mexican oregano (*Lippia graveolens* Kunth, synonym *Lippia berlandieri* Schauer), which contains polyphenols such as thymol and carvacrol, metabolites with proven antioxidant and antimicrobial activities that can be used to prevent the deterioration and contamination of packaged foods [14].

It is important to note that, although bio-based materials have potential for use in food packaging, their exact application will depend on various factors, such as compatibility with other ingredients, food regulations, and consumer preferences. Additional research and testing would be required to determine the feasibility and effectiveness of their use in this specific context.

2.4 CHARACTERISTICS OF SEMIDESERT PLANTS AS SOURCES OF BIO-BASED MATERIALS

2.4.1 ADAPTATIONS TO THE SEMIDESERT ENVIRONMENT

Plant species that thrive in deserts and semideserts have developed physical and physiological adaptations to survive in environments with scarce water and high temperatures. The key to their survival lies in their ability to conserve water and regulate their temperature [15]. Some of the most common adaptations include water conservation mechanisms, such as deep roots that seek groundwater, succulent leaves or stems that store water, and thick cuticles on leaves that reduce water loss by evaporation [16]. Plants also have adaptations to reduce exposure to extreme heat, such as small- or needle-shaped leaves that reduce the surface exposed to the sun and help prevent excessive water loss.

The architecture of plants often changes in semidesert environments, with low, branching forms that help reduce exposure to wind and water loss through evaporation. They may also have physiological adaptations such as the ability to close stomata during dry periods to reduce transpiration and conserve water [17]. To sustain their physical shape, desert and semidesert plants develop fibrous and sturdy structures, such as strong roots and thick trunks, which allow them to withstand strong winds and dry soils. The leaves of some species, such as *lechuguilla* (*Agave lechuguilla* Torr.), are of particular interest due to the large amount of fibers they possess, providing them with strength and the ability to retain water [18].

Another important adaptation is the ability of some desert plants to photosynthesize under extreme heat and lack of humidity, leading them to produce useful secondary metabolites for their survival, such as anti-oxidant compounds and protective pigments [19]. Other metabolites serve to protect the plant from pests and diseases that may be more common in desert environments [20]. Polyphenols such as thymol and carvacrol present in some desert plants, such as Mexican oregano (*L. graveolens* Kunth), have antimicrobial properties and help protect them from potential infections. The production of these compounds is a key adaptation for the survival of plants in desert environments, allowing them to withstand extreme conditions and protect themselves from competition, pathogens, predators, and environmental stressors [21], all of which influence the production of these metabolites.

2.4.2 AVAILABILITY AND SUSTAINABILITY OF SEMIDESERT PLANTS

The availability and sustainability of semidesert plants are crucial for the conservation of ecosystems in these areas, as uncontrolled exploitation can lead to soil degradation and loss of biodiversity, as useful plants are associated with other species and play a fundamental role in the ecological balance of the region [22].

The vast majority of species exploited in these regions correspond to wild populations, and their collection must be managed sustainably to ensure their long-term availability [23]. In this sense, it is important to promote conservation practices and sustainable management of semidesert plants. This involves implementing practices such as regulating collection, establishing protected areas or nature reserves, and promoting sustainable agriculture and forestry. This may include strategies such as responsible harvesting of plants and derived products, implementing measures to prevent overexploitation, promoting sustainable agricultural practices that encourage species diversity and soil protection, and incentivizing reforestation and restoration of degraded areas [24], all of which are regulated by current environmental legislation.

The exploitation of semidesert plants has been practiced for centuries as part of the traditional practices of local communities, first by the original native inhabitants and later by inhabitants who have colonized desert regions [25]. Currently, in Mexico, botanical species useful in various fields such as medicine, gastronomy, and the cosmetics industry are extracted from this habitat [26], and the materials obtained from them have led to the development of bio-based materials for various industrial sectors [27], including the development of packaging materials.

Examples of raw materials of plant origin obtained from semideserts in Mexico and of economic importance include oregano, *lechuguilla*, candelilla, prickly pear, aloe vera, damiana, creosote bush, and mesquite [28]. These plant-derived materials from semideserts have special properties that make them valuable for the production of various products, such as *lechuguilla* fiber, candelilla wax, and oregano essential oil [18, 29, 30].

Historically, efforts have been made to conserve and sustainably manage these species and their habitats in semidesert regions. However, it is essential to strengthen and expand these initiatives to ensure long-term conservation. This involves the active participation of local communities, especially those dependent on semidesert resources, and the implementation of appropriate policies and regulations that promote sustainable management of these resources. Additionally, the use of bio-based packaging materials derived from semidesert plants can have a positive impact on local communities that depend on these resources. The sustainable use of these plants can generate economic opportunities and improve the quality of life of communities by promoting the development of local value chains and job creation. Figure 2.2 shows a sample of this habitat.

FIGURE 2.2 Semidesert ecosystem in southeastern Coahuila, Mexico.

Source: Authors.

2.4.3 PROPERTIES OF MATERIALS EXTRACTED FROM THESE PLANTS

There are several properties of bio-based materials that can be utilized in various industrial applications, including the development of packaging for food preservation. These properties depend on the type of processing and the final molecular composition and structure of the materials obtained, which also depend on the botanical source from which they are extracted, as well as the biotic and abiotic factors in which these plants develop.

2.4.3.1 MICROCELLULOSE

An example of such materials is microcellulose, which is a particular form of cellulose consisting of very small fibers with a diameter ranging from a few nanometers (nanocellulose) to several micrometers. These fibers are known to be highly crystalline and to have impressive strength and rigidity relative to their size. Microcellulose is mainly obtained through isolation processes from native cellulose sources [31].

Cellulose sources for microcellulose production are varied and may include traditional materials such as wood and agricultural residues [32]. In fact, the use of more sustainable and less conventional sources to obtain cellulose is being explored due to environmental concerns associated with obtaining materials from sources such as wood, which may be related to deforestation issues and their negative impact on climate, hydrology, soil, and biodiversity [33]. One unconventional source of this material is the *lechuguilla* plant (*Agave lechuguilla* Torr.), which thrives abundantly in the arid and semiarid regions of northern and central Mexico (Figure 2.3).

FIGURE 2.3 Obtaining *lechuguilla* fiber (*Agave lechuguilla* Torr.).
Source: Authors.

The production of microcellulose from these materials generally involves an acid hydrolysis process in which cellulose fibers are treated with strong acids, such as sulfuric acid, under controlled conditions to obtain microscopic cellulose crystals [32]. This process selectively degrades the less crystalline regions of cellulose, leaving the highly crystalline parts intact. Microcellulose obtained in this way can be used to reinforce materials, in absorption applications, as a thickening agent, and in various technological and biomedical applications due to its unique properties.

2.4.3.2 STARCH

Some plants that thrive in desert and semidesert ecosystems are capable of storing starch in their tissues as energy reserves. Examples of these are the roots of *Agave* plants, such as different species and varieties of *Agave* (*Agave* spp.), sotol (*Dasylirion* spp.), and *Nolina* (*Nolina* spp.); however, many of them are scarce, making their commercial exploitation unfeasible, and some others, such as agaves, are used for other commercial purposes, such as the production of alcoholic beverages.

Other species capable of thriving in these ecosystems remain undervalued as potential unconventional and sustainable sources of starch, such as the buffalo gourd (*Cucurbita foetidissima* Kunth), in whose roots they store up to 52% of starch on a dry basis with properties similar to those of starches such as corn and cassava [34].

2.4.3.3 OTHER MATERIALS FROM THE SEMIDESERT WITH POTENTIAL USE IN FOOD PACKAGING

Another botanical material in the semidesert is the wax from the candelilla plant (*Euphorbia antisyphilitica*), used in various products [35], including food products, as its lustrous and moisture barrier properties favor preservation, for example, of fruits, prolonging their shelf life and good appearance by preventing or delaying dehydration. Combined with other biopolymers, this wax contributes improvements to the mechanical and barrier properties of the resulting biocomposites, which could favor its use in different applications, including the packaging of perishable and nonperishable foods.

On the other hand, mesquite gum (*Prosopis* spp.) has been proposed as it can act as a good barrier to moisture and oxygen [36], which is crucial, for example, for preserving the freshness and quality of dried fruits, such as dates. It has also been tested as an additive in packaging material formulation; when combined with other biopolymers, it optimizes the mechanical and barrier properties of these, which is important and useful when protecting food during transportation and storage, also aiding the antimicrobial properties reported in some studies.

Research continues on the physicochemical properties of bio-based materials originating from semidesert species, either to construct materials with new molecular architectures, or that could provide or reinforce favorable properties for their use in food packaging, as well as in various industrial applications where synthetic plastics have reigned for decades.

2.5 TYPES OF BIO-BASED MATERIALS FROM SEMIDESERT PLANTS FOR POSSIBLE USE IN FOOD PACKAGING

Bio-based materials obtained from semidesert plants in Mexico can be a sustainable alternative for use in food packaging. These materials have specific properties that make them suitable for this purpose, such as strength, flexibility, and a barrier against moisture and gases [37, 38]. Additionally, by using these bio-based materials in food packaging, the dependence on petroleum-derived plastic materials, which are highly polluting and nonbiodegradable, is reduced.

Some examples of bio-based materials from semidesert plants that could be useful in food packaging are *lechuguilla* fiber, candelilla wax, oregano essential oil, and buffalo gourd root starch. These materials can be used to manufacture biodegradable and compostable packaging, contributing to reducing plastic waste in the environment and promoting the transition to a circular and sustainable economy [39, 40]. The use of these materials in the packaging of processed foods can be beneficial both for product conservation and environmental care.

2.5.1 BIOPLASTICS

Bioplastics are plastic materials made from renewable sources, such as plants and microorganisms, instead of petroleum derivatives. These materials are a more sustainable and environmentally friendly alternative compared to conventional plastics [41].

By using bioplastics in food packaging, the dependence on nonrenewable resources is reduced, and the carbon footprint and generation of plastic waste are decreased. Additionally, bioplastics can offer specific properties that make it suitable for food packaging, such as strength, flexibility, and barrier against moisture and gases [1, 38, 42].

Some bioplastic materials from unconventional natural sources of semidesert origin have been proposed, such as buffalo gourd root starch (*C. foetidissima* Kunth) [8], nopal mucilage (*Opuntia* spp.) [43], or chia seed flour (*Salvia hispanica*) [44], whose physicochemical properties make them suitable for packaging certain processed foods, especially those with low moisture content.

Although they are still far from achieving or even approaching the physicochemical properties of conventional synthetic plastics, the use of bioplastics in the packaging of processed foods can contribute to the conservation and commercialization of these, as these materials help protect and preserve foods during their transportation and storage. Additionally, bioplastics can be biodegradable and compostable, meaning they naturally decompose in the environment without leaving toxic residues [45].

2.5.2 BIOCOMPOSITES

In addition to bio-based plastics, or bioplastics, and edible coatings, biocomposites have also been developed for the packaging of processed foods [1]. Biocomposites are materials that combine biodegradable polymers with other organic components, such as natural fibers and plant extracts, to improve their mechanical and barrier properties. These biocomposites can offer better resistance and durability compared to bioplastics alone while reducing the environmental impact caused by conventional plastics.

Some plant-derived materials from the semidesert that have been used in the proposed manufacturing of biocomposites include fiber from different agaves, oregano essential oil, candelilla wax, buffalo gourd starch, mesquite gum, and nopal mucilage. These materials have interesting properties, such as the ability to integrate with other biomaterials, such as starches, to form flexible films with antimicrobial and/or antioxidant activity, as well as being biodegradable, making them potentially suitable for packaging, transportation, and commercialization of processed foods [45].

Other proposals that have been presented include biodegradable materials combined with botanical raw materials extracted from the desert or semidesert, such as oregano essential oil, with which functionalized materials have been developed, such as polysuccinimide–starch–oregano oil, which have shown antimicrobial and antioxidant properties along with interesting mechanical and barrier properties [9].

Alongside the aforementioned cases, several other proposals support the development of bio-based materials as a sustainable alternative to address pollution caused by single-use plastic waste.

2.5.3 EDIBLE COATINGS

On the other hand, although they are not plastic materials in the full extent of the word, coating materials have been developed for edible coatings with the purpose of extending their shelf life, improving their appearance, and protecting them from external factors using plant-derived biopolymers from semidesert plants such as nopal and chia mucilages [13]. The use of these materials provides a barrier to gases and water vapor, preventing or delaying dehydration and the senescence phenomenon of whole raw foods, such as fruits.

The use of candelilla wax in the elaboration of coatings for the conservation of fresh and processed foods has also been studied. These coatings have barrier properties that prevent moisture loss and oxygen entry, helping to extend the shelf life of foods and maintain their freshness [47, 48]. These advances in packaging materials, such as bioplastics, edible coatings, and biocomposites, have demonstrated their potential in the conservation and commercialization of processed foods.

2.6 ADVANTAGES OF BIO-BASED MATERIALS FROM SEMIDESERT PLANTS TO DEVELOP FOOD PACKAGING

In the field of food packaging research, there is a growing interest in utilizing materials derived from semidesert plants as a bio-based alternative. These materials offer several advantages for food packaging since they are renewable and sustainable, reducing reliance on fossil-based materials, have a lower carbon footprint compared to traditional packaging materials, contributing to environmental conservation, they can provide better protection to food products, maintaining their quality and freshness for a longer shelf life, and they are biodegradable and compostable, reducing plastic waste and its negative impact on ecosystems [49].

2.6.1 BIODEGRADABILITY AND COMPOSTABILITY

Bio-based materials derived from semidesert plants have the advantage of being biodegradable and compostable. This means that they can break down naturally over time, reducing the amount of waste in the environment and turn into nutrient-rich soil. This feature is particularly advantageous as it supports circular economy practices, where waste is converted into valuable resources for agriculture or landscaping. This is especially important in the field of food packaging, where single-use plastics contribute to the growing plastic waste problem [50]. By utilizing bio-based materials that are biodegradable and compostable, food packaging can be designed to have a reduced environmental impact. Analysis to determine the potential of these materials to break down in different environments is crucial to ensure their effectiveness and sustainability [1]. Such studies include testing the materials in various composting conditions to verify their ability to biodegrade and integrate into the natural environment without causing harm. Also, such analysis can be carried out through respirometry test to measure the degradation rate of these materials in contact with specific bacteria or enzymes that are present in composting facilities [51]. As consumers are becoming increasingly environmentally conscious, products packaged in biodegradable and compostable materials may appeal to eco-minded customers. This can enhance brand reputation and attract consumers seeking more sustainable choices.

2.6.2 REDUCTION OF THE DEPENDENCE ON FOSSIL RESOURCES

Bio-based materials from semidesert plants offer the advantage of reducing the dependence on fossil resources for food packaging [1]. Current problems derived from indiscriminate use of fossil-based products include depletion of finite resources, environmental pollution from the extraction and production processes, and climate change impact from greenhouse gas emissions. The current oil extraction process may be considered controversial due to its significant environmental impact, including habitat destruction and water pollution. By shifting to bio-based materials, we can significantly reduce our reliance on fossil resources and mitigate these environmental issues [52].

2.6.3 POTENTIAL TO IMPROVE THE CIRCULAR ECONOMY AT SEMIDESERT REGIONS

Bio-based materials from semidesert plants have the potential to improve the circular economy in these regions. By utilizing local plants that are abundant in semidesert regions, we can create a sustainable loop where the plants are grown, harvested, processed into raw materials and packaging products, used for packaging food, and then composted or recycled back into the soil to support future plant growth. This closed-loop system reduces the need for external resources and minimizes waste generation, creating a more sustainable and circular economy. Producers of these types of crops receive low payments for these materials, providing them with additional income opportunities and contributing to the economic development of semidesert regions [53].

2.6.4 ENVIRONMENTAL BENEFITS

Using bio-based materials derived from semidesert plants in food packaging can offer several environmental benefits. These plants typically require less water and resources for cultivation compared to traditional crops, making them more sustainable options [54]. Besides that, these types of plants are usually wasted or underutilized resources, which makes producers receive really low incomes and the local economy suffers. Several studies have been performed using basic food sources such as corn, potato, and sugarcane to produce bioplastics, which is a good alternative, but its disadvantages include competition with the food industry and potential impact on land use and price increasing; that is why, the use of underutilized desert plants would positively impact both on the economic as well as in the environmental aspects [55]. It is important not to forget that the main benefits of using bio-based plastics from semidesert plants include reduced carbon footprint, conservation of resources, promotion of a circular economy, and mitigation of plastic waste and pollution.

2.6.5 CHALLENGES AND CONSIDERATIONS

There are several challenges and considerations that need to be taken into account when using bio-based materials from semidesert plants in food packaging.

2.6.5.1 BARRIER PROPERTIES AND DURABILITY

One challenge is ensuring that bio-based packaging materials have sufficient barrier properties to protect the food from external factors such as moisture, oxygen, and light. Biomaterials derived from semidesert plants may have different barrier properties and durability compared to conventional packaging materials, derived either from synthetic materials or from fossil resources. Research and development efforts are needed to enhance the barrier properties and durability of these materials to warrant they meet the necessary food safety requirements and prolong the shelf life of the packaged food [56]. Talking about the moisture barrier, it can be stated that bio-based materials present a challenge in maintaining an effective moisture barrier, as they may have higher moisture absorption rates compared to conventional packaging materials. Regarding gas permeability, bio-based materials may have different permeability rates for gases such as oxygen and carbon dioxide, which can affect the shelf life of food products, not only in a negative way, but also with the potential for positive effects, such as allowing certain gases to escape and extend the freshness of certain products. Additionally, the durability of bio-based materials needs to be considered, as they should be able to withstand transportation and handling without compromising their integrity. In this sense, studies have shown promising results in improving the barrier properties and durability of bio-based packaging materials from semidesert plants through various methods such as blending with other polymers or incorporating additives; it is important that these incorporated products would also come from natural resources in order to maintain the sustainability and biodegradability of the packaging materials [57]. Figure 2.4A shows a fracture micrography, and Figure 2.4B and C shows a surface micrography (at 100×, 1000×, and 3000×, respectively, and at 15 kV), for a biocomposite enhanced with microcellulose from *lechuguilla* fibers and functionalized with oregano essential oil, which tortuous structure allowed oxygen retention. Such effect was improved with the presence of oregano essential oil, which is recognized as a potent antioxidant, allowing a longer preservation of the products contained in this type of biocomposites with regard to the oxidation process [58].

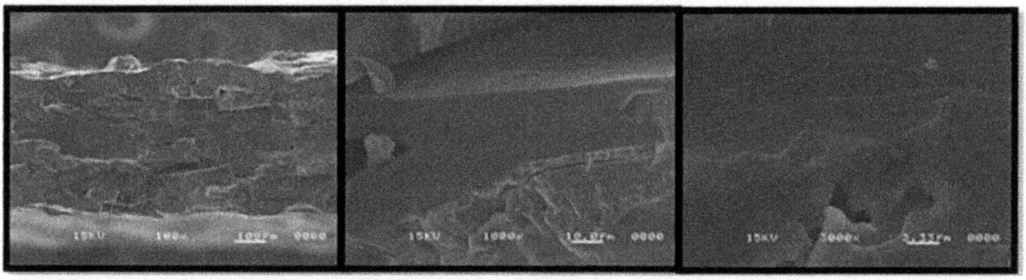

FIGURE 2.4 Scanning electron microscope (SEM) micrography. (A) Fracture 100×; (B) surface 1000×; (C) 3000× at 15 kV for a biocomposite enhanced with microcellulose from *lechuguilla* fibers and functionalized with oregano essential oil.

2.6.5.2 PRODUCTION COSTS AND SCALE-UP

Another challenge is the cost of producing bio-based materials from semi-desert plants on a large scale. The production costs of bio-based materials can be higher compared to conventional packaging materials, primarily due to the need for specialized equipment and processes. In addition, the availability and accessibility of semidesert plants as a raw material may also impact production costs, and the exploitation must be rational. However, with further research and advancements in technology, it is possible to optimize production processes and reduce costs. Furthermore, the scale-up of production is an important consideration, which requires strategic planning and investment in infrastructure and resources [59].

2.6.5.3 REGULATIONS AND FOOD SAFETY STANDARDS

There are regulatory considerations and food safety standards that must be met when using bio-based materials in food packaging [60]. These standards ensure that the materials are safe for contact with food and do not pose any health risks. Furthermore, there may be specific regulations regarding the labeling and marketing of bio-based packaging materials to ensure transparency and accurate information for consumers [61]. Also, the environmental impact of overexploiting plants should be carefully evaluated to prevent any negative consequences on biodiversity and ecosystem balance. The extraction and purification process to obtain the expected biomaterial from desert plants must be as eco-friendly as possible, according to the green chemical postulates [62].

2.6.5.4 CONSUMERS' EDUCATION AND AWARENESS

Another important factor is consumer education and awareness. Consumers need to be informed about the benefits and importance of using bio-based packaging materials from semidesert plants. This includes understanding the environmental advantages, such as reduced carbon footprint and conservation of finite resources. In addition to that, consumers need to be aware of proper disposal methods for bio-based packaging materials and the importance of recycling or composting them to reduce waste and promote a circular economy. Furthermore, educating consumers about the proper storage and handling of food products in bio-based packaging materials is crucial to maintain the quality and safety of the food product [63].

2.7 CASES OF STUDY AND APPLICATION EXAMPLES

Bio-based materials derived from semidesert plants offer a sustainable and eco-friendly alternative for developing different types of products generally used in the food industry. Some noteworthy cases and application examples are described below.

Packing such as trays, films, and containers for food products, these materials offer similar functionalities and durability to traditional plastics, including in some cases active functions to extend the shelf life of the food contained, rather than just being a passive package, while also being more sustainable and environmentally friendly. In this field, it can be mentioned as cactus (nopal) such as prickly pear cacti that have been utilized to create bioplastics with barrier properties suitable for food packaging products, in different presentations such as trays, cups, and films [64]. Another plant used is sisal that is often found in semidesert areas, and it yields fibers that can be processed into biodegradable films suitable for packaging fresh fruits and vegetables [65]. These films provide a natural barrier against moisture and help extend the shelf life. Continuing in the field of food preservation by preventing food dehydration, the use of jojoba wax, prickly pear seed oil, and mesquite seed gum adhesives has shown efficient results; for instance, jojoba plants are well adapted to semidesert climates and produce a natural liquid wax that has been used as a biodegradable coating for paper packaging, prickly pear seeds contain oil that can be extracted and processed into a natural coating material for paper-based food packaging, and ground mesquite pods have been used to create biodegradable wraps offering a natural and sustainable alternative to plastic cling films [66]. All these coatings provide a protective barrier, extend the shelf life of packaged foods, and reduce the reliance on synthetic coatings [67]. These wraps are compostable, breathable, and suitable for covering various food items. Not only humidity is an important barrier property, but the presence of other gases may also affect food's quality. It has been reported that *Acacia* gums, obtained from *Acacia* trees, produce gum Arabic, which can be used to create edible films; these biodegradable films are flexible, transparent, and can serve as a functional barrier [68]. Another action that is carried out to avoid gas transference when packing food is the proper sealing; studies using mesquite seed gum adhesives have been carried out. Mesquite seeds contain a natural gum that can be extracted and used as a bio-based adhesive in food packaging. This eco-friendly adhesive can be applied to laminated packaging materials, ensuring proper sealing and protective barrier properties. The resinous extract from creosote bushes, found in arid regions, can be processed into natural adhesives for food packaging. Creosote bush resins offer strong bonding properties and can be used in laminates, labels, and seals for sustainable packaging solutions [69].

Another important topic to be considered in food preservation is the spoilage caused due to microorganism action; some desert plants have evidenced antimicrobial properties such as Aloe vera plants, known for their gel-like substance, which can be processed into coatings for food packaging materials [70]. Aloe vera coatings have antimicrobial properties and can be applied to paper-based packaging to enhance product safety and shelf stability. Yucca plants contain saponins that can be extracted and incorporated into biodegradable plastics used for food packaging. Yucca extracts help improve the biodegradability of plastics and can also act as antimicrobial agents, enhancing food safety [71].

Another important application in food processing is maintaining or even improving food flavor; it exists several synthetic additives that have been used for this purpose, but nowadays consumers demand more natural products; so, this function can also be covered by biomaterials derived from desert plants; one of the most remarkable studies is the one based on gum Arabic encapsulation. Gum Arabic, derived from *Acacia* trees in semidesert regions, can be utilized for encapsulating flavors, colors, or nutrients in food packaging applications. This encapsulation technology helps protect sensitive food ingredients and enhances their flavor properties as well as their shelf stability [72].

Finally, but not least important, there is a complete market for disposable plates and cutlery, which indiscriminate usage has caused a severe impact on the environment. Different studies in this field have been carried out with the main challenge to overcome being that of obtaining similar mechanical properties to those obtained from fossil derivatives. Outstanding studies include *Agave* fiber packaging. *Agave* plants, commonly found in semidesert regions, provide strong and flexible fibers that can be used to create packaging materials such as biodegradable nets, food containers, and even biodegradable cutlery. Mesquite pod flour is obtained from mesquite trees, which thrive in semidesert regions [73]. This flour can be used to develop biodegradable food packaging materials such as bags, wraps, and containers, offering a sustainable alternative to traditional plastic packaging. Date palm trees are abundant in semidesert regions and provide strong fibers that can be used to create packaging materials such as trays, boxes, and pouches. These date palm fiber-based products are compostable and offer a sustainable option for food packaging [74].

2.8 CONCLUSION

Recapitulation of the benefits of bio-based materials from semidesert plants in food packaging is described below.

2.8.1 USING BIO-BASED MATERIALS DERIVED FROM SEMIDESERT PLANTS FOR FOOD PACKAGING OFFERS SEVERAL BENEFITS

1. Renewability: Semidesert plants are often hardy species that require minimal water and resources to grow, making them a sustainable source of bio-based materials. Unlike traditional petroleum-based plastics, which are derived from finite fossil

fuels, bio-based materials can be replenished through natural processes.

2. Biodegradability: Bio-based materials tend to be more biodegradable than conventional plastics, especially those derived from petroleum. This means that when disposed of properly, they break down into nontoxic components, reducing environmental pollution and waste accumulation.
3. Reduced carbon footprint: Production of bio-based materials typically generates fewer greenhouse gas emissions compared to petroleum-based plastics. Semidesert plants often grow in arid regions where land may not be suitable for agriculture, so utilizing them for bio-based materials can also help utilize land that might otherwise remain unused.
4. Customizability: Bio-based materials can be engineered to have specific properties suited for food packaging, such as moisture resistance, barrier properties, and strength. By selecting appropriate semidesert plant species and processing methods, it is possible to tailor these materials to meet the requirements of different food packaging applications.
5. Support for local economics: Cultivating semidesert plants for bio-based materials can create economic opportunities in regions where traditional agriculture may be challenging. This can help support local communities and diversify income sources.
6. Health and safety: Bio-based materials are generally considered safer for food packaging than certain conventional plastics that may contain harmful chemicals such as bisphenol A (BPA) or phthalates. Using plant-based materials reduces the risk of such chemicals leaching into food.
7. Marketing and consumer appeal: With increasing awareness and concern about environmental issues, using bio-based materials for food packaging can be a selling point for brands, appealing to environmentally conscious consumers who prefer sustainable products.

In conclusion, utilizing bio-based materials from semidesert plants for food packaging offers a range of benefits, including environmental sustainability, reduced carbon footprint, biodegradability, and support for local economies, while also providing customizable and safe packaging options for food products.

2.8.2 POTENTIAL FOR A MORE SUSTAINABLE AND ENVIRONMENTALLY FRIENDLY PACKING INDUSTRY

Undoubtedly, it is a major concern that the changes caused to the environment are due to the indiscriminate usage of food packing materials; that is why, consumers are actively seeking sustainable alternatives to traditional plastic packaging. Brands that adopt bio-based materials from semidesert plants can appeal to environmentally conscious consumers, enhancing their brand image and market competitiveness, along with governments and regulatory bodies that are increasingly implementing policies and regulations to encourage the use of sustainable materials and reduce plastic waste. Supportive policies, such as incentives for bio-based packaging and restrictions on single-use plastics, can further stimulate the growth of this industry. The raw material used to develop this growing industry is important; semidesert plants offer a plus, since these types of plants are often resilient to arid conditions, requiring minimal water and resources for growth. This makes them a sustainable source, whose cost is low and underestimated, so their application in this industry can promote local sourcing, benefiting communities in semiarid regions. This can create new economic opportunities for farmers and businesses in these areas, fostering rural development and resilience. Respectful engagement with these communities can not only provide valuable knowledge and expertise in plant cultivation and processing but also contribute to preserving cultural heritage and promoting social equity. So, continued research and innovation in bio-based materials science and engineering can drive advancements in the performance, cost-effectiveness, and scalability of semidesert plant-based packaging solutions. Plants will also be integrated into a circular economy model, where products are designed to be reused, recycled, or returned to the natural environment as nutrients. Through initiatives such as closed-loop recycling systems and composting programs, the life cycle of bio-based packaging materials can be optimized to minimize waste and maximize resource recovery. Collaboration between academia, industry, and government can accelerate progress in this field, focusing efforts on scaling up production. Overall, developing a sustainable and environmentally friendly industry based on bio-based materials from semidesert plants for food packaging has the potential to address pressing environmental challenges while fostering economic growth and innovation.

2.8.3 NEED TO CONTINUE RESEARCH AND INNOVATION IN THIS FIELD

As can be seen, there are multiple advantages of using bio-based materials, but there is still much to develop to comply with the challenges and disadvantages in the field.

Establishing a sustainable and efficient supply chain for semidesert plants can be complex, especially if they are sourced from remote or economically disadvantaged regions. Issues such as transportation logistics, storage facilities, and infrastructure development need to be addressed to ensure reliable and timely access to raw materials. Cultivation and land use are other topics since competition with other land uses, such as agriculture, forestry, or conservation efforts, may impact. Balancing the demand for land for different purposes while minimizing negative environmental and social impacts is essential; also, scaling up production to meet commercial demand requires significant investment in infrastructure, technology, and workforce capacity. Overcoming scale-up challenges while maintaining product quality and consistency is essential for market success. This topic is influenced by global market dynamics such as economic trends, trade policies, and consumers' preferences. Economic changes, including variability in commodity prices and transformations in consumption patterns, can influence the capacity of bio-based materials to emerge as sustainable alternatives. Consumers will always be looking for more in this sense; packaging materials must comply not only with a passive purpose, which is transportation, but also an active function needs to be comply at warranting that the food contained in it will be conserved in the best conditions; such property is based on that the packing product must have an adequate moisture resistance, which is crucial for maintaining the integrity of packaging materials, especially in humid or wet environments. It is also necessary to withstand various mechanical stresses during handling, transportation, and storage. Ensuring sufficient mechanical strength and durability is essential to prevent tearing, puncturing, or deformation that could compromise the packaged goods. Also, derived products should resist chemical degradation when exposed to substances commonly encountered in the packaging environment, such as oils, acids, or alkaline solutions. Ensuring chemical resistance is vital to maintaining the quality and safety of the packaged contents and preventing contamination, as well as ultraviolet (UV) resistance since exposure to UV radiation from sunlight can cause degradation and discoloration and temperature resistance to retain structural integrity and functional properties over a wide range of temperatures to prevent melting, warping, or brittleness under extreme heat or cold conditions.

It is also very important to ensure that derived products degrade at an appropriate rate without releasing harmful by-products or toxins, which can be challenging. Balancing biodegradability with other resistance properties is essential to maintain product performance while minimizing environmental impact, including the end-of-life management since bio-based packaging materials may be biodegradable or compostable under certain conditions, ensuring that proper end-of-life management remains a challenge. Establishing collection systems, composting facilities, and recycling infrastructure is necessary to maximize the environmental benefits of bio-based materials and prevent pollution.

KEYWORDS

- **bio-based materials**
- **microcellulose**
- **food packaging**
- **biodegradability**
- **synthetic plastics**

REFERENCES

1. Ncube, L.K., Ude, A.U., Ogunmuyiwa, E.N., Zulkifli, R., Beas, I.N. (2020). Environmental impact of food packaging materials: A review of contemporary development from conventional plastics to polylactic acid based materials. *Materials (Basel)*, 13, 4994. doi: 10.3390/ma13214994.
2. Guillard, V., Gaucel, S., Fornaciari, C., Angellier-Coussy, H., Buche, P., Gontard, N. (2018). The next generation of sustainable food packaging to preserve our environment in a circular economy context. *Frontiers in Nutrition*, 5, 121. doi: 10.3389/fnut.2018.00121.
3. Al-Tayyar, N.A., Youssef, A.M., Al-Hindi, R. (2020). Antimicrobial food packaging based on sustainable Bio-based materials for reducing foodborne pathogens: A review. *Food Chemistry*, 310, 125915. doi: 10.1016/j.foodchem.2019.125915.
4. Majid, I., Nayik, G.A., Dar, S. M., Nanda, V. (2018). Novel food packaging technologies: Innovations and future prospective. *Journal of the Saudi Society of Agricultural Sciences*, 17, 454–462. doi: 10.1016/j.jssas.2016.11.003.

5. Chamchong, M., Sangsom, V., Charoeamkitti, N. (2012). Computer-based on-line assessment of sterilizing value and heat distribution in retort for canning process. doi: 10.5772/34617.
6. Han, J.H., Lee, D.S., Min, S.C., Chung, M.S. (2012). Eco-design of food and beverage packaging. *Emerging Food Packaging Technologies*, 361–379. doi: 10.1533/9780857095664.4.361.
7. Peelman, N., Ragaert, P., Meulenaer, B.D., Adons, D., Peeters, R., Cardon, L., Impe, F.V., Devlieghere, F. (2013). Application of bioplastics for food packaging. *Trends in Food Science & Technology*, 32, 128–141. doi: 10.1016/j.tifs.2013.06.003.
8. Hernández-Centeno, F., Hernández-González, M.D.C., Tirado-Gallegos, J.M., Rodríguez-González, C., Rios-Velasco, C., Rodríguez-Hernández, A.M., Peña, H.Y.L.L., Zamudio-Flores, P.B. (2024). Biobased films from unconventionally sourced starch (*Cucurbita foetidissima* Kunth) and oregano essential oil (*Lippia berlandieri* Schauer): A look at their physicochemical properties. *MRS Advances*, 9, 523–530. doi: 10.1557/s43580-024-00824-1.
9. Hernández-González, M., Berúmen, C.M.P., Ruíz, H.S., Salazar, C.R., Paz, J., Olivas-Armendáriz, I., Martel-Estrada, S., González, C.A.R. (2017). Polysuccinimide functionalized with oregano's essential oil extracts, an antimicrobial extended release bio-material. *Materials Letters*, 191, 73–76. doi: 10.1016/j.matlet.2017.01.035.
10. González-López, H., Méndez-González, J., Peña, H.L.L., Hernández-Centeno, F., González-Uribe, D. (2019). Relaciones de producción en *Lippia graveolens* Kunth con variables ambientales al norte de zacatecas. *Árido-Ciencia*, 6, 13–17.
11. López, H.D.G., Manzo, S.V., Uribe, D.U.G. (2022). Tabla de rendimiento de biomasa para Euphorbia antisyphilitica al norte de Zacatecas. Revista Mexicana de Ciencias Agrícolas, 13, 443–453. doi: 10.29312/remexca.v13i3.2908.
12. López, H.D.G., Uribe, D.U.G. (2022). Estimación del rendimiento de biomasa y fibra de *Agave lechuguilla* Torr. al norte de Zacatecas. *Revista mexicana de ciencias forestales*, 14, 97–117. doi: 10.29298/rmcf.v14i75.1289.
13. Villavicencio-Gutiérrez, E.E., Quiroz, D.C., Hernández-Ramos, A., Martínez-Burciaga, O.U. (2021). Manejo forestal sustentable de los recursos no maderables en el semidesierto del norte de México. *Revista Mexicana de Ciencias Forestales*. doi: 10.29298/rmcf.v12iespecial-1.1083.
14. Aguilar-Sánchez, R., Munguía-Pérez, R., Reyes-Jurado, F., Navarro-Cruz, A.R., Cid-Pérez, T.S., Hernández-Carranza, P., Beristain-Bauza, S.D.C., Ochoa-Velasco, C., Avila-Sosa, R. (2019). Structural, physical, and antifungal characterization of starch edible films added with nanocomposites and Mexican oregano (*Lippia berlandieri* Schauer) Essential Oil. *Molecules*, 24, 2340. doi: 10.3390/molecules2412340.
15. Trejo-Calzada, R., Pedroza-Sandoval, A., Arreola-Ávila, J.G., García-González, F. (2020). Native plants to arid areas: A genetic reservoir for drought-tolerant crops. *IntechOpen eBooks*. doi: 10.5772/intechopen.86485.
16. Plant Adaptations. (2006, June 25). Retrieved, from: http://museum2.utep.edu/chih/NHCD/plantadaptations.html.
17. Abobatta, W.F. (2019). Drought adaptive mechanisms of plants—a review. *Advances in Agriculture and Environmental Science*, 2, 42–45. doi: 10.30881/aaeoa.00021.
18. Pando-Moreno, M., Pulido, R., Castillo, D., Jurado, E., Jiménez, J. (2008). Estimating fiber for lechuguilla (*Agave lechuguilla* Torr., Agavaceae), a traditional non-timber

forest product in Mexico. *Forest Ecology and Management*, 255, 3686–3690. doi: 10.1016/j.foreco.2008.02.053.
19. Xue, W., Li, X.Y., Lin, L.S., Wang, Y.J., Li, L. (2011). Effects of elevated temperature on photosynthesis in desert plant *Alhagi sparsifolia* S. *Photosynthetica*, 49, 435–447. doi: 10.1007/s11099-011-0054-7.
20. Godoy, F., Olivos-Hernández, K., Stange, C., Handford, M. (2021). Abiotic stress in crop species: Improving tolerance by applying plant metabolites. *Plants (Basel),* 10, 186. doi: 10.3390/plants10020186.
21. Soliman, S., Mohammad, M.G., El-Keblawy, A.A., Omar, H., Abouleish, M., Madkour, M., Elnaggar, A., Hosni, R.M. (2018). Mechanical and phytochemical protection mechanisms of *Calligonum comosum* in arid deserts. *PLoS One*, 13, e0192576. doi: 10.1371/journal.pone.0192576.
22. Reid, W.V., Mooney, H.A., Capistrano, D., Carpenter, S.R., Chopra, K., Cropper, A., Dasgupta, P., Hassan, R., Leemans, R., May, R.M., Pingali, P., Samper, C., Scholes, R., Watson, R.T., Zakri, A.H., Zhao, S. (2006). Nature: The many benefits of ecosystem services. *Nature*, 443, 27–28. doi: 10.1038/443749a.
23. Vilela, A., Bolković, M.L., Carmanchahi, P., Cony, M., de Lamo, D., Wassner, D. (2009). Past, present and potential uses of native flora and wildlife of the Monte Desert. *Journal of Arid Environments*, 73, 238–243. doi: 10.1016/j.jaridenv.2007.10.013.
24. Zhang, Y.M., Wu, H. Q., Bai, Y., Wang, H., Li, J.L., Li, Y.X. (2013). The Impact of biodiversity in sustainable agroecosystems. *Advanced Materials Research*, 4017–4020. doi: 10.4028/www.scientific.net/amr.726-731.4017.
25. Rivera-Hernández, J.E., Flores-Hernández, N., Vargas-Rueda, A.F., Salinas, G.A., Basáñez, M.D.J.C., García-Albarado, J.C. (2018). Flora y vegetación de la región semiárida de Acultzingo-Maltrata, Veracruz, México. *Acta Botánica Mexicana.* doi: 10.21829/abm126.2019.1433.
26. Palma-Tenango, M., Miguel-Chávez, R.S., Soto-Hernández, R. (2017). Aromatic and medicinal plants in Mexico. *InTech eBooks*. doi: 10.5772/66507.
27. Arellano-Perusquia, A., Bañuelos-Hernández, B., Luz, U., Froylán, A., Agustín, A.C., Antonio, G. M., Junquera-Martínez, S., Cruz-Hernández, A. (2019). Mexican indigenous species with agroecological Uses. *IntechOpen eBooks*. doi: 10.5772/intechopen.83400.
28. Blanckaert, I., Vancraeynest, K., Swennen, R., Espinosa-García, F.J., Piñero, D., Lira-Saade, R. (2007). Non-crop resources and the role of indigenous knowledge in semi-arid production of Mexico. *Agriculture Ecosystems & Environment*, 119, 39–48. doi: 10.1016/j.agee.2006.06.015.
29. Calamaco, Z.G., Montfort, G.R., Marszalek, J.E., González, G.A. (2023). Revisión sobre el orégano mexicano Lippia graveolens HBK. (Sinonimia Lippia berlandieri Schauer) y su aceite esencial. *Investigación y Desarrollo en Ciencia y Tecnología de Alimentos*, 8, 861–871. doi: 10.29105/idcyta.v8i1.109
30. Rojas, R., Tafolla-Arellano, J.C., Martínez-Ávila, G.C.G. (2021). *Euphorbia antisyphilitica* Zucc: A source of phytochemicals with potential applications in industry. *Plants*, 10, 8. doi: 10.3390/plants10010008.
31. Tonoli, G.H.D., Teixeira, E.M., Corrêa, A.C., Marconcini, J.M., Caixêta, L.A., Pereira-da-Silva, M.A., Mattoso, L.H. (2012). Cellulose micro/nanofibres from Eucalyptus kraft pulp: Preparation and properties. *Carbohydrate Polymers*, 89, 80–88. doi: 10.1016/j.carbpol.2012.02.052.

32. Elanthikkal, S., Gopalakrishnapanicker, U., Varghese, S., Guthrie, J.T. (2010). Cellulose microfibres produced from banana plant wastes: Isolation and characterization. *Carbohydrate Polymers*, 80, 852–859. doi: 10.1016/j.carbpol.2009.12.043.
33. Amaya Vergara, M.C., Cortés Gómez, M.P., Restrepo, M.C., Manrique Henao, J., Pereira Soto, M.A., Gañán Rojo, P.F., Castro Herazo, C.I., Zuluaga Gallego, R. (2018). Novel biobased textile fiber from Colombian agro-industrial waste fiber. *Molecules,* 23, 2640. doi: 10.3390/molecules23102640.
34. Hernández-Centeno, F., López-De la Peña, H.Y., Hernández-González, M., Rodríguez-González, C.A., Tirado-Gallegos, J.M., Rios-Velasco, C., Zamudio-Flores, P.B. (2020). Physicochemical, thermal, rheological and morphological characteristics of flour and starch from a non-conventional source: *Cucurbita foetidissima* Kunth roots. *Journal of Food Measurement and Characterization*, 14, 1976–1985. doi: 10.1007/s11694-020-00444-x.
35. Hernández-Centeno, F., López-De La Peña, H.Y., Rojas, R. (2018). Candelilla (*Euphorbia antisiphylitica* Zucc): Geographical distribution climate and edaphology. In: *Food Process Engineering and Quality Assurance* (pp. 473–492). Apple Academic Press.
36. Salehi, F. (2019). Characterization of new biodegradable edible films and coatings based on seeds gum: A review. *Journal of Packaging Technology and Research*, 3, 193–201. doi: 10.1007/s41783-019-00061-0.
37. Colín-Chávez, C., Valdez, H.S., Féliz, A.R., Peralta, E., Corona, A.R.S., Corona, M.G., Turrado-Saucedo, J. (2021). Papermaking as potential use of fibers from Mexican *Opuntia ficus*-indica waste. *Biotecnia*, 23, 141–150. doi: 10.18633/biotecnia.v23i1.1315.
38. Gante, C.R.D., Rocha, J.A., Collins, C.P.S. (2019). Using native plants of the Northeast of Mexico for developing active antimicrobial food packaging films. *IntechOpen eBooks*. doi: 10.5772/intechopen.80779.
39. Rosenboom, J.G., Langer, R., Traverso, G. (2022). Bioplastics for a circular economy. *Nature Reviews Materials*, 7, 117–137. doi: 10.1038/s41578-021-00407-8.
40. Silva, F.A.G.S., Dourado, F., Gama, M., Poças, F. (2020). Nanocellulose bio-based composites for food packaging. *Nanomaterials (Basel),* 10, 2041. doi: 10.3390/nano10102041.
41. Ortíz, S.P. (2023). Are bioplastics the solution to the plastic pollution problem?. *PLoS Biology,* 21, e3002045. doi: 10.1371/journal.pbio.3002045
42. Perera, K.Y., Jaiswal, A.K., Jaiswal, S. (2023, June 20). Biopolymer-based sustainable food packaging materials: Challenges, solutions, and applications. https://www.mdpi.com/2304-8158/12/12/2422/pdf?version=1687255268.
43. Sandoval, D.C.G., Sosa, B.L., Martínez-Ávila, G.C.G., Fuentes, H.R., Abarca, V.H.A., Rojas, R. (2019). Formulation and characterization of edible films based on organic mucilage from Mexican Opuntia ficus-indica. *Coatings*, 9, 506. doi: 10.3390/coatings9080506.
44. Dick, M., Pagno, C.H., Costa, T.M.H., Gomaa, A., Subirade, M., Rios, A.D.O., Flôres, S.H. (2015). Edible films based on chia flour: Development and characterization. *Journal of Applied Polymer Science*, 133. doi: 10.1002/app.42455.
45. Cuenca, P.S., Ramallo, L.A., Albani, O.A. (2020). Películas y materiales compuestos basados en almidón de mandioca acetilado de alto grado de sustitución. *Revista de Ciencia y Tecnología*, 33, 85–93. doi: 10.36995/j.recyt.2020.33.011.

46. Cazón, P., Velázquez, G., Ramírez, J.A., Vázquez, M. (2017). Polysaccharide-based films and coatings for food packaging: A review. *Food Hydrocolloids*, 68, 136–148. doi: 10.1016/j.foodhyd.2016.09.009
47. León-Zapata, M.Á.D., Sáenz-Galindo, A., Rojas-Molina, R., Rodríguez-Herrera, R., Jasso-Cantú, D., Aguilar, C.N. (2015). Edible candelilla wax coating with fermented extract of tarbush improves the shelf life and quality of apples. *Food Packaging and Shelf life*, 3, 70–75. doi: 10.1016/j.fpsl.2015.01.001.
48. Ruiz-Martínez, J., Aguirre-Joya, J.A., Rojas, R., Vicente, A., Aguilar-González, M.Á., Rodríguez-Herrera, R., Álvarez-Pérez, O.B., Torres-León, C., Aguilar, C.N. (2020). Candelilla wax edible coating with *Flourensia cernua* bioactives to prolong the quality of tomato fruits. *Foods*, 9, 1303. doi: 10.3390/foods9091303.
49. Shershneva, E.G. (2022). Biodegradable food packaging: benefits and adverse effects. *IOP Conference Series: Earth and Environmental Science*, 988, 022006. doi: 10.1088/1755-1315/988/2/022006.
50. Oliveira, A.S.B.D., Melo, N.R.D. (2019). Market and sustainability of food packaging: A review. *Boletim do Centro de Pesquisa de Processamento de Alimentos*, 36. doi: 10.5380/bceppa.v36i1.57846.
51. Pitombo, L.M., Ramos, J.C., Quevedo, H.D., Carmo, K.P.D., Paiva, J.M.F.D., Pereira, E.A., Carmo, J.B.D. (2018, January 1). Methodology for soil respirometric assays: Step by step and guidelines to measure fluxes of trace gases using microcosms. https://pubmed.ncbi.nlm.nih.gov/30003050.
52. Johnston, J., Lim, E., Roh, H. (2019). Impact of upstream oil extraction and environmental public health: A review of the evidence. *Science of the Total Environment*, 657, 187–199. doi: 10.1016/j.scitotenv.2018.11.483.
53. Dahiya, S., Katakojwala, R., Ramakrishna, S., Mohan, S.V. (2020). Biobased products and life cycle assessment in the context of circular economy and sustainability. *Materials Circular Economy*, 2. doi: 10.1007/s42824-020-00007-x.
54. Siracusa, V., Rocculi, P., Romani, S., Rosa, M.D. (2008). Biodegradable polymers for food packaging: A review. *Trends in Food Science & Technology*, 19, 634–643. doi: 10.1016/j.tifs.2008.07.003.
55. Nguyen, T.K., That, N.T.T., Nguyen, N.T., Nguyễn, H.T.L. (2022). Development of starch-based bioplastic from jackfruit seed. *Advances in Polymer Technology*, 2022, 1–9. doi: 10.1155/2022/6547461.
56. Jacob, J., Lawal, U., Thomas, S., Valapa, R.B. (2020). Biobased polymer composite from poly(lactic acid): Processing, fabrication, and characterization for food packaging. *Elsevier eBooks*, 97–115. doi: 10.1016/b978-0-12-818795-1.00004-6.
57. Jabeen, N., Majid, I., Nayik, G.A. (2015). Bioplastics and food packaging: A review. *Cogent Food & Agriculture*, 1, 1117749. doi: 10.1080/23311932.2015.1117749.
58. Orozco Plancarte, M.D., Hernández González, M., López De la Peña, H.Y., Hernández Centeno, F. Evaluación de las propiedades antioxidantes, mecánicas y de barrera de un material biobasado (polisuccinimida-almidón) funcionalizado con extractos de aceite esencial de orégano (Lippia berlandieri Schauer).
59. Mirhashemi, M.S., Mohseni, S., Hasanzadeh, M., Pishvaee, M.S. (2018). Moringa oleifera biomass-to-biodiesel supply chain design: An opportunity to combat desertification in Iran. *Journal of Cleaner Production*, 203, 313–327. doi: 10.1016/j.jclepro.2018.08.257.

60. Petersen, K., Nielsen, P.V., Bertelsen, G., Lawther, M., Olsen, M.B., Nilsson, N.H., Mortensen, G. (1999). Potential of biobased materials for food packaging. *Trends in Food Science & Technology*, 10, 52–68. doi: 10.1016/s0924-2244(99)00019-9.
61. Misko, G.G. (2020, February 21). Biobased plastics and the sustainability puzzle. Retrieved from https://www.packaginglaw.com/special-focus/biobased-plastics-and-sustainability-puzzle#:~:text=%E2%80%9CBiobased%E2%80%9D%20means%20related%20to%20or,with%20the%20aid%20of%20microorganisms.
62. Balandrin, M.F., Klocke, J.A., Wurtele, E.S., Bollinger, W.H. (2023, September 29). Natural plant chemicals: Sources of industrial and medicinal materials: https://www.science.org/doi/10.1126/science.3890182?cookieSet=1.
63. Shogren, R., Wood, D., Orts, W., Glenn, G. (2019). Plant-based materials and transitioning to a circular economy. *Sustainable Production and Consumption*, 19, 194–215.
64. Kudanga, T., Aruwa, C.E. (2021). Industrial Applications of *Opuntia* spp.(Nopal, Fruit and Peel). *Opuntia spp.: Chemistry, Bioactivity and Industrial Applications* (pp. 841–875).
65. Castro, S., Matos, F.R., Mercuri, L.P., Santos, A.V. (2023). Synthesis and evaluation of the incorporation of sisal fiber cellulose in the polymeric matrix of starch from jackfruit seed (*Artocarpus heterophyllus* Lam.) using thermogravimetry. *Journal of Thermal Analysis and Calorimetry*, 148, 97–105.
66. Aguirre-Joya, J.A., Álvarez, B., Ventura, J.M., García-Galindo, J.O., De León-Zapata, M. A., Rojas, R., Saucedo, S., Aguilar, C.N. (2016). Edible coatings and films from lipids, waxes, and resins. *Edible Food Packaging: Materials and Processing Technologies*, 36, 121–152.
67. Fei, T., Wang, T. (2017). A review of recent development of sustainable waxes derived from vegetable oils. *Current Opinion in Food Science*, 16, 7–14.
68. Khezerlou, A., Zolfaghari, H., Banihashemi, S.A., Forghani, S., Ehsani, A. (2021). Plant gums as the functional compounds for edible films and coatings in the food industry: A review. *Polymers for Advanced Technologies*, 32, 2306–2326.
69. Uzodinma, E.O., Mbaeyi-Nwaoha, I.E., Onwurafor, E.U. (2020). Suitability of bacterial fermentation and foil packaging of condiment from African mesquite (*Prosopis africana*) seeds for nutritional retention and commercialization. *African Journal of Microbiology Research*, 14, 340–348.
70. Kumar, S., Kalita, S., Das, A., Kumar, P., Singh, S., Katiyar, V., Mukherjee, A. (2022). Aloe vera: A contemporary overview on scope and prospects in food preservation and packaging. *Progress in Organic Coatings*, 166, 106799.
71. Gutiérrez-García, G.J., Quintana-Romero, L.A., Morales-Figueroa, G.G., Esparza-Romero, J., Pérez-Morales, R., López-Mata, M.A., Juárez, J., Sánchez-Escalante, J.J., Peralta, E., Quihui-Cota, L., Soto-Valdez, H. (2021). Effect of *Yucca baccata* butanolic extract on the shelf life of chicken and development of an antimicrobial packaging for beef. *Food Control*, 127, 108142.
72. Al-Hamayda, A., Abu-Jdayil, B., Ayyash, M., Tannous, J. (2023). Advances in microencapsulation techniques using Arabic gum: A comprehensive review. *Industrial Crops and Products*, 205, 117556.

73. Dholakia, P., Sabharwal, P. K., Vandana. (2022). Seed gums: Sources, applications, and recent trends in edible films. In: (Eds.), *Edible Food Packaging: Applications, Innovations and Sustainability* (pp. 263–276). Springer Nature: Singapore.
74. Amroune, S., Bezazi, A., Dufresne, A., Scarpa, F., Imad, A. (2021). Investigation of the date palm fiber for green composites reinforcement: Thermo-physical and mechanical properties of the fiber. *Journal of Natural Fibers*, 18, 717–734.

CHAPTER 3

New Trends in Edible Coating-Based Starch: A Sustainable Alternative for Postharvest Preservation of Fruits and Vegetables

GISELA MARTÍNEZ-MENDOZA,[1] NDAHITA DE DIOS-AVILA,[1] JUAN MANUEL TIRADO-GALLEGOS,[2] VERA TORRES-ARMENDARIZ,[1] NORA AIDEÉ SALAS-SALAZAR,[1] CLAUDIO RÍOS-VELASCO,[3] and MARÍA JANETH RODRÍGUEZ-ROQUE[1*]

[1]*Faculty of Agrotechnological Sciences, Autonomous University of Chihuahua, Chihuahua, Mexico*

[2]*Faculty of Zootechny and Ecology, Autonomous University of Chihuahua, Chihuahua, Mexico*

[3]*Centro de Investigación en Alimentación y Desarrollo—CIAD, AC, Unidad Cuauhtémoc, Chihuahua, México*

*Corresponding author

ABSTRACT

Over the past few years, edible films and coatings have been evaluated to preserve the quality and extend the shelf life of fresh fruits and vegetables. In general, applying edible packages over fresh products during the post-harvest storage allows maintaining or increasing the quality by delaying the physiological changes, microbial spoilage, and oxidative reactions. Biopolymers are considered a sustainable alternative to petroleum-derived plastics, as they are renewable and degradable. Starch is a biopolymer

Sustainable Food Packaging Materials and Technology. Juan Manuel Tirado-Gallegos, Juan Daniel Rivaldi Chávez, and José Juan Buenrostro-Figueroa (Eds)

DOI: 10.1201/9781779641434-3

frequently used in coating formulation due to its availability, biodegradability, low cost, and high filmogenic capacity. Films and coatings-based starch regulate the gas exchange/humidity, control the senescence, and extend the shelf life of fruits and vegetables. However, the functionality can be enhanced by using active packaging technologies, for instance, by incorporating biological compounds that modify the interactions among the packaging-product-external space. Plant extracts and their derivatives contain many bioactive compounds with important antimicrobial and antioxidant properties. These extracts have been evaluated in the formulation of coatings; their addition prevents the deterioration of fresh products, spoilage of microorganisms, and prolongs their shelf life. Considering the new trends for the postharvest conservation of horticultural products, this chapter summarizes the most recent research on the potential use of bioactive compounds from plants in the formulation of edible films and coatings for the preservation of physicochemical, organoleptic, and nutritional properties of fresh products.

3.1 INTRODUCTION

Fruits and vegetables are alive even after harvest and change their quality in response to storage conditions as they are metabolically active. During postharvest, the respiration, ethylene production, and carbon dioxide (CO_2) generation are constant, promoting physicochemical changes such as the weight loss, degradation of organic compounds (proteins, lipids, and carbohydrates), leading to the deterioration of the quality of fresh produce [1, 2]. For this reason, the food industry is constantly searching for methods to reduce postharvest losses and increase the shelf life of these perishable commodities. In this sense, the use of refrigeration and controlled or modified atmospheres, the thermal and nonthermal treatments, the use of preservatives, the genetic modification, and the packaging are effective strategies for their preservation [3].

Overall, food packaging is considered a system that protects food from microbial, chemical, and environmental contaminants, ensuring the quality, safety, and, therefore, prolonging its shelf life [4, 5]. Saklani et al. [6] indicate that the main advantage of packaging is the easier transportation of products, the safety, and maintenance of their nutritional quality until consumption.

In the past, petroleum-based plastics were considered the main materials used for fruit and vegetable packaging. However, their use has important limitations; for instance, the expensive material cost, the high-energy consumption, and the environmental pollution [6, 7]. In response to this problem and to the growing environmental awareness, the search for sustainable alternatives has been driven. Considerable progress has been made in the development of edible packaging and coatings as an environmentally friendly alternative to extending the shelf life of fresh produce [6, 8–10].

Among the most widely used materials for forming edible coatings (ECs), polysaccharides (i.e., chitosan, alginate, pectin, cellulose, and starch) have been extensively investigated for their potential in plant food preservation. Starch is an outstanding option due to its remarkable properties, biodegradability, and biocompatibility. In addition, it does not present toxicity to living organisms [11]. The biocompatibility of starch facilitates the incorporation of bioactive compounds to develop specific properties [12]; these substances can be added as pure compounds, essential oils, or plant extracts [13].

This chapter addressed the most recent advances and new trends in ECs as a promising and sustainable alternative for the postharvest preservation of fruits and vegetables. It is mainly focused on polysaccharide-based coatings, such as starch, the application techniques for fruits and vegetables, the benefits of the addition of bioactive compounds, and a wide range of examples where they were applied.

3.2 ECS FOR THE POSTHARVEST PRESERVATION OF FRUITS AND VEGETABLES

The edible coatings are considered a simple and safe postharvest technique to maintain the quality of fresh produce such as fruits and vegetables [14]. Generally, the coating is a thin layer of edible material directly applied to the product surface that provides protection against adverse environmental conditions by acting as a physical barrier [5, 7]. Among the advantages of using ECs, it is essential to highlight that their application can extend the shelf life of fresh fruits and vegetables by reducing the respiration rate, maintaining firmness, providing gloss, preventing weight loss, protecting light-sensitive and bioactive compounds (e.g., soluble solids, vitamins, anthocyanins, phenolic compounds, among others), reducing bacterial growth and maintaining organoleptic properties [13] (Figure 3.1).

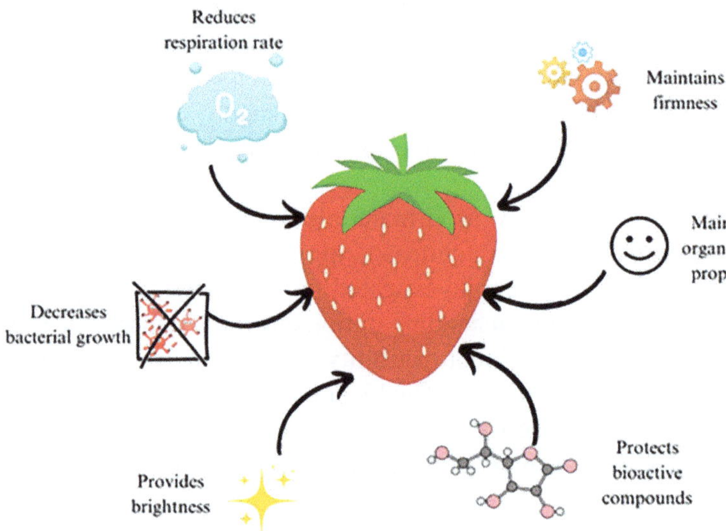

FIGURE 3.1 Main functions of edible coatings when applied in the postharvest of fruit and vegetable products.

Various packaging materials from alternative sources have been recently studied as alternatives to those derived from petroleum, as they have economic viability and are obtained from renewable sources [15]. Table 3.1 shows the main biomolecules used to formulate ECs and briefly describes their characteristics.

In a recent study, Dadzie et al. [16] evaluated the effect of a beeswax (lipid) coating on the quality of eggplant (*Solanum melongena* L.) fruits. They showed that this coating minimized changes in weight and firmness and delayed color changes during cold storage of eggplant. On the other hand, Eshetu et al. [17] formulated coatings from beeswax in combination with chitosan (polysaccharide) to maintain the quality of mango (*Mangifera indica* L.) fruits. Their results showed that the coating made with a combination of beeswax and chitosan effectively delayed fruit ripening, maintained fruit quality for approximately 21 days, and that the chitosan reduced the incidence of diseases and showed antimicrobial activity.

TABLE 3.1 Characteristics and Biomaterials Used in Edible Coating Formation to Preserve Fruits and Vegetables During the Postharvest

Biomaterials	Characteristics	Examples
Lipids	Capacity to form continuous coatings Good barrier properties against moisture loss Improves appearance and polishes surfaces High hydrophobicity	Carnauba wax Candelilla wax Beeswax
Proteins	Capacity to form continuous coatings Excellent barrier to O_2 Reduces loss of aromatic compounds and oils Improves appearance Hydrophilic nature	Soy proteins Whey milk Wheat gluten Corn zein
Polysaccharides	Capacity to form continuous coatings Efficient barrier to O_2 Colorless appearance High compatibility with other substances, such as plasticizers and bioactive additives Grease free Hydrophilic nature Significantly reduces dehydration, oxidative rancidity, and surface browning	Chitosan Alginate Pectin Cellulose and derivatives Starch and derivatives

Source: Adapted with permission from Al-Tayyar et al. [2], Kumar et al. [7], and Perez et al. [5].

Regarding the development of protein-based coatings for fresh produce preservation, Dehghani et al. [18] proposed the use of fish gelatin (protein) in combination with bitter almond gum (polysaccharide) as suitable materials to prolong the shelf life of tomato (*Solanum lycopersicum* L.) for up to 28 days of storage at 20°C. They reported that this formulation effectively preserves the nutritional quality and sensory attributes of tomato by retarding changes in firmness, lycopene content, and titratable acidity.

Although it is possible to obtain ECs from different materials and features for the postharvest preservation of fruits and vegetables, polysaccharide-based materials (i.e., chitosan, alginate, pectin, cellulose, starch, and their derivatives) have been the most explored sources. One of the most outstanding characteristics of these types of biopolymers is their high natural availability; additionally, they have excellent film-forming properties and high compatibility with other substances such as antioxidant compounds, natural preservatives, antimicrobial agents, among others. A detailed information on this biopolymer type is provided in the next section of this chapter book.

3.3 COATINGS BASED ON POLYSACCHARIDES

From a sustainability point of view, biodegradable polysaccharides are among the most abundant biopolymer in nature, as they are derived from biomass, which represents an inexhaustible source of raw material compared to petroleum derivatives [19]. They are formed by long polymer chains linked by glycosidic bonds resulting in a hydrophilic structure that is one of the most important features to take into consideration in the formulation of ECs [14].

As previously described in Table 3.1, coating based on polysaccharides also possesses important features standing up to their capability to form continuous coatings, they have a colorless appearance and are grease-free, they reduce dehydration, oxidative rancidity, and surface browning. Moreover, they are considered a good barrier against gases, aromas and oxygen compared to coatings formulated with lipids and proteins [20], extending the shelf life of fresh produce.

Another important attribute of these types of biopolymers is their compatibility with other substances, which broadens their industrial potential. For instance, they can be used as carriers of active ingredients, such as antidarkening, antioxidants, antimicrobial, pigments, nutrients, flavorings, among other bioactive substances [15], conferring functional properties of the coatings.

On the other hand, among the most reported natural polysaccharides in the formulation of ECs are pectin, chitosan, cellulose, and starch and their derivatives (Table 3.2) [21–25]. According to Salgado et al. [26] chitosan-based ECs have been shown to maintain the freshness of fruits and vegetables for a longer time. They are also considered an effective method for replacing synthetic fungicides during postharvest stages of fruits and vegetables due to their antimicrobial properties. Among cellulose derivatives, carboxymethyl cellulose has been evaluated in the formulation of ECs. It is an efficient alternative to prolong the shelf life of perishable products by delaying the ripening process, especially in nonclimacteric fruits. However, the carboxymethyl cellulose use is limited due to its high production cost [5]. Regarding pectin, Lazaridou and Biliaderis [27] pointed out that this polymer can produce ECs due to its gelling and emulsifying capacity. The pectin application in fresh produce for preservation purposes are attributed to its permeability to gases (O_2, CO_2, and volatile compounds) and oils, the reduction of oxidative rancidity, and inhibition of microbial growth.

Starch is the most widely used polysaccharide in the formulation of ECs due to its easy processing, low cost, abundance, biodegradability, null toxicity, and its ability to form a continuous matrix [19]. Although all the biopolymers mentioned above maintain the freshness of coated fruits and vegetables, starch-based coatings can significantly reduce weight loss, maintain firmness/color, reduce nutrient loss (vitamins, bioactive compounds, etc.), reduce respiration rate, and inhibit the activity of some enzymes [20]. All these features make starch-based coatings with unique physical, chemical, and functional characteristics.

3.4 COATING BASED ON STARCH

Starch is synthesized by plants as an energy reserve and is composed of amylose and amylopectin, which have different structures and functionalities [29]. Starch is the most widely used biopolymer among the polysaccharides for producing edible films and coatings. It is considered one of the most promising and sustainable polymers of natural origin for producing environmentally friendly packaging materials applied to food, especially fresh produce such as fruits and vegetables [19]. Additionally, starch is easily processed with a low cost, it is also abundant, biodegradable, and can form continuous matrices.

This biopolymer is commercially obtained from various sources, such as corn, potato, wheat, and cassava; however, starch can be gotten from fruit by-products, such as avocado (*Persea americana* Mill.), mango (*M. indica* L.), jackfruit (*Artocarpus heterophyllus* Lam.), and tamarind (*Tamarindus indica* L.) seeds, among others [10, 30].

According to Pedreiro et al. [4], starch-based edible films and coatings are characterized by being tasteless and colorless; their compact structure and low solubility contribute to improve their barrier properties against oxygen. The latter feature has a favorable impact on the preservation of

TABLE 3.2 The Main Advantages and Disadvantages of Polysaccharide-Based Coatings

Polysaccharides	Characteristics		References
	Advantages	Disadvantages	
Pectin	Nontoxic, biodegradable, and biocompatible Permeability properties Barrier to gases (O_2, CO_2, and ethylene) Hydrophilic nature tends to reduce lipid absorption Antimicrobial properties	Low barrier against moisture Multilayer coatings and emulsions are necessary	Lazaridou and Biliaderis [27]
Carboxymethyl cellulose (cellulose derivative)	Obtained from the carboxylation of cellulose Forms colorless coatings Fat-free Water soluble Barrier against gases (O_2, CO_2) and aromas Highly adhesive (coating/fruit surface)	Low moisture barrier High production cost	Abdul Khalil et al. [28]
Chitosan	Obtained from the deacetylation of chitin Nontoxic, biodegradable, and biocompatible Form continuous coatings Antimicrobial, antioxidant, and antifungal activity Acts as a gas barrier	Low mechanical resistance Low solubility neutral and alkaline pH	Salgado et al. [26]
Starch	Renewable, biodegradable, and low-cost production Continuous gel-forming capability Transparent, odorless, and tasteless Excellent gas barrier Creates a controlled atmosphere It can be used in its native or modified forms High mechanical strength Compatible with a wide variety of compounds	Low solubility in cold water Hydrophobic nature Low water vapor barrier capacity	Oyom et al. [14]

fresh food. Although these coating types present some limitations in terms of their mechanical properties and water vapor permeability, they can be overcome with the application of additives such as plasticizers. It has been reported that plasticizers provide flexibility and a barrier against gases, improving the mechanical properties of starch-based coatings [7]. In this sense, Sapper and Chiralt [31] showed that the permeability to water vapor and gases, migration of compounds, physical/mechanical protection capacity, and impact on product appearance (maintaining color and gloss) are the most important functional properties for the starch application as ECs on fruits and vegetables. Table 3.3 briefly describes the requirements of a starch-based coating.

Nawab et al. [32] used coatings based on mango seed starch (2%) and various plasticizers, such as sorbitol, glycerol, and a combination of both, to extend the shelf life of tomatoes. Their results indicate that using the starch-based coatings delayed the ripening process for up to 20 days by significantly reducing weight loss, nutritional changes (soluble solids content, titratable acidity, ascorbic acid content), and firmness. On the other hand, they indicated that using sorbitol as a plasticizer improved surface property, maintaining the quality of the climacteric fruit. In a second study, Nawab et al. [22] evaluated this same formulation in developing heat-sealable biofilms for preserving chili powder compared to a polyethylene-based film (commercial). The most efficient formulation for preserving dry products was mango seed starch/glycerol, which preserved color and pungency for six months. The authors reported that these results could be attributed to the biobags' low oxygen permeability.

Regarding the application of starch-based coatings on nonclimacteric fruits, Nogueira et al. [33] evaluated the use of arrowroot starch (*Maranta arundinacea* L.) at different concentrations (1–5%) in combination with glycerol as a plasticizer. Although the formulations were able to form homogeneous, transparent, and flexible films, only those obtained at a concentration of 0%, 2% and 4% starch were evaluated as potential ECs of plum (*Prunus domestica* L.) during postharvest storage at 25 and 5°C for 35 days. The results indicated that plums coated with 2% starch gave the fruit a higher gloss without showing changes in the texture. In addition, it is essential to note that storage temperature significantly influences the shelf life of this fruit since the fruit stored at 25°C had a shelf life of only seven days compared to those stored at 5°C (35 days). This behavior was also attributed to the respiration rate, which was affected by storage temperature.

TABLE 3.3 Requirements for Starch-Based Coating Forming Agents for Preserving Fruits and Vegetables in the Postharvest

Requirement Type	Property	Impact on Coating
Surface	↑ Wettability	Enhances the ability of coatings to wet, adhere, and spread on the food surface
	↑ Adhesion and extensibility	Improves the coating functions while enhancing the product appearance and attractiveness
Tensile	↑ Mechanical resistance	Prevents coating fracture and protects against physical/mechanical damage during storage
Mass transfer	↓ Water vapor permeability	Prevents moisture loss or gain
		Delays changes in product texture, flavor, and appearance
	↓ O_2 permeability	Reduces the respiration rate and the production of ethylene
		Delays the senescence
	≠ Moderate CO_2 permeability	Prevents anaerobic respiration and avoids unwanted metabolic changes

Source: Adapted with permission from Sapper and Chiralt [31].

However, after storage (at 5°C) of coated plums, there was an increase in pH, soluble solids, and anthocyanin content but decreased acidity.

3.5 APPLICATION METHODS OF EDIBLE COATING ON FRUITS AND VEGETABLES

Edible coatings are directly applied on the food surface, forming thin layers that are considered an integral part of the final product [9]. A significant feature of ECs is their capability to improve the appearance, functional, and preservation of food. The application technique must ensure the desired beneficial effect on the coated area of the food [1]. For this reason, there are different ways of applying biopolymers onto fruits and vegetables, but it depends on the physicochemical, rheological, and thickness properties of the coating [4].

Among the most used techniques for coating fruits and vegetables are dipping, spraying, electrospraying, and brushing. In these procedures, the filmogenic solution is placed on the surface of the food, and later, the adhesion of EC to the product takes place [7] (Figure 3.2).

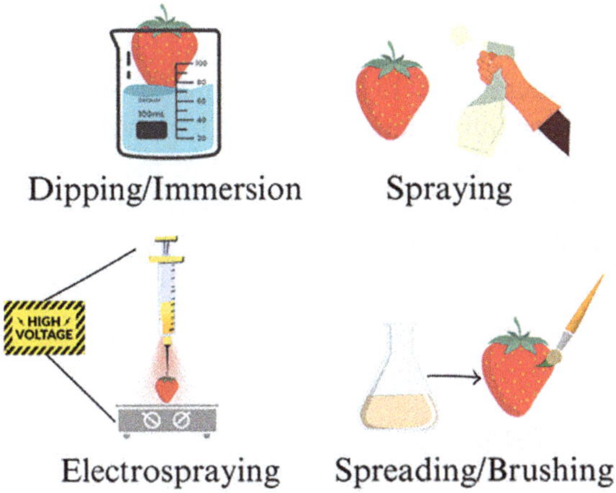

FIGURE 3.2 Scheme of application methods for edible coatings

3.5.1 DIPPING/IMMERSION

Dipping is the most widely used method for coating fruits and vegetables because of its ease application, low cost, and capability to achieve adequate coverage on irregular surfaces [34]. This technique consists of the complete immersion of the food in the filmogenic solution for a determined time, which depends on the characteristics of the solution as well as the food to be coated (i.e., 30 s to 5 min). Subsequently, food is removed from the solution to eliminate the excess, and it is subjected to the drying process, which can be done with specialized equipment or by leaving it at room temperature. The result is the formation of a thin layer onto the product surface [35, 36].

Key factors affecting the effectiveness of dipping/immersion coating application include the type of solvent used, the temperature and viscosity of the filmogenic solution, speed, duration of immersion, frequency, and extraction [1]. However, this method has certain disadvantages, such as the dilution of the coating solution, the possible growth of microorganisms in the dip tank, and the accumulation of residual materials [37].

3.5.2 SPRAYING

Coating by spraying is a conventional method used mainly in the food industry. It consists of atomizing small particles on the surface of the food using an atomizer or spray nozzle to apply liquids [35]. These sprayed particles adhere homogeneously to the product surface, forming a semipermeable membrane [5]. The advantages of this technique are obtaining a homogeneous coating with greater thickness control, it can be applied in multilayers, and contamination of the film-forming solution is avoided compared to the immersion technique [37]. However, within the requirements for this technique, it is important to consider the nozzle size, temperature, viscosity of the solution, atomization pressure (approx. 60–80 psi), and the time of application [8].

3.5.3 ELECTROSPRAYING

Electrospraying is a common technique for liquid food formulation. It involves hydrodynamic electrostatic atomization, producing efficient nanoparticles or microparticles [1, 38]. This technique allows adjusting the droplet size and density of the coating by controlling the flow and viscosity of the filmogenic solution [39]. This process uses a high-voltage electrostatic field to charge a filmogenic solution passed through an atomizing nozzle. The electrostatic charge causes the solution to fragment into tiny droplets. Finally, these nanoparticles are attracted to a surface that contains the food [1, 38]. The advantages of this method in comparison to spraying include more precise droplet size control, uniform application, and high substrate coverage efficiency, resulting in less solution waste [40, 41].

3.5.4 BRUSHING

Brushing is a simple technique recommended for solutions with high viscosity, making it suitable for small-scale food production. It consists of applying the coating on the feed surface using a brush [37]. Generally, the fundamental aspects of the coating's distribution are the amount of wetting and how quickly it spreads. The quality and properties of filmogenic solution, as well as the surface and geometry of food, considerably influence the application; besides, the human factor (who is applying the EC by brushing) significantly impacts the quality and homogeneity of the coating thickness [39].

3.6 BIOACTIVE STARCH-BASED COATINGS

In recent decades, various strategies have been investigated to extend the shelf life of fruits and vegetables. These products are characterized by high perishability, making shelf life extension a significant challenge [11]. As a result of these efforts, the ECs have emerged. As previously mentioned, starch-based coatings have been widely explored as a method of postharvest preservation of fresh produce. However, recent research highlights the use of bioactive coatings for the same purposes [4, 11, 12, 35, 42–44].

Active food packaging technologies include incorporating chemical or biological substances, which can change the interactions between the packaging and the food [45]. Using these bioactive edible coatings applied on fruits and vegetables is a promising alternative to increase shelf life [46–48]. Combining biopolymers and bioactive compounds created active packaging with improved characteristics, such as antioxidant and antimicrobial properties [43]. The main advantages of this type of packaging lie in the gradual and controlled release of the active agents on the coated surface. This ensures a constant concentration of the active compound in the food, helping to counteract the effect of prolonged storage use [44]. Moreover, adding bioactive substances to coatings can improve their physicochemical and antimicrobial properties, such as water/steam barriers, tensile strength, color, solubility, oxidation resistance, antimicrobial activity, and organoleptic characteristics [12, 49].

The bioactive compounds are usually added as either pure compounds, essential oils, or plant extracts [13]. Plant extracts and essential oils added in coating formulations on fruits and vegetables have been shown to retard the respiration rate and reduce the microbial spoilage on foods, as well as to improve the mechanical functions of active ECs [50].

According to do Nascimento et al. [51] and Yadav et al. [52], coatings formulated with starch and bioactive compounds stand out for their barrier properties against microorganisms, antioxidant capacity, UV protection, oxygen regulation, and water vapor barriers. In addition, they can suppress the proliferation of microorganisms and extend the shelf life of packaged fresh produce [11].

Incorporating antioxidant compounds in the formulation of ECs, such as essential oils [11] and phenolic compounds [4], showed a synergic interaction among active EC and the covered products by trapping or neutralizing oxidizing compounds or directly releasing the antioxidants in the products. This contributes to preserving the nutritional value of foods without altering their sensory properties since the presence of oxidizing compounds results in the deterioration of essential nutrients such as proteins, lipids, and carbohydrates. All these underline the importance of active coatings in preserving food quality [13].

From another perspective, adding antimicrobial agents in the formulation of active coatings improves their capability to protect against pathogenic microorganisms and preserves food quality for a longer period [49]. Antimicrobial agents are divided in two main categories according to their chemical composition: organic and inorganic. Organic antimicrobials include essential oils [53] and vegetable extracts [54]; while among the inorganic are zinc oxide, titanium dioxide, and silver [55]. However, natural antimicrobial compounds are often preferred instead of synthetic additives because they are healthier [48]. However, it is important to determine the minimum concentration of organic or inorganic antimicrobial agents needed to achieve an efficient protection against spoilage microorganisms and avoid undesirable results. Active packaging may also help to reduce the adverse effects on the cell membrane of fruits and vegetables, alterations in intracellular biomolecules, regulation of metabolic processes, and oxidative stress [56].

Table 3.4 summarizes the most recent research focused on starch-based active coatings, which were functionalized with bioactive compounds and applied onto fruits and vegetables for postharvest preservation.

3.7 CONCLUSIONS AND FUTURE TRENDS

Starch-based edible coatings are a promising alternative for maintaining the quality and extending the shelf life of fruits and vegetables due to their capability to form continuous coatings, improving the food's appearance, protecting against physical or mechanical damage, decreasing weight loss/moisture, and reducing the respiration rate. Currently, there is a trend to incorporate ingredients that, in addition to their adequate surface, tensile, and mass transfer properties, prevent oxidation, protect against microbial attack, and increase the nutritional quality of the food. This is the case of active coatings. Among the ingredients incorporated in the formulation of active coatings, stand out pure compounds (e.g., thymol and carvacrol), essential oils (e.g., rosemary, lavender, cinnamon, thyme, among others), and

TABLE 3.4 The Main Effect of Starch-Based Active Coatings on Fruits and Vegetables

Coating Material	Food Coated	Coating Application Methods	Main Effects on Fruit and Vegetables	Reference
High amylose corn starch, two combinations of thymol and carvacrol	Mango (*Mangifera indica* var. Ataulfo) and papaya (*Carica papaya* L. var. Maradol). The samples were stored at 20°C for 18 days	Dipping	↑ Antifungal effect against *Colletotrichum gloeosporioides*, the causal agent of anthracnose ↓ Delayed softening, respiration rate, and ethylene production ≠ Resistance to water vapor contributed to decrease the degree of dehydration ↓ Decreased the permeability to O_2 and CO_2 ↔ Avoided color changes of the fruit surface	Ochoa et al. [46]
Ginger rhizome starch, keratin (obtained from chicken feather waste), and polyphenol extract of avocado pear peel	Ripe tomato fruits, which were infected with *Aspergillus flavus* and *Aspergillus niger* after 6 days of storage at room temperature (25 ± 2°C)	Dipping	↓ Decreased the weight loss by increasing the concentration of polyphenol extract from avocado pear peel ↑ Increased the titratable acidity by increasing the extract loading, ascorbic acid, and lycopene contents ↑ Inhibited the incidence of infection caused by *A. flavus* and *A. niger*, as well as the severity of the disease	Oluba et al. [47]
Fruit starch and stem bark phenolic extract of *Spondias purpurea* L.	Minimally processed mangoes (MPM) stored at 12 ± 1°C and 88% ± 1 of RH (relative humidity) for 10 days	Dipping	↔ Coatings functionalized with phenolic extracts maintained external appearance and exhibited antimicrobial activity ↓ Decreased moisture loss, respiration rate, and oxidative processes, thereby contributing to an extension of product shelf life ↔ Changes in browning index were minimal, suggesting that there is antioxidant activity upon adding Stem bark phenolic extract of *S. purpurea* ↓ The ripening process of MPM was delayed, making it an excellent barrier to respiration. Decreased the degradation of complex carbohydrates to simpler ones	Rodrigues et al. [48]

TABLE 3.4 *(Continued)*

Coating Material	Food Coated	Coating Application Methods	Main Effects on Fruit and Vegetables	Reference
Cassava starch (*Manihot esculenta* Crantz), added with clove (*Syzygium aromaticum*) or cinnamon (*Cinnamomum* sp.) essential oils	Papaya (*Carica papaya* L. Hawaiian variety) stored at 25°C, for 12 days	Dipping	↓ The formulations inhibited the halos for *Escherichia coli*, *Salmonella enterica*, *Colletotrichum gloeosporioides*, and *Penicillium* sp. ↓ Reduced total soluble solids content and weight loss compared to the control, making these treatments effective in preserving these quality attributes ↑ Higher retention rate of vitamin C and carotenoid content during storage	do Nascimento et al. [51]
Yam (*Dioscorea rotundata* L.) starch, soy lecithin (emulsifying agent), and three essential oils (lime, fennel, and lavender)	Strawberries (*Fragaria vesca* L.) were stored at 25°C and 85% RH (using a supersaturated KCl solution) for 15 days	Spraying	↑ Improved the antioxidant capacity of the coatings, especially the lavender essential oil (38 mL/mg DPPH and 8 mL/mg ABTS) ↓ Decreased water vapor permeability and oxygen permeability because of the addition of essential oils ↓ Significantly reduced weight loss (22.8%) due to improved water vapor permeability ↑ The addition of essential oils inhibited the growth of *Aspergillus niger* until day 8 of storage	Gómez et al. [57]
Moth bean (*Vigna aconitifolia*) starch, and aqueous extract Basil leaves (*Ocimum tenuiflorum*)	Eggplants (*S. melongena*) were stored at 25 ± 1°C, 50–55% RH to monitor shelf life and quality	Dipping	↔ Maintained color changes because the coating acted as a barrier against gas migration, creating a modified atmosphere ↑ Improved fruit firmness up to 16 days of storage resulted from decreased metabolic processes	Kumar et al. [58]

TABLE 3.4 *(Continued)*

Coating Material	Food Coated	Coating Application Methods	Main Effects on Fruit and Vegetables	Reference
Mango kernel seed starch and lemongrass essential oil	Guava fruit (*Psidium guajava* L.) is stored at 23 ± 2 °C and a RH of 85 ± 5% for 9 days	Dipping	↑ Improved the shelf life of guava fruit while maintaining its quality attributes ↓ Reduced physiologic weight loss and oxidation rate ↔ Maintained physicochemical characteristics such as acidity, total soluble solids, textural properties, phenolic content, and sensory profiles ↓ Reduced the yeast and mold microbial load during the storage period ↔ The retention of quality attributes was correlated with higher phenolic content and less count of yeast and mold	Yadav et al. [52]
Water-chestnut (*Trapa natans*) starch, rosemary essential oil, and Tween-20 (emulsifying agent)	Apples (*Malus domestica* var. golden delicious) were stored for 120 days and analyzed at an interval of 30 days	Dipping	↔ Anthocyanin content, ascorbic acid, total phenolic content, and antioxidant activity were maintained ↓ Decreased weight loss ↓ Reduced microbial growth of *Penicillium expansum*, resulting in potential postharvest biocontrol ↑ More protection to fruit against low-temperature damage, thereby protecting it against chilling injury	Bashir et al. [59]
Pregelatinized corn starch with a cellulose nanofiber (pickering emulsifier) and basil essential oil (*Ocimum basilicum*)	Turkish Tango mandarins (*Citrus reticulata* × *C. sinensis*) stored at room temperature for 12 days	Spraying	↓ Reduced the changes caused by biochemical degradation and minimized color changes ↔ Maintained color stability in terms of color density ↓ Decreased citric acid losses	Wigati et al. [54]

vegetable extracts (e.g., extract basil leaves, stem bark phenolic compound) which conferred barrier properties against pathogenic microorganisms, antioxidant capacity, protection against UV radiation, as well as improve the gas permeability (O_2, CO_2, and water vapor). Another advantage is that active coatings can preserve the physicochemical, organoleptic, and nutritional characteristics of coated products due to the controlled release of the bioactive compounds added to the edible coating. Despite to the current research on active packaging, future studies must explore new sources of active ingredients (e.g., food waste) and formulations, recommended doses of bioactive compound, as well as the effects between the ingredients and the coated product to promote the synergistic effects and reduce those antagonistic. All these efforts may contribute to the Food and Agriculture Organization's Sustainable Development Goal number 12 "*Responsible Production and Consumption*" by reducing food losses and waste through the production and supply chain, where the postharvest is included and, therefore, the application of edible coatings on fruits and vegetables.

KEYWORDS

- **bioactive compounds**
- **biopolymer**
- **edible films and coatings**
- **quality**
- **shelf life**
- **starch**
- **plant extracts**

REFERENCES

1. Monteiro, F.A.R., Fonseca, J. d.M., Trevisol, T.C., Fagundes, C., Valencia, G.A. (2019). Active, eco-friendly and edible coatings in the post-harvest—A critical discussion. *Polymers for Agri-Food Applications,* 433–463.
2. Al-Tayyar, N.A., Youssef, A.M., Al-Hindi, R.R. (2020). Edible coatings and antimicrobial nanoemulsions for enhancing shelf life and reducing foodborne pathogens of fruits and vegetables: A review. *Sustainable Materials and Technologies*, 26, e00215.

3. Brasil, I.M., Siddiqui, M.W. (2018). Chapter 1—Postharvest quality of fruits and vegetables: An overview. In: Siddiqui, M. W. (Eds.), *Preharvest Modulation of Postharvest Fruit and Vegetable Quality* (pp. 1–40). Academic Press.
4. Pedreiro, S., Figueirinha, A., Silva, A.S., Ramos, F. (2021). Bioactive edible films and coatings based in gums and starch: Phenolic enrichment and foods application. *Coatings*, 11, 1393.
5. Perez, V.A., Barciela, P., Carpena, M., Prieto, M.A. (2023). Edible coatings as a natural packaging system to improve fruit and vegetable shelf life and quality. *Foods,* 12, 3570.
6. Saklani, P., Nath, S., Das, S.K., Singh, S.M. (2019). A review of edible packaging for foods. *International Journal of Current Microbiology and Applied Sciences*, 8, 2885–2895.
7. Kumar, L., Ramakanth, D., Akhila, K., Gaikwadd, K.K. (2021). Edible films and coatings for food packaging applications: A review. *Environmental Chemistry Letters*, 20, 875–900.
8. Atieno, L., Owino, W.O., Ateka, E.M., Ambuko, J.L. (2019). Influence of coating application methods on the postharvest quality of cassava. *International Journal of Food Science,* 2019, 2148914.
9. Díaz, M.E., Castro, M.R. (2021). Edible films and coatings as food-quality preservers: An overview. *Foods*, 10, 249.
10. Hernández, G.S.E., Balois, M.R., Palomino, H.Y.A., López, G.G.G., Berumen, V.G., Bautista, R.P.U., Alejo, S.G. (2020). Novel edible coating of starch-based stenospermocarpic mango prolongs the shelf life of mango "Ataulfo" fruit. *Journal of Food Quality,* 2020, 1320357.
11. Cui, C., Ji, N., Wang, Y., Xiong, L., Sun, Q. (2021). Bioactive and intelligent starch-based films: A review. *Trends in Food Science & Technology*, 116, 854–869.
12. Falcão, L.d.S., Coelho, D.B., Veggi, P.C., Campelo, P.H., Albuquerque, P.M., de Moraes, M.A. (2022). Starch as a matrix for incorporation and release of bioactive compounds: Fundamentals and applications. *Polymers*, 14, 2361.
13. Cruz, M.R.G., Rayas, A.A.A., González, R.R.M., Zambrano, Z.M.L., Aguilar, T.J.E., Liceaga, A.M. (2023). Application of polysaccharide-based edible coatings on fruits and vegetables: Improvement of food quality and bioactivities. *Polysaccharides*, 4, 99–115.
14. Oyom, W., Zhang, Z., Bi, Y., Tahergorabi, R. (2022). Application of starch-based coatings incorporated with antimicrobial agents for preservation of fruits and vegetables: A review. *Progress in Organic Coatings*, 166, 106800.
15. Basumatary, I.B., Mukherjee, A., Katiyar, V., Kumar, S. (2022). Biopolymer-based nanocomposite films and coatings: Recent advances in shelf life improvement of fruits and vegetables. *Critical Reviews in Food Science and Nutrition*, 62, 1912–1935.
16. Dadzie, R.G., Ampofo-Asiama, J., Quaye, B., Kizzie-Hayford, N., Amoako-Andoh, F., Abano, E.E., Amoah, R.S. (2023). Effect of beeswax coating on the quality of eggplant fruit during cold storage and optimization of coating conditions. *Journal of Food Quality.* 2023, 5879085.
17. Eshetu, A., Ibrahim, A.M., Forsido, S.F., Kuyu, C.G. (2019). Effect of beeswax and chitosan treatments on quality and shelf life of selected mango (*Mangifera indica* L.) cultivars. *Heliyon,* 5, e01116.

18. Dehghani, S., Hosseini, E., Rousta, E. (2022). Shelf life extension of tomato (*Solanum lycopersicum* L.) using an edible coating of bitter almond gum-fish gelatin conjugates. *Progress in Organic Coatings*, 170, 106980.
19. Pelissari, F.M., Ferreira, D.C., Louzada, L.B., dos Santos, F., Corrêa, A.C., Moreira, F.K.V., Mattoso, L.H. (2019). Chapter 10 - Starch-based edible films and coatings: An eco-friendly alternative for food packaging. In: Silva Clerici, M.T.P., Schmiele, M. (Eds.), *Starches for Food Application* (pp. 359–420). Academic Press.
20. Wu, J., Zhang, L., Fan, K. (2024). Recent advances in polysaccharide-based edible coatings for preservation of fruits and vegetables: A review. *Critical Reviews in Food Science and Nutrition,* 64, 3823–3838.
21. Martínez, K., Ortiz, M., Albis, A., Gilma Gutiérrez Castañeda, C., Valencia, M.E., Grande Tovar, C.D. (2018). The effect of edible chitosan coatings incorporated with Thymus capitatus essential oil on the shelf life of strawberry (Fragaria x ananassa) during cold storage. *Biomolecules*, 8, 155.
22. Nawab, A., Alam, F., Haq, M.A., Haider, M.S., Lutfi, Z., Kamaluddin, S., Hasnain, A. (2018). Innovative edible packaging from mango kernel starch for the shelf life extension of red chili powder. *International Journal of Biological Macromolecules*, 114, 626–631.
23. Botin, F.C., Pellá, M.G., Silva, O.A., Raimundo, K.F., Caetano, J., Linde, G.A., Colauto, N.B., Dragunski, D.C. (2020). Shelf life of guavas coated with biodegradable starch and cellulose-based films. *International Journal of Biological Macromolecules*, 152, 272–279.
24. Moreira, E.d.S., Silva, N.M.C.d., Brandão, M.R.S., Santos, H.C., Ferreira, T.A.P.d.C. (2022). Effect of modified starch and gelatin by-product based edible coating on the postharvest quality and shelf life of guava fruits. *Food Science and Technology*, 42, e26221.
25. Soto, M.L., Palou, L., Argente, S.M., Ramos, L.M.A., Pérez, G.M.B. (2021). Optimization of antifungal edible pregelatinized potato starch-based coating formulations by response surface methodology to extend postharvest life of 'Orri' mandarins. *Scientia Horticulturae*, 288, 110394.
26. Salgado, C.M.d.l.P., Salgado, C.J., García, H.A.B., Calderón, D.G., Gómez, V.H., Oliver, E.R., Fernández, M.M.C., Yáñez, F.J. (2021). Chitosan as a coating for biocontrol in postharvest products: A bibliometric review. *Membranes*, 11, 421.
27. Lazaridou, A., Biliaderis, C.G. (2020). Edible films and coatings with pectin. In: Kontogiorgos, V. (Eds.), *Pectin: Technological and Physiological Properties* (pp. 99–123). Springer International Publishing: Cham.
28. Abdul Khalil, H., Banerjee, A., Saurabh, C.K., Tye, Y., Suriani, A., Mohamed, A., Karim, A., Rizal, S., Paridah, M. (2018). Biodegradable films for fruits and vegetables packaging application: Preparation and properties. *Food Engineering Reviews*, 10, 139–153.
29. Lauer, M.K., Smith, R.C., (2020). Recent advances in starch-based films toward food packaging applications: Physicochemical, mechanical, and functional properties. *Comprehensive Reviews in Food Science and Food Safety*, 19, 3031–3083.
30. Chacon, W.D.C., Gaviria, Y.A.R., de Melo, A.P.Z., Verruck, S., Monteiro, A.R., Valencia, G.A.J.S.S. (2024). Physicochemical properties and potential food applications of starches isolated from unconventional seeds: A review. *Starch*, 76, 2200228.

31. Sapper, M., Chiralt, A. (2018). Starch-based coatings for preservation of fruits and vegetables. *Coatings*, 8, 152.
32. Nawab, A., Alam, F., Hasnain, A. (2017). Mango kernel starch as a novel edible coating for enhancing shelf life of tomato (*Solanum lycopersicum*) fruit. *International Journal of Biological Macromolecules*, 103, 581–586.
33. Nogueira, G.F., Leme, B.d.O., Santos, G.R.S.d., Silva, J.V.d., Nascimento, P.B., Soares, C.T., Fakhouri, F.M., de Oliveira, R.A. (2021). Edible films and coatings formulated with arrowroot starch as a non-conventional starch source for plums packaging. *Polysaccharides*, 2, 373–386.
34. Soto, M.L., Pérez, G.M.B., Martínez, B.V., Palou, L. (2023). Postharvest application of potato starch edible coatings with sodium benzoate to reduce sour rot and preserve mandarin fruit quality. *Coatings*, 13, 296.
35. Pandey, V.K., Islam, R.U., Shams, R., Dar, A.H. (2022). A comprehensive review on the application of essential oils as bioactive compounds in Nano-emulsion based edible coatings of fruits and vegetables. *Applied Food Research*, 2, 100042.
36. Suhag, R., Kumar, N., Petkoska, A.T., Upadhyay, A. (2020). Film formation and deposition methods of edible coating on food products: A review. *Food Research International*, 136, 109582.
37. Youssef, A., Assem, F., El-Sayed, S., Salama, H., Abd El-Salam, M. (2017). Utilization of edible films and coatings as packaging materials for preservation of cheeses. *Journal of Packaging Technology and Research*, 1, 87–99.
38. Khan, M.K.I., Nazir, A., Maan, A.A. (2017). Electrospraying: A novel technique for efficient coating of foods. *Food Engineering Reviews*, 9, 112–119.
39. Pham, T.T., Nguyen, L.L.P., Dam, M.S., Baranyai, L. (2023). Application of edible coating in extension of fruit shelf life: Review. *AgriEngineering*, 5, 520–536.
40. Jiang, Y., Yu, L., Hu, Y., Zhu, Z., Zhuang, C., Zhao, Y., Zhong, Y. (2019). Electrostatic spraying of chitosan coating with different deacetylation degree for strawberry preservation. *International Journal of Biological Macromolecules*, 139, 1232–1238.
41. Tavassoli, K.E., Gamage, M.V., Dumée, L.F., Kong, L., Zhao, S. (2022). Edible films and coatings for shelf life extension of mango: A review. *Critical Reviews in Food Science and Nutrition*, 62, 2432-2459.
42. Feng, M., Yu, L., Zhu, P., Zhou, X., Liu, H., Yang, Y., Zhou, J., Gao, C., Bao, X., Chen, P. (2018). Development and preparation of active starch films carrying tea polyphenol. *Carbohydrate Polymers*, 196, 162–167.
43. Majeed, T., Dar, A.H., Pandey, V.K., Dash, K.K., Srivastava, S., Shams, R., Jeevarathinam, G., Singh, P., Echegaray, N., Pandiselvam, R. (2023). Role of additives in starch-based edible films and coating: A review with current knowledge. *Progress in Organic Coatings*, 181, 107597.
44. Oliveira Filho, J.G.d., Braga, A.R.C., Oliveira, B.R.d., Gomes, F.P., Moreira, V.L., Pereira, V.A.C., Egea, M.B. (2021). The potential of anthocyanins in smart, active, and bioactive eco-friendly polymer-based films: A review. *Food Research International*, 142, 110202.
45. Tunma, S. (2017). Starch based nanocomposites in active packaging for extended shelf life of fresh fruits. *Walailak Journal of Science and Technology (WJST)*, 15, 273–281.

46. Ochoa, V.C.E., Pérez, P.J.C., Varillas, T.J.M., Navarro, C.A.R., Hernández, C.P., Munguía, P.R., Cid, P.T.S., Avila, S.R. (2021). Starch edible films/coatings added with carvacrol and thymol: In vitro and in vivo evaluation against Colletotrichum gloeosporioides. *Foods*, 10, 175.
47. Oluba, O.M., Obokare, O., Bayo-Olorunmeke, O.A., Ojeaburu, S.I., Ogunlowo, O.M., Irokanulo, E.O., Akpor, O.B. (2022). Fabrication, characterization and antifungal evaluation of polyphenolic extract activated keratin starch coating on infected tomato fruits. *Scientific Reports*, 12, 4340.
48. Rodrigues, F.A.M., dos Santos, S.B.F., Lopes, M.M.d.A., Guimarães, D.J.S., de Oliveira Silva, E., de Souza Filho, M.d.S.M., Mattos, A.L.A., da Silva, L.M.R., de Azeredo, H.M.C., Ricardo, N.M.P.S. (2021). Antioxidant films and coatings based on starch and phenolics from *Spondias purpurea* L. *International Journal of Biological Macromolecules*, 182, 354–365.
49. Kola, V., Carvalho, I.S. (2023). Plant extracts as additives in biodegradable films and coatings in active food packaging. *Food Bioscience*, 54, 102860.
50. Saxena, A., Sharma, L., Maity, T. (2020). Chapter 34 - Enrichment of edible coatings and films with plant extracts or essential oils for the preservation of fruits and vegetables. In: Pal, K., Banerjee, I., Sarkar, P., Kim, D., Deng, W.-P., Dubey, N.K., Majumder, K. (Eds.), *Biopolymer-Based Formulations* (pp. 859–880). Elsevier.
51. do Nascimento, A., Toneto, L.C., Lepaus, B.M., Valiati, B.S., Faria-Silva, L., de São José, J.F.B. (2023). Effect of edible coatings of cassava starch incorporated with clove and cinnamon essential oils on the shelf life of papaya. *Membranes*, 13, 772.
52. Yadav, A., Kumar, N., Upadhyay, A., Singh, A., Anurag, R.K., Pandiselvam, R. (2022). Effect of mango kernel seed starch-based active edible coating functionalized with lemongrass essential oil on the shelf life of guava fruit. *Food Reviews International*, 14, 103–115.
53. Villar, C.K.A., Ruiz, P.G.A., Fuertes, R.C.M., Biofilms based on chitosan. (2021). Edible films based on chitosan and essential oil of *Minthostachys Mollis* (Muña) with antioxidant and antimicrobial properties. *Revista de la Sociedad Química del Perú*, 87, 309–320.
54. Wigati, L.P., Wardana, A.A., Jothi, J.S., Leonard, S., Van, T.T., Yan, X., Tanaka, F., Tanaka, F. (2023). Preserving mandarin quality during ambient storage using edible coatings of pregelatinized corn starch pickering emulsions and essential oil. *Food Bioscience*, 53, 102710.
55. Mousavi Khaneghah, A., Hashemi, S.M.B., Limbo, S. (2018). Antimicrobial agents and packaging systems in antimicrobial active food packaging: An overview of approaches and interactions. *Food Bioproducts Processing*, 111, 1–19.
56. Mohd Nizam, N.H., Mohammad Rawi, N.F., Mhd Ramle, S.F., Abd Aziz, A., Abdullah, C.K., Rashedi, A., Mohamad Kassim, M.H. (2021). Physical, thermal, mechanical, antimicrobial and physicochemical properties of starch based film containing aloe vera: A review. *Journal of Materials Research and Technology*, 15, 1572–1589.
57. Gómez, C.P., Figueroa, L.K.J., Hernández, F.J., Cortés, R.M., Ortega, T.R. (2021). Effect of different essential oils on the properties of edible coatings based on yam (*Dioscorea rotundata* L.) starch and its application in strawberry (*Fragaria vesca* L.) preservation. *Applied Science*, 11, 11057.

58. Kumar, R., Ghoshal, G., Goyal, M. (2021). Effect of basil leaves extract on modified moth bean starch active film for eggplant surface coating. *LWT-Food Science and Technology*, 145, 111380.
59. Bashir, O., Amin, T., Hussain, S.Z., Naik, H.R., Goksen, G., Wani, A.W., Manzoor, S., Malik, A.R., Wani, F.J., Proestos, C. (2023). Development, characterization and use of rosemary essential oil loaded water-chestnut starch based nanoemulsion coatings for enhancing post-harvest quality of apples var. Golden delicious. *Current Research in Food Science*, 7, 100570.

CHAPTER 4

Natural Bio-Based Polymers as Active Coating Barrier for Food Contact Paper Packaging

JUAN DANIEL RIVALDI CHÁVEZ,[1,2,3*] NURIA GÓMEZ-HERNÁNDEZ,[2] ÚRSULA FILLAT-LATORRE,[2,4] PRISCILLA VERGARA,[2] JOSE MARIA CARBAJO,[2] FÉLIX GARCÍA-OCHOA,[3] and JUAN CARLOS VILLAR[2]

[1]*Industrial Applications Department, Faculty of Chemical Sciences, Universidad Nacional de Asunción, San Lorenzo, Paraguay*

[2]*Cellulose and Paper Group, Forest Product Department, Forest Research Centre (ICIFOR-CSIC), Madrid, Spain*

[3]*Chemical and Materials Engineering Department, Faculty of Chemical Sciences, Universidad Complutense de Madrid, Madrid, Spain*

[4]*CELBIOTECH Paper Engineering Research Group, Universitat Politècnica de Catalunya (UPC), Terrassa, Barcelona, Spain*

*Corresponding author

ABSTRACT

Selection of packaging materials plays a crucial role in preserving the nutritional value and quality of foods. Currently, the increasing concerns regarding nonsustainable packaging based on single-use plastic promote intense research for obtaining biocompatible, compostable, and recyclable materials. Effective approaches to provide adequate moisture, gases, and oil/fat barriers, as well as antioxidant and antimicrobial properties for paper

Sustainable Food Packaging Materials and Technology. Juan Manuel Tirado-Gallegos, Juan Daniel Rivaldi Chávez, and José Juan Buenrostro-Figueroa (Eds)

DOI: 10.1201/9781779641434-4

packaging, are being studied extensively. Natural polymers are sustainable materials with the ability to provide improved properties and capacity as coating paper for preserving the quality and shelf life of food. Barrier coatings containing innocuous biopolymers are sustainable alternatives to traditional food packaging. Mixtures of natural polymers (composites) have the potential to replace current plastic-paper and plastic-paperboard coatings. The main drawback in coating formulation is that natural polymers show hydrophilicity and high polarity that mainly promotes a low water vapor barrier. In addition, the hydrophobicity of some polymers affects the adequate dispersion and barrier properties of the coating. Biodegradability, recyclability, and cost-effectiveness are issues that need more studies to improve the sustainability of the paper packaging industry. This chapter aims to provide a comprehensive overview of advances in active paper coating for food packaging, with a special focus on barrier properties, methods of production, and recyclability.

4.1 INTRODUCTION

Currently, food packaging represents most of the total packaging sector, with materials like paper, aluminum foil, metallized films, and certain types of plastics. The type of material used in packaging greatly affects its barrier properties. Plastic is one of the best-performing materials due to its low production cost and mechanical and barrier properties. However, these materials are of fossil origin, with low recyclability or unrecycled, affecting the environmental quality and health [12]

Paper is used extensively in packaging applications because of its versatility, low cost, nontoxicity, high recyclability, and biodegradability [74, 80]. The Confederation of European Paper Industries (CEPI), which brings together producer associations from 18 European nations, reported a total paper and board production of 85 million tons in 2022 [13]. Paper packaging consumption will increase with the increasing demand for renewable, sustainable, and competitive materials, which represent a revolution compared to traditional packaging.

The sustainability strategies of the paper industry are oriented toward the substitution of plastics and ephemeral plasticizing agents considered problematic. In this regard, the European Union (Directive EU 2019/904) promotes the prevention and reduction of the impact of certain single-use plastic products on the environment, encouraging the use of suitable and more sustainable alternative products, suitable for multiple uses and that, once converted into waste, can be prepared for reuse and recycling.

Paper consists of microfibrils of cellulose, a linear homopolymer formed by cellobiose units consisting of a D-glucose dimer obtained by condensation of β (1–4) glycosidic bonds. Cellulose structure presents amorphous and crystalline regions with numerous free OH groups in the glucose structure. Due to the high porosity and hydrophilicity of the cellulose, paper packaging absorbs water and offers low resistance to the transmission of gases and water vapor, which reduces its applicability to preserve and extend the shelf life of foods [35, 46]. The uncoated paper and board packaging lack gas, liquid, and grease barrier properties, which are required for preserving perishable foodstuffs such as meat, fish, dairy, and fatty foods.

The food industry requests sustainable packaging materials with barrier properties, sufficient mechanical properties as well as antioxidant and antimicrobial capacity to preserve the food characteristics. Traditionally, polyfluoroalkyl additives are used in the paper industry to achieve barrier properties [31]. However, it has been demonstrated that these substances are associated with endocrine and immunological alterations in humans, in addition to the environmental problems caused by their difficult degradation, constituting priority ecological challenges [18]. Thus, the search for new alternatives that allow the substitution of single-use plastics and plasticizers that are safe for human health stimulates new scientific research.

Since the 2000s, biopolymers for active coating have been receiving strong global attention for their potential as a barrier. Packaging barrier properties refer to the ability of packaging materials to prevent the passage of gases, moisture, light, and other external factors that can affect the quality and shelf life of packaged products [59]. Surface sizing is one of the most traditional procedures used in the paper industry as a barrier. This treatment aims to fill the pores between the fibers with polymeric solutions, typically starch with a hydrophobic solution (acrylic copolymers), impart improved properties to the paper, such as stiffness, surface resistance, printability, and certain hydrophobicity. Several factors influence packaging barrier properties, including thickness, composition (multilayer of different materials), temperature and moisture, mechanical stress, and the application of coatings and surface treatments [1, 14]. In addition, physical and mechanical properties, as well as the geometry of the package, are critical in keeping the product fresh, frozen, dehydrated, or aseptic.

Paper coating with biopolymers, such as starch, chitosan, wax, lignin, and others, constitutes an alternative to generate "active" papers with advanced and biocompatible properties for the preservation of the quality of food products. A combination of biopolymers applied as a coating can reduce water-vapor permeability and wettability, promote resistance to oils and greases, reduce gas transmission, and impregnate antimicrobial and antioxidant properties, among others. This chapter will present current solutions to develop sustainable paper coatings for food packaging, their barrier properties and characteristics, methods of industrial coating, and aspects of biodegradability and recyclability.

4.2 BARRIER PROPERTIES

Barrier properties refer to the capacity of certain materials used as packaging to resist the entry of moisture, light, oxygen, carbon dioxide, aromatic compounds, microorganisms, grease, and oil inside the packaging, protecting and preserving the food quality and characteristics such as texture, smell, color, and nutritional properties (see Figure 4.1). The barrier properties offered by paper and paperboard packaging are weaker than those of plastic-based packaging [80].

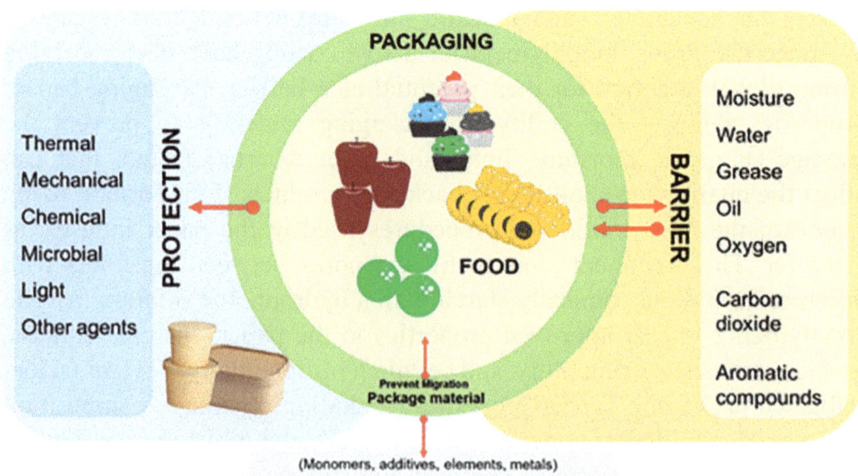

FIGURE 4.1 Packaging requirements for food products.

Various polymers such as ethylene vinyl alcohol, polyethylene, and polypropylene (PP) have been used to reinforce the barrier properties of paper-based packaging. However, these plastics as coating generates a negative environmental impact because of low recyclability and nonbiodegradability. Barrier properties depend on the nature and characteristics of the polymers used as coating (chemical and superficial structure, thickness, and crystallinity), the characteristics of the product to be protected, the storage conditions (e.g., humidity and temperature), and the characteristics of the permeant (e.g., molecular size, polarity). Among the permeants, oxygen and water vapor are the most important since they cross through the pores of the cellulose and can affect the level of oxidation and moisture of the packed food, respectively. For fresh products, it is recommended to have low water vapor permeability (WVP) to maintain the humidity and texture of the product, while it is desirable to have low oxygen permeability (OP) to reduce lipid oxidation or to retard respiration of some products [61, 103].

The OP is determined by the analysis of the oxygen transmission rate, defined as the amount of oxygen that permeates at a determined permeant gradient through the paper sample of fixed thickness per area and unit time—ASTM D3985-81 [4]. The OP is calculated as follows:

$$\mathrm{OP} = \frac{F\,T}{A\,\Delta P}\left(\frac{\mathrm{cm}^3\,\mu\mathrm{m}}{\mathrm{day\,m}^2\,\mathrm{kPa}}\right) \tag{4.1}$$

where F is oxygen flow, mL/min; T is thickness of the sample, μm; A is area, m^2; and ΔP is the pressure across the sample, kPa.

The WVP is measured as water vapor transmission rate (WVTR) and consists of a measure of the passage of water vapor through a paper sample of fixed thickness at specified conditions of temperature and relative humidity (ASTM E96), calculated as:

$$\mathrm{WVP} = \frac{\Delta m T}{\Delta t\, A\,\Delta P}\left(\frac{\mathrm{g}\,\mu\mathrm{m}}{\mathrm{day\,m}^2\mathrm{kPa}}\right) \tag{4.2}$$

where Δm is weight loss of water, g; T is thickness of the sample, μm; Δt is time of assay, s; A is the area of paper sample, m^2; and ΔP is pressure, kPa.

Another important parameter to be considered in coated paper is the surface wettability of a paper, which is determined by optical methods according to Test Method T 458 cm^{-14} [92], and it is expressed in terms of static or dynamic water contact angle. This method estimates the free energy and hydrophobicity of the paper and coated paper surface, which depends on the interactions of hydrogen bonding forces, polarity, and dispersion forces [42]. A contact angle higher than 90° indicates a repulsion of liquid by the paper surface; lower values (< 90°) indicate attraction of the liquid by the surface or higher wettability of the paper. Furthermore, the porosity of the paper surface is a key factor to determine the wettability since a small diameter of pores increases the resistance to the passage of liquids through the paper. The water absorption can also be estimated by the Cobb method, which consists of measuring the amount of water absorbed by a paper sample of a specific area after 60 s (Cobb60) or 180 s (Cobb180) of contact according to IS0 535 (ISO, 2023). The water absorbency is expressed in g/m^2.

The oil-resistant packaging paper is undisputedly necessary for serving, wrapping, or packaging processed food products. Grease resistance testing Technical Association of the Pulp and Paper Industry (TAPPI) T-559-cm02 [91] is the most accepted method to measure the repellence of paper and paperboard materials to grease, oil, and waxes. This technique evaluates greaseproof papers treated with perfluorinated compounds but is applied to paper substrates coated with different compounds. Mixtures of castor oil, toluene, and *n*-heptane, with different concentrations, viscosity, and surface tension, are numbered according to their aggressiveness. The factor increases with increasing content of heptane and decreasing content of castor oil from 1 (least aggressive) to 12 (most aggressive). Each mixture is dropped on the paper surface and removed (wiped off) after 15 s, observing the formation of a darkened area. Darkening mark indicates that the pores were filled with the liquid mixture, and the test of repellency was not successful. This procedure continues until determining the highest kit number corresponding to a nondarkening area, indicating an effective repellency [42]. For packaging indicated for food applications, it is recommended a minimum resistance level of 5 [63].

4.3 LIMITATIONS OF SINGLE PAPER-BASED PACKAGING AND PREPARATION OF COATED PAPERS

Paper and cardboard are versatile, economical, and environmentally friendly materials used in packaging to protect different products. However, paper-based packaging has some limitations, such as lack of strength that compromises the integrity of the packaged product, moisture sensitivity that makes this material less suitable to maintain resistance of the package in humid environments, and limited barrier properties against greases, oils, odors, and gases compared to other materials like plastics or metal [93]. This makes it less suitable for packaging products such as fresh and processed foods that require extended shelf life or protection from external elements [22]. Although cellulose is inherently biodegradable, most paper packages have limited biodegradability and recyclability due to the coating applied to improve its barrier properties. Most of the coatings are petroleum-derived compounds such as paraffin or plastics (i.e., PP, epoxy, polyurethane, plastisol, latex), which are difficult to remove in recycling processes or in the final waste management [44]. Also, the printability of the paper-based packaging is limited compared to other materials like plastics, which usually permit printing high-resolution graphics.

Regarding cost, paper is considered a cost-effective packaging material; however, specialized treatments or coatings to enhance its properties may increase the cost. Despite these limitations, advancements in paper technology and the development of composite materials using biodegradable polymers are continuously expanding the capabilities of paper-based packaging. Bio-based polymers may replace most of the nonrenewable materials used for coating paper, even increasing barrier properties, and promoting environmental balance and biocompatibility. Biopolymers are molecules with high molecular weights and viscosity that difficult the manipulation and coater runnability. Hydrophilic polymers show a low water vapor barrier, whereas hydrophobic polymers prevent an adequate dispersion in the paper surface [55].

For coating paper, a complex issue that should be considered is the solid content and preparation methods. Also, some parameters affecting rheological behavior, such as viscosity, and other conditions or properties, as temperature, pH, and co-adjutants of dispersion, must be considered to obtain stabilized and homogeneous suspensions [56, 78].

In general, for obtaining continuous coating by surface sizing, a solid concentration between 10 and 15 wt% is needed. A lower concentration of solids can require high energy consumption for drying. However, a lower solid content can affect the uniformity of the coating surface. For nonsoluble compounds, such as nanocellulose, the homogenous dispersion within the substrate or base paper represents a bottleneck for fast application and quality of the coated product. Thus, dispersing agents are necessary for an adequate distribution of the nanofibers; however, they should be compatible with the substrate and with the packed product [67, 85].

Another aspect to be considered is that the polysaccharides and proteins used as coating are highly hydrophilic and can be exposed to conditions that favor microbial contamination during preparation, application, and handling. Besides, biopolymer coatings can be dispersed or removed from the paper surface when applied alone, reducing the barrier and mechanical properties of the coated paper. By incorporating hydrophobic compounds such as lignin and natural waxes, or by chemical modification of natural polymers (e.g., acetylated starch, hydroxyprolylated, and cationic starch), it is possible to increase the retention of hydrophilic biopolymers on substrate surfaces [80]. Blends with other biodegradable and biocompatible polymers (polyvinyl alcohol, polyglycolic acid) or metallic ions can also improve the coating fixation and coated paper characteristics.

4.4 BIOPOLYMERS FOR COATING

Paper has poor barrier properties and is commonly associated with nonbiodegradable materials (plastic/aluminum/resin) to achieve moisture resistance and other functional properties [30]. Biopolymers offer a promising alternative for food paper packaging due to their biodegradability and repulpability. The last term means that the paper coated with polymers can be rewetted and the fibers recovered (at least 80% of the fiber based on the total weight). Barrier coatings containing innocuous or generally recognized as safe (GRAS) substances such as starch, chitosan, gelatin, alginate, soy protein, amino acids, natural waxes, microbial polymers (e.g., xanthan, polyhydroxyalkanoates (PHA), functionalized nanocellulose, and degradable polymers from biobased sources (e.g., polylactic acid (PLA), polyglycolic acid (PGA)) are sustainable alternatives to synthetic paper coating used in food packaging. These biopolymer coatings and their blends are effective to oxygen, water vapor, and oil barriers, which alone or in combination with substances such as lysozyme, zinc chloride ($ZnCl_2$), zinc oxide (ZnO), magnesium oxide (MgO), among others, can prevent the growth or adherence of selected microorganisms on the packaging and food [43, 74].

Bio-based polymers in food paper packaging can help to reduce environmental impact, promote sustainability, and meet increasing consumer demand for eco-friendly packaging solutions. However, it is essential besides mechanical and barrier properties to consider factors such as cost, performance, regulatory compliance, and end-of-life disposal when selecting biopolymer materials for specific packaging applications. Figure 4.2 shows the main biopolymer studied for paper coating applications for food packaging.

4.4.1 STARCH

Starch is a polysaccharide (glucan) traditionally used as a retention aid, bonding agent, and surface size and coating agent in the papermaking industry [68]. This biopolymer is classified as amylose (linear branchless), which consists of a molecule composed of glycosidic units linked by α-(1→4) bonds) and amylopectin (branched form) with branches of glucose bound to the linear glucose chain by α-(1→6) glycosidic bonds. This polysaccharide can be extracted from starchy crops such as potato, tapioca, wheat, corn, and rice. Polymers derived from starch, such as thermoplastic starch or starch blends, are renewable and biodegradable molecules that can be used as coatings, films, or composites with paper fibers to improve mechanical properties, moisture resistance, and antimicrobial activity [75].

4.4.2 NANOCELLULOSE/CELLULOSE NANOFIBERS (CNF)

The application of nanomaterials as fillers in paper coating, such as nanoclays, pigments, minerals, and most recently nanocellulose, is a strategic alternative to improve paper and cardboard functionalities. Nanocellulose refers to biopolymers obtained from cellulosic biomass and microbial source with structural dimensions on the nanometric scale. Bacterial cellulose is obtained from cultivation of various genera of bacteria, including *Komagataeibacter, Gluconacetobacter; Aerobacter, Azotobacter, Agrobacterium, Pseudomonas, Achromobacter, Alcaligenes, Rhizobium,* and *Sarcina* [70]. The extraction of cellulose from vegetable biomass requires physical (ultrasonic, microwave radiation, and high shear mechanical techniques) and chemical (acid/basic hydrolysis,

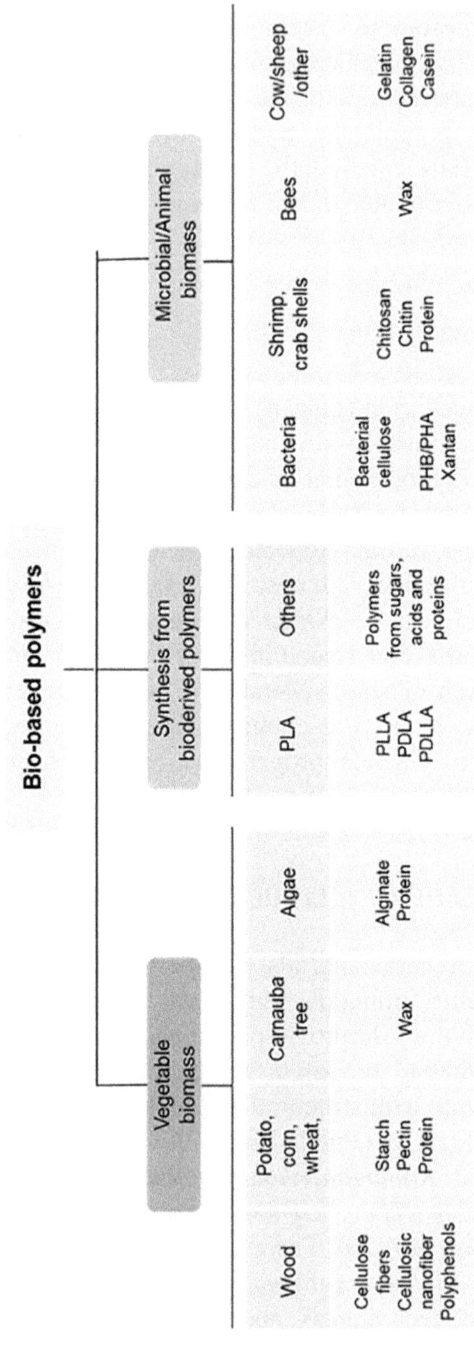

FIGURE 4.2 Bio-based polymers applied as a coating for food packaging.

TEMPO-mediated oxidation/carboxylation, and/or enzymatic hydrolysis) treatments [26, 96]. According to the intensity of the treatments, the cellulose obtained can result in CNF, microfibrillated cellulose [96]. On the other hand, crystalline nanocellulose can be obtained by acid hydrolysis. The use of vegetable or microbial nanocellulose can enhance mechanical and thermal properties and increase the gas, oil, and grease barrier in papers and cardboards, improving their biodegradability and recyclability [26, 60, 65]. Nanocellulose can be functionalized or nonfunctionalized, the first one for obtaining natural films with antimicrobial or antioxidant activity, and the other can be used to improve and reinforce the mechanical properties.

The application method and drying process of biopolymers for coating have relevance to the barrier properties of a coated paper. Fillat et al. [26] analyzed the influence of two metering systems (rod and blade) and two drying methods (hot polished surface and hot air jet) on the barrier properties of a commercial recycled paper coated with a 2% carboxymethylated nanofibrils (CNFs) (solids) suspension. CNFs mechanically prepared by super mass collider (m-CNFs), freeze-dried carboxymethylated nanofibrils (c-CNFs), and freeze-dried TEMPO nanofibrils (T-CNFs) were used and compared with cationic starch as a coating. Paper coated with a monolayer of c-CNFs allowed to reduce air permeance of the paper applied by a wound rod and blade micrometer, with a smaller extent in paper coated with T-CNFs and m-CNFs. Multilayer c-CNFs coating also showed the lowest air permeance (0.01 μm/Pa·s) and the best surface aspect. Air permeance is associated with the film formation and depends on chemical properties, particle size, and rheological properties of the polymers used, as well as by drying conditions. Furthermore, the density and compactability of the film is influenced by the mean porous diameter and hydrogen bonds in the nanocellulose structure that affects air permeability [60, 93].

Studies conducted by Gómez et al. [32] showed that c-CNF as a coating applied with similar metering systems and drying processes successfully improved the barrier properties of the commercial paper. WVP value at 90% RH and 20°C decreased slightly when a monolayer of CNFs was applied. However, by increasing the CNF coat weight on the surface through multilayers, the WVP increased, probably due to the damage promoted on the fibrillar and porous structure of the coated paper by consecutive wetting and drying cycles. cCNF coating allowed to reduce the air permeance, decrease the wettability rate of the samples (about 60% compared to reference paper), and increase the resistance to oil–grease

(grease resistance >1800 s and TAPPI UM 557 kit test No. >10). In general, paper coated with CNFs showed a contact angle lower than 62° due to the hydrophilicity of the nanocellulose structure, associated with the large number of hydroxyl groups.

Besides gas permeability and grease resistance, the addition of nanocellulose to wood pulp can improve the mechanical properties of papers. The tensile index, burst index, and tear index values were almost doubled by increasing the wet bacterial cellulose content (1–5 wt%) in sheets produced from softwood pulp [28]. The period of dispersion and consistency of the wet bacterial cellulose suspension are important keys to be considered. Thus, dispersion time higher than 3 mins for a 0.3% suspension at 11,000 rpm can shorten fibers and affect the properties of the sheets. Higher consistency of fiber could result in a good sheet formation.

4.4.3 CHITOSAN

Chitosan is a linear polysaccharide derived from chitin, a natural biopolymer found in the shells of crustaceans (including crab, shrimp, and lobster) composed of residues of glucosamine and N-acetylglucosamine. This biopolymer is soluble in water and acidic aqueous media, depending on its molecular weight and acetylation degree (high degree of acetylation prevents its solubilization) [3]. It exhibits antimicrobial properties and can be incorporated into paper coatings or films to extend the shelf life of food products. Many studies have been conducted to apply chitosan coating using several coating and drying techniques in different papers. Good film formation and surface closure were observed in scanning electron microscopy images when a coating weight of up to 2.4–6 g/m^2 was coated with rod and dried under ambient conditions [53, 54] and using an applicator roll in a laboratory film press unit and drying by infrared and hot air [55]. In these studies, an important decrease in air permeance was measured, reaching values below 0.001 nm/Pa s and an airflow close to 0 mL/min; good grease repellency (kit No. 12) and grease repellency up to 1500 s were achieved.

Some authors reported a decrease in WVTR from 30% to 70% [8, 55, 81], whereas no improvement was achieved by Kjellgren 2006. When comparing water absorption in different studies, some decrease in Cobb (water absorption capacity test) and an increase in contact angle are produced in soft-sized or paper without sizing. On the other hand, no improvements are found when the initial base papers show low water absorption or high contact angle [8, 9, 55, 81, 90, 99]. Mechanical properties such as tensile strength or tensile index were improved by chitosan coating [55, 99]. Antimicrobial activity was observed when chitosan is impregnated in paper and bacterial cellulose [11], but it seems that bactericidal properties decrease when chitosan is in solid form [90].

4.4.4 PROTEINS/AMINO ACIDS

Plant and animal-derived proteins are complex macromolecules which contain amine (–NH_2) and carboxyl (–COOH) functional groups suitable to react with other natural polymers to obtain a biodegradable coating for improving the mechanical strength and barrier properties of papers. Wheat gluten (WG), calcium caseinate, and corn gluten meal have been reported as good gas barriers and good capacity to form films [30]. The surface paper preparation for coating is an important factor for the penetration of the biopolymer. Sodium caseinate is a compound derived from casein, a protein obtained from mammalian milk and main subproduct of cheese production. It is obtained by microfiltration, ultrafiltration, coagulation, or precipitation. Films produced from caseinate are stable to variations of pH, temperature, and salt concentrations, with good mechanical properties. Besides, caseinate shows low OP associated with its high structural cohesion and reduction of porosity because of its polarity [51].

WG coating was prepared in a sodium sulfate solution (as a reducing agent) and used to coat treated paper with calcium carbonate and starch and untreated paper. A coating of 20.7 g/m^2 of WG promoted a strong interaction with the untreated cellulose structure (composite-like structure) reducing WVP, gas permeability (oxygen and carbon dioxide), and creating an oil barrier [33].

4.4.5 LIGNIN

Lignin is a widely distributed and abundant biopolymers in nature; it is obtained from vegetable biomass and as a byproduct of the paper industry and biofuel production. Its valorization, beyond the energy value, could substantially improve the economics and environmental performance of the paper industries and future lignocellulosic biorefineries. This biopolymer consists of a complex network of phenolic compounds mainly composed of synapyl, *p*-coumaroyl alcohols, and coniferyl phenylpropanoid monomers [79]. Currently, the market for lignin and its derivatives is growing at a rate of 2.28% per year, from $599 million in 2014 to $704 million in 2022 [49]. Lignin can be obtained by solubilization from lignocellulosic fraction, followed by separating the solid residue from the spent liquor (Kraft, Organosolv, soda, and sulphite pulping methods), and processes of selective hydrolysis of polysaccharides and separation of the solid fraction rich in lignin [27]. The fractionated lignin obtained will depend largely on the biomass raw material and the extraction method used, so its selection will depend on the desired final application [49, 105].

The complex structure of lignocellulosic biomass is one of the main obstacles to its conversion into chemical products. Therefore, a pretreatment stage is required in order to separate and maximize the recovery of each of the main components [29]. Lignin structure can be chemically modified to enhance its compatibility with paper fibers and used as a renewable barrier coating in food packaging applications.

Different forms of lignin applications for producing active coating for paper include the use of lignin by itself, as part of thermoplastics, blended with other substances, and as a precursor for thermoset resin [83]. Modifications of paper surface by combining water-soluble Kraft lignin (cationic) and lignosulphonate (anionic) allowed for promoting lower air permeability and surface energy (decrease in surface polarity) [77]. The electrostatic attraction between both compounds created insoluble films with strong hydrophobic properties. Most biopolymers have limited use as paper coating due to the high solubility in water, being easily removed by moisture or by liquid contact [23, 67].

Formulations based on starch and Kraft lignin showed capacity for reducing the water solubility of the carbohydrate in dry coatings [45]. However, it was reported that partial migration of lignin to the solution occurred in presence of starch at low pH. Adding zirconium carbonate as a cross linker reduced the migration of lignin from the coating and increased the water contact angle (hydrophobicity) of the base paper.

Xie et al. [102] reported the immobilization of Ag in lignin-polyurethane (PU) mixture for paper coating that resulted in improved mechanical properties, thermal stability, and antibacterial properties when applied on paper packaging. Inhibition of *E. coli* and *S. aureus* was observed for formulations containing Ag^+ and less than 40% of lignin (<1 cfu/mL). Furthermore, no inhibition zone (microbial halo) was observed in Petri plate agars containing paper coated with the mixture, indicating that the Ag was firmly anchored in lignin-PU coating. Despite the synthetic origin of the PU, there are studies of biodegradation of this polymer and its potential combination with lignin for coatings [40, 41].

4.4.6 ALGAE POLYMERS

Polymers derived from algae, such as alginate or carrageenan, offer biodegradability and barrier properties suitable for food packaging [20, 84, 108]. Alginate is a water-soluble biopolymer extracted from the cell wall of brown algae and traditionally used in food, pharmaceutical, and packaging industries [20]. In the papermaking industry, it is used to improve smoothness in sheets, enhance grease–oil resistance, increase strength, and improve ink holdout and gloss. Furthermore, this biopolymer, alone or in combination with other biopolymers, improves the flow properties, runnability, and water retention of paper coating during application and metering [55]. The last property is important to preserve the visual aspect, avoiding binder migration and mottle on the paper surface during the application and drying process. The combination of sodium alginate with other biopolymers, such as hydroxypropyl methylcellulose (a safety polysaccharide), promoted by hydrogen bonds within the hydroxyl and hydroxypropyl groups and the oxygen atom of alginate, results in films with a good barrier to grease and oxygen [108].

4.4.7 LIPIDS AND NATURAL WAX

Paraffin waxes obtained from petroleum, coal, distillates of wood or shale oil have been traditionally used as a coating of papers to improve their barrier properties against water, for packaged foods [50, 66]. However, these compounds are being substituted as coatings for food or food packaging due to environmental concerns and their negative effects on human health. Lipids of long-chain fatty acids and natural waxes can be used as coating to reduce gas and WVP, wettability, and to increase oil resistance in paper for food packaging applications. Coated papers with wax present some disadvantages compared with other polymers used for coating, such as low dispersion or homogeneity on the coated surface, brittleness, and cracks.

Natural waxes such as carnauba wax, beeswax, and rice bran wax appear as potential alternatives to paraffin. Recently, carnauba wax, a powder obtained from the outer surface of leaves of the Brazilian palm tree (*Copernicia prunifera*), has attracted a lot of interest in recent years as a feasible biopolymer for coatings [16, 23]. This natural wax is rich in aliphatic esters, diesters of cinnamic acids, aromatic compounds, and free alcohols. The esters are predominantly composed of carbon chains in the range of C46 and C60 and cinnamic acid in the form of *para*-hydroxyl and *para*-methoxy substituted [21, 36]. Carnauba wax is considered a GRAS additive and is widely used in the food industry as an acidity regulator, anticaking carrier, mass agent, lubricant, formulation aid, within other several applications—CODEX Alimentarius/INS 903 [101]. Carnauba wax is used to produce films for covering and preserving fresh fruits because of its hydrophobic and antimicrobial properties. Ethanolic extract of carnauba wax powder containing phenolic gallic acid, catechin, and chlorogenic acid showed inhibitory capacity against dermatophyte fungi *Trichophyton rubrum* and *Microsporum canis*. Also, it has important antioxidant properties, requiring concentrations of 365 and 317 μg/mL to scavenge 50% of 2,2-diphenyl-1-picrylhydrazyl or 2,2′-azino-bis (3-ethylbenzothiazoline-6-sulfonic acid) radicals (EC_{50}), respectively [19].

In general, the industrial process for coating papers requires a high temperature (> 100°C) and a controlled dry rate to avoid fast evaporation and defects on the coated surface. Using carnauba wax alone as a coating is unfeasible because of its relatively low melting point (82–86°C) and high cost. The mixture with other biopolymers (i.e., gelatin, starch, and among others) can promote a higher melting point and thermal stability during the coating process due to the interactions between the polar groups such as hydroxyl, carbonyl, and amino groups [16, 23].

A bilayer coating (15.0 g/m^2) of a mixture of different proportions of cationic starch and carnauba wax on kraftliner paper was prepared by Santos et al. [23]. In this work, the cationic starch (4.0 wt%) was gelled and mixed with glycerol (20 wt%) as a plasticizer, and the wax emulsion (10%, w/v) was prepared with the addition of a high content of polysorbate 80 as an emulsifier. This mixture promoted coating with high hydrophobicity verified by the high-water contact angle (100°), mainly associated to the roughness on the coating surface. However, the wettability of the coating was higher than that of the control, probably associated with the major accessibility of water to the hydroxyl groups promoted by the emulsifier.

Regarding the oil resistance using the oil kit test (TAPPI), coated paper with a blend of starch to wax of 3:1 showed a higher level of resistance to oil (Level/ Kit N° 9) than that coating containing a higher proportion of wax (1:3 starch to wax), which achieved level 5. This level is the limit or minimum safe oil resistance required for paper to be applied to packed food. According to the authors, the low resistance could be associated with the major content of nonpolar C–C bonds and a lower proportion of polar C–O bonds, as well as with the lower surface energy of the paper coating with starch and wax. Nonetheless, other parameters such as roughness of the surface should be considered since an increasing roughness associated with a noncontrolled dry process could increase the oil absorption of the coated paper. Furthermore, during controlled drying rates, cationic starch can form a film on the paper surface that increases the oil resistance by reducing or inhibiting the oil permeation into porous [63].

Another natural lipid that can be used as a coating for paper and offers an excellent barrier to WVP is that secreted by the wax glands of *Apis mellifera* and *Apis cerana*. The hydrophobicity and certain antimicrobial activity are conferred by long-chain saturated esters, alcohols, and acids. Food-grade linseed oil, previously epoxidized and acrylated, was combined with beeswax (10 wt%) to obtain a coating that reduced WVTR and improved the WCA (111°), preserving high values (> 90°) even after 50 cycles of abrasive treatments. The roughness imparted and the composition of the acrylated linseed oil-beeswax composite promoted high hydrophobicity and thermal stability of the coated paper [97].

4.4.8 OTHER POLYMERS

Other biodegradable and biocompatible polymers used as coating paper include PHAs), (PLA, and its amorphous copolymer poly(L-*co*-D,L-lactide), and PGA [17, 39, 88]. These polyesters are used to form a film structure on papers and other substrates, promoting a barrier toward oil–grease, moisture, and air, as well as the mechanical resistance of materials. As thermoformable, polyesters can be processed by different methods such as extrusion, thermoforming, injection molding, and fiber processing, making them easy industrial manipulation [15]. Table 4.1 shows studies of active coatings applied to paper and cardboard with potential applications in food packaging.

TABLE 4.1 Active Coating Applied to Paper and Paperboard with Potential Applications in Food Packaging

Biopolymer/Blends	Substrate	Function	Coat	Properties and Barrier Capacity	Reference
Whey protein, beeswax, and sucrose	Solid bleached sulfate paperboard, grammage not specified.	Wettability, grease repellence.	0.144 g (5 × 5 cm) paperboard, 57.6 g/m^2.	Increased water resistance and decreased WVP. High oil-barrier properties	[46]
Alginate and chitosan	100% primary-fiber furnish (mixture of hardwood and softwood), 72 g/m^2. 100% secondary or recycled fibers with no surface treatment, 129 g/m^2.	Wettability, air permeability, grease repellence, WVP.	6.0–6.1 g/m^2.	Water-vapor transmission rate was reduced by at least 35%. Reduction of air permeability. The permeation of deuterated *n*-alkanes was reduced by up to 50%. Grease resistance for alginate (Kit N° 7–12).	[55]
Nanofibrillated cellulose, (NFC) from *Eucaliptus* bleached kraft pulp paper.	Unbleached Pinus Kraft paper, 60 g/m^2.	Air permeability; water absorption, thermal stability, mechanical properties.	20 g/m^2 (1 mm surface deposition of NFC, 28 g/m^2 (2 mm surface deposition of NFC).	Reduction of water absorption between 55 and 58%. Reduction of air permeability. Thermal stability of the paper with NFC coatings met requirements for food packaging. Increase in burst index, tear index, and tensile index.	[60]
Pineapple peel cellulose, cellulose nanofibers, and carnauba wax	Cellulose nanocomposite.	Oxygen permeability, WVP, and thermal stability, mechanical properties.	Grammage not specified.	Wax increases hydrophobicity with water contact angle over 100. Material exhibited ultraviolet resistance and antioxidation properties. Reduce WVTR and oxygen permeability. Improve mechanical properties and thermal stability of the cellulose nanocomposite. Cellulose nanocomposite delayed postharvest deterioration and maintained storage quality of cherry tomatoes.	[108]

TABLE 4.1 *(Continued)*

Biopolymer/Blends	Substrate	Function	Coat	Properties and Barrier Capacity	Reference
Polylactic acid (PLA) with xanthan gum (thickening agent)	Industrial paper, 50 g/m^2.	WVP, air permeability.	5-15 g/m^2 (PLA-xanthan gum emulsion).	Reduction of WVP and WVTR when increasing the coating film. PLA coating weights gives better barrier properties to air permeability.	[1]
Mechanical, carboxymethylated, and TEMPO cellulose nanofibrils (CNF)	Recycled paper, 116–122 g/m^2.	Air permeability	12–100-µm-thick (wet CNF film), theoretical coat 0.24 to 10 g/m^2.	Decrease in air permeance. CNF coating was translucent, uniform and with gloss values like paper coated with plastics.	[26]
Mechanical, carboxymethylated, and TEMPO CNF	Recycled paper, 115 g/m^2.	WVP, wettability, air permeability, grease repellence	6.6–36.6 g/m^2 (wound rod coater). 7.0–33.4 g/m^2 (blade coater).	Air permeance decreased by the application of carboxymethylated CNF. Paper wettability rate decreased by 60%. Mechanical and TEMPO CNFs coatings provided no grease barrier properties; carboxymethylated CNF increased grease resistance (Kit N°>10).	[32]
Chitosan, bacterial nanocellulose, and ZnO	Kraft paper, 90 g/m^2.	Antibacterial activity, mechanical properties	12-µm-thick, grammage not specified.	High antibacterial response (assigned to the combination of chitosan and ZnO, independent of its shape and size) and boosted mechanical properties.	[43]
Chitosan and xylan	Paper made from unbleached pulp, 50 g/m^2.	WVP, wettability, air permeability, and grease repellence.	5 g/m^2.	Improvements in barrier properties for WVTR (30 g/m^2.day) and grease resistance (KIT N°8). Inhibition of *Bacillus sp, Aspergillus niger,* and *Penicillium sp.*	[110]
Chitosan	Paper from bamboo fibers-nanofibrilated cellulose, 84 g/m^2.	Mechanical properties		Improvement in the mechanical properties and wettability of the coated paper.	[90]

TABLE 4.1 *(Continued)*

Biopolymer/Blends	Substrate	Function	Coat	Properties and Barrier Capacity	Reference
Wheat gluten	Bleached pulp paper treated on two sides with calcium carbonate and starch, 108 g/m^2. Untreated paper, 36.5 g/m^2.	WVP, gas permeability, oil repellence	20.7–25.0 g/m^2.	Improvement of gas barrier properties of paper by reducing oxygen, carbon dioxide, and air permeability. WVP was decreased by 45–56%. Significant reduction in oil wettability.	[33]
Cationic starch and carnauba wax	Kraftliner paper, 86 g/m^2.	Water absorption Grease resistance.	15.0 g/m^2.	Mixture promotes high hydrophobicity and water contact angle (100°). Addition of carnauba wax to the cationic starch decreased the resistance to oil. Decrease water absorption and wettability.	[23]
Sodium alginate, hydroxypropyl methylcellulose-Layer 1	Paper, grammage not specified.	WVP, wettability, air permeability, and grease resistance	1–5 g/m^2.	Grease resistance (Kit N° 12), water resistance (Cobb value: 4.23 g/m^2), reduced air permeability, low WVTR, and improved mechanical properties.	[108]
Polyvinyl butyral and hydrophobic silica nanoparticles-Layer 2					
Kraft lignin and starch	Machine-glazed paper, 59.9 g/m^2.	WVP	9.4–9.7 g/m^2 (with a pilot coating machine with paper moving at 400 m/min).	WVTR of the pilot-coated paper was decreased significantly	[45]

4.5 OTHER POTENTIAL ACTIVE SUBSTANCES FOR PAPER COATING

Paper surfaces can be appropriately modified to impart a determined function or property for improving paper characteristics. The modification or functionalization of cellulose is favored by the presence of free OH groups that can be linked with different active molecules, such as biopolymers, vegetable oils, antioxidants, and others. The use of bio-based polymer formulations with these substances to coat paper surfaces could be a feasible solution to manufacture sustainable food packaging with barrier properties and a low carbon footprint [38, 80]. These substances interact with the packaged product or its environment to extend shelf life, improve safety, or enhance quality. Some of these bio-based polymers could potentially act on the foodstuffs environment by removing oxidants/contaminants or providing antimicrobial resistance [6].

Active substances include absorbers of oxygen, ethylene, and moisture, lysozyme (antioxidant) and metallic ions (antimicrobial) among others. Oxygen absorbers that prevent or inhibit oxidation reactions, including iron-based compounds, ascorbic acid, and enzymes like glucose oxidase, can be applied in coatings. Cherpinski et al. [17] reported the production of highly efficient oxygen-scavenging multilayered coatings of poly(3-hydroxybutyrate) and polycaprolactone polymer-containing palladium nanoparticles over a paper from bleached Kraft eucalyptus pulp, reducing the oxygen concentration (%volume) by 40–80% after 50 h. Also, it has been reported that paper coated with ethylene absorbers slow down the ripening process. Activated carbon impregnated with potassium permanganate was used as an ethylene absorber and fixed in paper a mixture of cassava starch and gelatin [2]. The paper was used as a corrugated sheet of corrugated cardboard showing faster ability to oxidize the ethylene from the atmosphere of storage when compared with untreated activated carbon corrugated cardboard. In another work, the corrugated cardboard covered with the same ethylene absorber showed that mango fruit quality was preserved after short-term storage [100].

Antimicrobial agents have been incorporated into coatings to produce papers for protecting foods. Most of the antimicrobial compounds prevent the surface adherence of microorganisms on the paper packaging, reducing the growth of bacteria, molds, and yeast in the storage products. Antimicrobial agents in active packaging include silver nanoparticles, organic acids, lysozymes, essential oil, and bacteriophages [58, 87]. For example, a bioactive paper developed by Lai et al. [58] showed highest antimicrobial activity of the alginate–nanosilver particle coating with bacteriophage incorporated into the alginate matrix. The secondary paper (not intended for direct contact with food) coated with microencapsulated citronella oil showed antimicrobial activity against *E. coli* and *Saccharomyces cerevisiae* after 24 h of cultivation. After 48 h of incubation, low growth of the microorganism was observed compared with the blank (culture plates without coated paper) [87]. Paper coated with chitosan solution (2% w/v) promoted an inhibition zone in agar plates inoculated with *E. coli* [106].

Lysozyme is an active alkaline protein with antimicrobial activity against bacteria based on the capacity to disrupt the peptidoglycan of the cell wall. This protein acts by breaking the 1,4-β-glucoside bonds between *N*-acetylmuramic acid and *N*-acetylglucosamine. Also, this active protein reduces the oxidation of lipids and other proteins by scavenging free radicals and other reactive oxygen species that contribute to oxidation [6, 107]. This enzyme is usually extracted from avian eggs and mammalian milk and can be hydrolyzed by proteases naturally present in fresh foods that can limit its use in packaging. Nonetheless, new strategies are being developed to produce lysozyme coating by combination with polyelectrolytes and other polymers (e.g., protein, pectin, alginate) for a controlled delivery [107]. Barbiroli et al. [6] reported the incorporation of lysozyme and lactoferrin into paper containing carboxymethyl cellulose that allowed noncovalent binding of the positively charged proteins to the paper matrix.

Furthermore, discoloration, flavor-off, and loss of nutritional value in packaged products are results of oxidation reactions in food. This effect can be inhibited or delayed by a coating containing antioxidants. Antioxidants scavenge free radicals and stabilize lipid molecules, helping to maintain product quality and extend shelf life. Battisti et al. [109] reported the production of paper coated with gelatin crosslinked with transglutaminase enzyme (using glycerol as plasticizer) and citric acid as antimicrobial and antioxidant agent. Beef packaged with the coated paper showed lower microbial concentration after 4 days of storage and stability to lipid oxidation, compared with initial conditions.

4.6 FABRICATION STRATEGIES OF ACTIVE BARRIER COATING ON PAPER

The industrial coating is applied to a wide variety of commercial papers to protect and prevent damage of the surface and the contained product.

The paper-based packaging would have new functionalities that contribute to extending the shelf life of food. Therefore, the design of these innovative active packaging introduces complexity in the manufacture because the solution must be economically feasible to implement industrial and its recyclability and/or biodegradability must be guaranteed.

For the implementation of bioactive polymers in the paper-based food packaging manufacture, the active barriers should be achieved with low amounts of these bioactive compounds, because the container cost must be minimal to avoid raising the price of perishable foodstuff, and the price of these additives is currently high [55, 71]. The coating suspension (bioactive + additives) is usually applied to one sheet side and then is dried and finished; as a result, the coating layer must exhibit uniform thickness and enough strength to withstand the packaging manufacturing stresses, keeping its integrity. In addition, the barrier components must not migrate into the food, which could cause spoilage or modify its organoleptic characteristics.

The runnability of biopolymer suspensions in different scenarios involved in paper surface coating is a key factor to obtain quality barriers, coupled with the base paper properties and operational variables of coating technology [52, 86]. Coating involves different phases, first a surplus of coating formulation is applied on the paper surface, subsequently it is metered, and then the wet film is dried and finally finished to achieve the surface structure targeted [62].

Typically, barrier properties are imparted by applying polymer coatings on the surface of papers and/or paperboards by different methods, such as extrusion coating, dispersion coating, dip coating, curtain coating, size press coating, bar coating, and blade coating, depending on the coating and base paper characteristics. However, the last three methods are the most widely employed in the manufacture of packaging paper and board. The typical industrial paper coating method is presented in Figure 4.3.

Coating by extrusion (Figure 4.3a) mostly refers to the method that extrudes a suspension or solution of polymers onto the surface of a moving base paper. This method is suitable to cover the paper with thermoplastic polymers; however, nonthermoplastic polymers, such as polysaccharides, proteins, and lipids, have severe difficulties to be applied by this technique, being required dissolving or dispersing the polymer in a compatible thermoplastic solvent. The major drawback of extrusion is the high coat weight necessary to achieve the required properties. This renders the

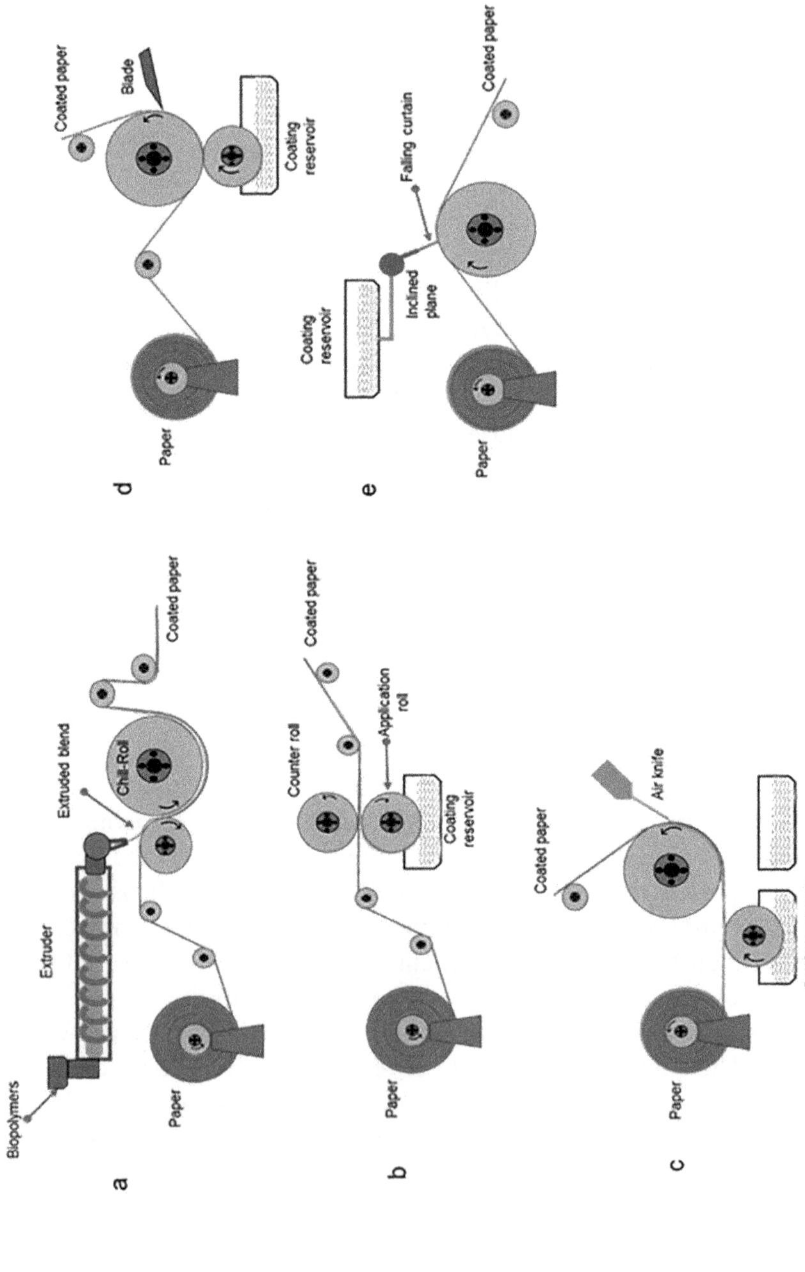

FIGURE 4.3 Typical industrial paper coating processes. (a) Extrusion coating; (b) size press coating; (c) air knife coating; (d) blade coating; and (e) curtain coating.

application of this type of biopolymer economically unfeasible, as they are often expensive [7].

The size-press coating (Figure 4.3b) is suited to apply a coating film with low-suspension solid concentration, which is absorbed into the paper, then the paper web is compressed in the roll nip, and the coating suspension penetrates the external sheet structure, and finally the film split takes place. This technology is well established in the paper industry and is usually used to size paper with starch-based suspension. Laboratory equipment is also available to apply sizing emulsion on paper surfaces and has been used to apply biopolymers [34]. However, several authors have reported that the low solid content of coating dispersions leads to uneven coverage of the paper surface and does not provide the required barrier properties [50, 95]. Other methods applied in industry include the air knife coating (Figure 4.3c) and blade coating (Figure 4.3d).

Typically, paper coating methods for laboratory scale include the dip coating, bar coating and curtain coating. Size press coating, dip coating, curtain coating, bar coating, and blade coating achieve the full surface coverage of paper surface with low coating weights [80]. Therefore, low amounts of active biopolymers could be dispersed in the coating formulation to get the desired barrier properties. Dip coating involves dipping the paper into an aqueous coating solution, soaking both paper sides, and then the coating surplus is removed in different ways and dried. This easy application method is widely used at laboratory scale [89]. However, the lack of control in the coating thickness applied on each side and the impregnation of the whole fibrous structure, consuming an unnecessary amount of suspension, are two drawbacks that consequently limit its industrial application.

The curtain coating technology (Figure 4.3e) has become increasingly interesting for the industry since it allows higher coating weight, uniformity, and better gas barrier properties. This method consists in creating an uninterrupted curtain of coating suspension that falls over paper surface while being transported on a conveyor belt or calendar rolls at a regulated speed through the curtain to ensure uniform coating [53, 80].

In contrast, coating stations are used to cover the paper with coating color solutions with the solids concentrations as high as possible (50–70%). Also, they can be applied to a functional coating with solids content between 10% and 30% [62]. Among the different metering systems, rod and blade are suitable to coat paper surfaces with low solid content (less than 10%), assuming the coating slurry has an allowable viscosity. In this coating process, the rheological features of the coated substance are crucial. Spray coating is another possibility to improve the barrier in papers based on the spraying of polymers through tiny nozzles; however, the concentration of solids in the polymeric solution must be considered to avoid clogging of dispersion holes.

After coating, a critical factor is how the drying of the wet biopolymer coating is carried out; the dehydration rates of the slurry affect the integrity and quality of the final coated layer. In addition, biopolymer suspensions usually have a low solid content (10–15%), which leads to a high-water removal at this stage, and this limits the industrial application, as drying of the coating is an energy-intensive process, and therefore a very expensive process [48, 55, 71].

The importance of the drying process in coating quality is well known in papermaking; several methods are used to dry wet coating, such as contact with a hot surface, hot air, and infrared, being hot air and oven drying are used preferentially in a laboratory scale. However, very few studies are available about how the drying method affects the coating's quality; The results of these studies reveal that drying must be fast, to avoid deep penetration of biopolymers into the paper structure, keeping them on the surface, especially on porous papers. On the other hand, an evaporation rate of coating too fast leads to the breakdown of the coating layer, as the viscosity of the coating increases more rapidly [26, 32, 48, 55, 57]. Therefore, it is necessary to determine the most suitable conditions for each type of polymer-based coating to achieve the target barrier properties.

Finishing operations in papermaking, such as smoothing and calendaring (process of smoothing and compressing paper), determine the structural properties of coated paper. Surface roughness defines the porous structure of coated paper, such that rougher paper has a larger surface pore size than smoother paper.

4.7 BIODEGRADABILITY AND RECYCLABILITY

Typically, the cellulose-based packaging is laminated with flexible aluminum and plastic foils to reduce the wettability, the water vapor, and the oxygen transmission. Consequently, most paper-based packaging with barrier properties contains unsustainable and nonbiodegradable materials [10]. Paper and board are known as biodegradable and highly recycled materials that contribute to the concept of circularity and environmental equilibrium. Nowadays, paper and paperboard present a recycling rate of 82.5% against 39.7% for plastic, 74.8% for metallic, and 74.9% for glass [25]. However, most paper-based packaging for food contains plastics as coatings that are nonrepulpable or are of limited repulpability, recyclability, biodegradability, and compostability [47, 73]. Usually, the reutilization or recycling of non-natural polymers (containing paraffin, plasticizers, and others) releases substances that negatively affect the environment (per- and polyfluoroalkyl substances, benzene, bisphenol, styrene, aldehydes, and phthalates). Today, the main challenge of the coated paper and paperboard is the compostability and repulpability of these materials to produce recycled products [37, 47].

In order to maintain the sustainability and circularity of the paper/board value chain, it is important to ensure that new products based on paper and board are recyclable by the paper industry. To achieve this goal, new coatings based on biopolymers of easy recyclability and biodegradability are being developed. Nevertheless, information about the recyclability and biodegradability of these products is scarce.

The biodegradability of papers can be investigated by the soil burial method. This method consists of paper samples of defined dimensions (4 cm × 4 cm), weighed and placed about 10 cm beneath the soil during 50 days at 24°C to allow natural degradation. During this period, a paper sample is removed every 10 days and dried at 40°C to determine the weight loss until the constant weight of each sample, expressing the biodegradability rate as a percentage of weight loss [8, 104]. The requirements for packaging recoverable through composting and biodegradation are described in EN 13432: 2001 (UNE, 2019).

A paper of 80 g/m^2 obtained from corn husk and sugarcane bagasse coated with chitosan and bees-wax-chitosan (with improved mechanical and barrier properties) was analyzed regarding its biodegradability by the soil burial method. Uncoated paper showed a faster rate of degradation (within 30 days), followed by paper coated with chitosan, and beeswax-chitosan that achieved complete degradation in about 40–50 days [8]. The low rate of biodegradation of the coating papers depends upon several factors, such as the accessibility of the soil microorganisms and synthesis of enzymes for degrading the coating polymers, humidity, temperature, erosion, and porosity of the sample within others. Nazhad et al. [72] tested the biodegradation of polyester-based coated boards (polyester-based resin and PLA) and found that the materials were biodegradable, with PLA-coated layers degraded faster. Similar results were found by Helanto et al. [39] when studying the influence of minerals in PLA-coated paperboard, which also found low recyclability in these boards due to problems in repulpability.

The repulpability of paper and cardboard is the ability of these materials (fibers) to be redispersed in water, and it is a fundamental requirement of an efficient recycling process. However, the repulpability of coated paper has certain limitations because wax and some polymers cannot be easy to release from the fibers and can cause undesirable staining and defects on the surface during the recycling paper production. The variables to be considered for the repulpability of a coating paper are the water temperature, chemical treatments (basic/acid treatment), pH, specific characteristics of the pulper (mill for cellulose disaggregation), and nature of the coating used [69]. In general, for the repulpability test, 50 g of samples of 2.5 cm × 2.5 cm is placed in a dish containing 950 g of distilled water (23°C), mixed for 15–17 min at 3000 rpm. The slurry formed is diluted to 1% consistency, mixed for 4 min, and an aliquot of 200 mL is spread into a glass tray for visual and microscopic observation of fiber aggregates (fiber bundles), undispersed coating pieces, colored particles, or other noncellulose constituents [39]. The recyclability of fiber-based packaging can be estimated based on the mass percentage of rejected material (noncellulosic materials retained in a sieve) of the total mass of repulped material. The harmonized test to quantify the recyclability of paper and cardboard products in standard paper and board recycling mills was recently published by the CEPI [13]. A fiber-based packaging is acceptable as recyclable if the total mass of rejected material is lower than 20% [82]. The recyclability can be estimated as follows:

$$\mathrm{TR} = \mathrm{CR} + \mathrm{FR} \times a \tag{4.3}$$

$$Y = 100\% - \mathrm{TR} \tag{4.4}$$

where TR is the total reject (%); CR is the coarse reject rate (%); FR is the fine reject rate (%); a is the correction factor; and Y is the yield as mass percentage of material that can be reused (%).

Polyesters such as PLA represent an alternative to traditional nondegradable polymers used as coatings, because of their potential to replace wax coatings to produce hydrophobic coating papers. Nechita and Roman [73] describe the polysaccharides used in coating for food packaging papers, describing the recyclability as one of the advantages of these coatings. On the other hand, nanofibrillar cellulose-based products are biodegradable and compostable [98]. Paper and board packaging sustainability is well rated not only by consumers but also by scientific facts [76]. In general, consumers judge packaging material by criteria of circular economy, natural-looking material, and design.

4.8 FUTURE TRENDS AND OPPORTUNITIES

Biopolymer active coating paper is a technology that offers several advantages for food packaging, such as its relatively low cost, mechanical resistance, barrier to grease, oil, water, vapor, and gas (under certain conditions), and antibacterial and antioxidant capacities, in addition to improving the appearance and printability of the paper, recyclability, and biodegradability. However, the current and future viability of active coated paper products will depend on the easy preparation of the coating, the homogeneity and overall quality, and the possibility of industrial mass production that allows cost reduction. Actually, laboratory research shows great potential of the biopolymers as coating material, although most studies are still not conclusive or suitable for industrial applications, with the exception of starch which has been used for decades.

Furthermore, the expenses and cost of biopolymers are still high in comparison with petroleum-based polymers. It is necessary to explore and develop new blends and composites to obtain low-cost coatings with desirable properties. Also, biopolymer coatings show relatively low stiffness and strength when compared with some non-natural polymers. Nonetheless, it could be solved by a combination of biopolymers and composites (ceramics, metallic ions, and nanocomposites) to achieve a cheap coating with adequate mechanical and barrier properties.

In the future, the biopolymer active paper coatings technology can also be implemented for other uses in food production and conservation, such as encapsulated oil or flavor, which can be released from the paper packaging after thermal activation. Further studies should analyze in detail rheological properties of the biopolymeric solutions or polymeric blends used as coating, since the viscosity depends on the nature and concentration of these compounds. Moreover, significant influence on the preparation, application, and surface homogeneity of the coating are important issues. Novel coating technologies should be created to ensure barrier properties that preserve food characteristics and to facilitate the composting, repulpability, and recycling of paper and cardboard.

4.9 CONCLUSION

Petroleum-based polymers and wax are being replaced by natural biopolymers to obtain active and sustainable coating paper for food packaging. Extensive studies are being developed worldwide in this field to create new sustainable coatings with barrier and functional properties. However, the expansion and future of this field will depend on the characteristics and applicability of the biopolymer-based coating paper, production costs, functionality, compatibility of the emerging materials, and the potential for scale-up. The recyclability and compostability of natural coatings in paper and paperboard materials are advantages over plastic packaging, since this aspect is of increasing importance for the market and final consumers. Food packaging with new functionality and barrier properties has not yet been totally explored. New and sustainable coatings for the paper and cardboard industry are expected to be developed in the future.

ACKNOWLEDGMENT

Juan D. Rivaldi gratefully acknowledges the postdoctoral fellowship support from Programa Nacional de Becas Don Carlos Antonio López – BECAL-Paraguay (Contract N° 278/2022).

KEYWORDS

- **biopolymer**
- **coating paper**
- **food-grade packaging**
- **active barrier**
- **biodegradable**

REFERENCES

1. Abdenour, C., Eesaee, M., Stuppa, C., Chabot, B., Barnabé, S., Bley, J., Tolnai, B., Guy, N., Nguyen-Tri, P. (2023). Water vapor and air barrier performance of sustainable paper coatings based on PLA and xanthan gum. *Materials Today Communications*, 36, 106626.
2. Aprilliani, F., Warsiki, E., Iskandar, A. (2018). Kinetic studies of potassium permanganate adsorption by activated carbon and its ability as ethylene oxidation material. *IOP Conference Series: Earth and Environmental Science*, 141, 012003.
3. Aranaz, I., Alcántara, A.R., Civera, M.C., Arias, C., Elorza, B., Heras Caballero, A., Acosta, N. (2021). Chitosan: An overview of its properties and applications. *Polymers*, 13, 3256.
4. American Society for Testing and Materials—ASTM. Designation D3985-81—Standard Method of Test for Oxygen Gas Transmission Rate Through Plastic Film and Sheeting Using a Coulometric Sensor, 2018.
5. American Society for Testing and Materials ASTM Designation E96-66, Standard Method of Test for Water Vapor Transmission of Materials in Sheet Form, 2022.
6. Barbiroli, A., Bonomi, F., Capretti, G., Iametti, S., Manzoni, M., Piergiovanni, L., Rollini, M. (2012). Antimicrobial activity of lysozyme and lactoferrin incorporated in cellulose-based food packaging. *Food Control*, 26, 387–392.
7. Basak, S., Dangate, M.S., Samy, S. (2024). Oil-and water-resistant paper coatings: A review. *Progress in Organic Coatings*, 186, 107938.
8. Bhardwaj, A., Sharma, N., Alam, T., Sharma, V., Sahu, J.K., Hamid, H., Bansal, V., Alam, M.S. (2023). Development and characterization of chitosan and beeswax–chitosan coated biodegradable corn husk and sugarcane bagasse-based cellulose paper. *Waste and Biomass Valorization*, 14, 1625–1636.
9. Bhardwaj, S., Bhardwaj, N.K., Negi, Y.S. (2020). Effect of degree of deacetylation of chitosan on its performance as surface application chemical for paper based packaging. *Cellulose*, 27, 5337–5352.
10. Brouwer, M.T., Thoden van Velzen, E.U., Ragaert, K., ten Klooster, R. (2020). Technical limits in circularity for plastic packages. *Sustainability*, 12, 10021.
11. Cabañas-Romero, L.V., Valls, C., Valenzuela, S.V., Roncero, M.B., Pastor, F.J., Diaz, P., Martínez, J. (2020). Bacterial cellulose–chitosan paper with antimicrobial and antioxidant activities. *Biomacromolecules*, 21, 1568–1577.
12. Campanale, C., Massarelli, C., Savino, I., Locaputo, V., Uricchio, V.F. (2020). A detailed review study on potential effects of microplastics and additives of concern on human health. *International Journal of Environmental Research and Public Health*, 17, 1212.
13. CEPI - Confederation of European Paper Industries, 2022 https://www.cepi.org/wp-content/uploads/2022/10/Cepi-recyclability-laboratory-test-method_FINAL.pdf.
14. Chen, X., Xiao, B., Yang, Y., Jiang, Y., Song, X., Chen, F., Wang, W., Wu, J., Meng, Y. (2024). A novel paper-based composite film with enhanced oxygen and water vapor barrier properties. *Progress in Organic Coatings*, 186, 108042.
15. Cheng, H.Y., Yang, Y.J., Li, S.C., Hong, J.Y., Jang, G.W. (2015). Modification and extrusion coating of polylactic acid films. *Journal of Applied Polymer Science*, 132, 42472.

16. Cheng, Y., Zhai, X., Wu, Y., Li, C., Zhang, R., Sun, C., Wang, W., Hou, H. (2023). Effects of natural wax types on the physicochemical properties of starch/gelatin edible films fabricated by extrusion blowing. *Food Chemistry*, 401, 134081.
17. Cherpinski, A., Szewczyk, P.K., Gruszczyński, A., Stachewicz, U., Lagaron, J.M. (2019). Oxygen-scavenging multilayered biopapers containing palladium nanoparticles obtained by the electrospinning coating technique. *Nanomaterials*, 9, 262.
18. Chow, Y.N., Foo, K.Y. (2023). Insights into the per-and polyfluoroalkyl substances-contaminated paper mill processing discharge: Detection, phytotoxicity, bioaccumulative profiling, and health risk verification. *Journal of Cleaner Production*, 384, 135478.
19. da Silva Andrade, L.B., da Silva Julião, M.S., Cruz, R.C.V., Rodrigues, T.H.S., dos Santos Fontenelle, R.O., da Silva, A.L.C. (2018). Antioxidant and antifungal activity of carnauba wax powder extracts. *Industrial Crops and Products*, 125, 220–227.
20. Dave, P.K., Rao, T.R., Thakkar, V.R. (2023). Development of films and coatings from alginates. In: *Biopolymer-Based Films and Coatings* (pp. 101–120), CRC Press.
21. de Freitas, C.A.S., de Sousa, P.H.M., Soares, D.J., da Silva, J.Y.G., Benjamin, S.R., Guedes, M.I.F. (2019). Carnauba wax uses in food–a review. *Food Chemistry*, 291, 38–48.
22. Deshwal, G.K., Panjagari, N.R., Alam, T. (2019). An overview of paper and paper based food packaging materials: Health safety and environmental concerns. *Journal of Food Science and Technolgy*, 56, 4391–4403.
23. dos Santos, A.D.A., Matos, L.C., Mendonça, M.C., do Lago, R.C., dos Santos Muguet, M.C., Damásio, R.A.P., Ponzecchi, A., Soares, J.R., Sanadi, A.R., Tonoli, G.H.D. (2023). Evaluation of paper coated with cationic starch and carnauba wax mixtures regarding barrier properties. *Industrial Crops and Products*, 203, 117177.
24. European Union – EU 2019. https://eur-lex.europa.eu/legal-content/ES/TXT/?uri=CELEX%3A32019L0904
25. Eurostat 2022. Recycling rates for packaging waste. Information on: https://ec.europa.eu/eurostat/databrowser/view/ten00063__custom_10465666/default/table?lang=en.
26. Fillat, Ú., Vergara, P., Villar, J.C., Gómez, N. (2023). Structural properties of coated papers with cellulosic nanofibres using different metering systems and drying technologies. *Progress in Organic Coatings*, 179, 107543.
27. Galkin, M.V., Samec, J.S. (2016). Lignin valorization through catalytic lignocellulose fractionation: A fundamental platform for the future biorefinery. *ChemSusChem*, 9, 1544–1558.
28. Gao, W.-H., Chen, K.-F., Yang, R.-D., Yang, F., Han, W.-J. (2011). Properties of bacterial cellulose and its influence on the physical properties of paper, *BioResources*, 6, 144–153.
29. Garcia-Ochoa, F., Vergara, P., Wojtusik, M., Gutiérrez, S., Santos, V.E., Ladero, M., Villar, J.C. (2021). Multi-feedstock lignocellulosic biorefineries based on biological processes: An overview. *Industrial Crops and Products*,72, 114062
30. Gastaldi, E., Chalier, P., Guillemin, A., Gontard, N. (2007). Microstructure of protein-coated paper as affected by physico-chemical properties of coating solutions. *Colloids and Surfaces A: Physicochemical and Engineering Aspects*, 301, 301–310.
31. Glenn, G., Shogren, R., Jin, X., Orts, W., Hart-Cooper, W., Olson, L. (2021). Per-and polyfluoroalkyl substances and their alternatives in paper food packaging. *Comprehensive Reviews in Food Science and Food Safety*, 20, 2596–2625.

32. Gómez, N., Vergara, P., Fillat, Ú., Villar, J.C. (2024). Effect of metering systems and drying methods on the barrier properties of paper coated with multiple layers of cellulose nanofibres. *Progress in Organic Coatings*, 189, 108323.
33. Guillaume, C., Pinte, J., Gontard, N., Gastaldi, E. (2010). Wheat gluten-coated papers for bio-based food packaging: Structure, surface and transfer properties. *Food Research International*, 43, 1395–1401.
34. Ham-Pichavant, F., Sèbe, G., Pardon, P., Coma, V. (2005). Fat resistance properties of chitosan-based paper packaging for food applications. *Carbohydrate Polymers*, 61, 259–265.
35. Haridevamuthu, B., Raj, D., Chandran, A., Murugan, R., Seetharaman, S., Dhanaraj, M., Almutairi, B.O., Arokiyaraj, S., Arockiaraj, J. (2024). Sustainable food packaging: Harnessing biowaste of Terminalia catappa L. for chitosan-based biodegradable active films for shrimp storage. *Carbohydrate Polymers*, 329, 121798.
36. Harron, A.F., Powell, M.J., Nunez, A., Moreau, R.A. (2017). Analysis of sorghum wax and carnauba wax by reversed phase liquid chromatography mass spectrometry. *Industrial Crops and Products*, 98, 116–129.
37. He, Z., Chowdhury, A., Tong, L., Reynolds, M., Ni, Y. (2019). Cellulose paper-based strapping products for green/sustainable packaging needs. *Paper and Biomaterials*, 4, 54.
38. Helanto, K., Matikainen, L., Talja, R., and Rojas, O.J. (2019). Bio-based polymers for sustainable packaging and biobarriers: A critical review, *BioResources*, 14, 4902–4951.
39. Helanto, K., Talja, R., Rojas, O.J. (2022). Mineral-filled biopolyester coatings for paperboard packaging materials: Barrier, sealability, convertability and biodegradability properties. *Nordic Pulp & Paper Research Journal*, 37, 212–221.
40. Howard, G.T. (2002). Biodegradation of polyurethane: A review. *International Biodeterioration & Biodegradation*, 49, 245–252.
41. Howard, G.T., Blake, R.C. (1999). Growth of Pseudomonas fluorescens on a polyester-polyurethane and the purication and characterization of a polyurethanase-protease enzyme. *International Biodeterioration & Biodegradation*, 42, 213–220.
42. Hubbe, M.A, Pruszyński, P. (2020). Greaseproof paper products: A review emphasizing ecofriendly approaches. *Bioresources*, 15, 1978–2004.
43. Jabłońska, J., Onyszko, M., Konopacki, M., Augustyniak, A., Rakoczy, R., Mijowska, E. (2021). Fabrication of paper sheets coatings based on chitosan/bacterial nanocellulose/ZnO with enhanced antibacterial and mechanical properties. *International Journal of Molecular Sciences*, 22, 7383.
44. Jasmani, L., Ainun, Z. M.A., Adnan, S., Ibrahim, R., Sapuan, S.M., Ilyas, R.A. (2021). Sustainable paper-based packaging. *Bio-based Packaging: Material, Environmental and Economic Aspects*, 225–244.
45. Javed, A., Rättö, P., Järnström, L., Ullsten, H. (2021). Lignin-containing coatings for packaging materials—Pilot trials. *Polymers*, 13, 1595.
46. Jeong, S., Yoo, S. (2020). Whey protein concentrate-beeswax-sucrose suspension-coated paperboard with enhanced water vapor and oil-barrier efficiency. *Food Packaging and Shelf life*, 25, 100530.
47. Kathuria, A., Zhang, S. (2022). Sustainable and repulpable barrier coatings for fiber-based materials for food packaging: A review. *Frontiers in Materials*, 9, 929501.

48. Khairunnisa-Atiqah, M.K., Salleh, K.M., Ainul Hafiza, A.H., Nyak Mazlan, N.S., Mostapha, M., Zakaria, S. (2022). Impact of drying regimes and different coating layers on carboxymethyl cellulose cross-linked with citric acid on cotton thread fibers for wound dressing modification. *Polymers*, 14, 1217.
49. Khan, R.J., Lau, C.Y., Guan, J., Lam, C.H., Zhao, J., Ji,Y., Wang, H., Xu, J., Lee, D.-J., Leu, S.Y. (2022). Recent advances of lignin valorization techniques toward sustainable aromatics and potential benchmarks to fossil refinery products. *Bioresource Technology*, 346, 126419.
50. Khwaldia, K., Arab-Tehrany, E., Desobry, S. (2010). Biopolymer coatings on paper packaging materials. *Comprehensive Reviews in Food Science and Food Safety*, 9, 82–91.
51. Khwaldia, K., Perez, C., Banon, S., Desobry, S., Hardy, J. (2004). Milk proteins for edible films and coatings. *Critical Reviews in Food Science and Nutrition*, 44, 239–251.
52. Kimpimäki, T., Savolainen, A.V. (1997). Barrier dispersion coating of paper and board. In: (Eds.), *Surface Application of Paper Chemicals*. Springer: Dordrecht.
53. Kjellgren, H., Gällstedt, M., Engström, G., Järnström, L. (2006). Barrier and surface properties of chitosan-coated greaseproof paper. *Carbohydrate Polymers*, 65, 453–460.
54. Kjellgren, H., Engström, G. (2006). Paper and paper coating performance chemicals: Influence of base paper on the barrier properties of chitosan-coated papers. *Nordic Pulp & Paper Research Journal*, 21, 685–689.
55. Kopacic, S., Walzl, A., Hirn, U., Zankel, A., Kniely, R., Leitner, E., Bauer, W. (2018). Application of industrially produced chitosan in the surface treatment of fibre-based material: Effect of drying method and number of coating layers on mechanical and barrier properties. *Polymers*, 10, 1232.
56. Kukkamo, V., Laari, J. (2020). Importance of rheology and viscosity in paper and board coating. *Japan Tappi Journal*, 74, 260–264.
57. Kumar, V., Ottesen, V., Syverud, K., Gregersen, Ø.W., Toivakka, M. (2017). Coatability of cellulose nanofibril suspensions: Role of rheology and water retention. *BioResources*, 12, 7656–7679.
58. Lai, T.T., Pham, T.T.H., van Lingen, M., Desaulniers, G., Njamen, G., Tolnai, B., Jabrane, T., Moineau, S., Barnabé, S. (2022). Development of antimicrobial paper coatings containing bacteriophages and silver nanoparticles for control of foodborne pathogens. *Viruses*, 14, 2478.
59. Lavoine, N., Desloges, I., Khelifi, B., Bras, J. (2014). Impact of different coating processes of microfibrillated cellulose on the mechanical and barrier properties of paper. *Journal of Materials Science*, 49, 2879–2893.
60. Lengowski, E.C., Franco, T.S., Viana, L.C., Bonfatti Júnior, E.A., de Muñiz, G.I. (2023). Micro and nanoengineered structures and compounds: Nanocellulose. *Cellulose*, 30, 10595–10632.
61. Lin, D., Zhao, Y. (2007). Innovations in the development and application of edible coatings for fresh and minimally processed fruits and vegetables. *Comprehensive Reviews in Food Science and Food Safety*, 6, 60–75.
62. Linnonmaa, J., Trefz, M. (2000). Pigment coating techniques. Pigment coating and surface sizing of paper. In: Gullichsen, J., Paulapuro, H. (Eds.), *Papermaking Science and Technology. Book. 11* (pp. 24–28). Fapet Oy: Helsinki.

63. Long, Z., Wu, M., Peng, H., Dai, L., Zhang, D., Wang, J. (2015). Preparation and oil-resistant mechanism of chitosan/cationic starch oil-proof paper. *BioResources*, 10, 7907–7920.
64. Long, Z., Wu, M., Peng, H., Dai, L., Zhang, D., & Wang, J. (2015). Preparation and oil-resistant mechanism of chitosan/cationic starch oil-proof paper. *BioResources*, 10(4), 7907–7920.
65. Lourenço, A.F., Martins, D., Dourado, F., Sarmento, P., Ferreira, P.J., Gamelas, J.A. (2023). Impact of bacterial cellulose on the physical properties and printing quality of fine papers. *Carbohydrate Polymers*, 314, 120915.
66. Mansoori, G.A., Barnes, H.L., Webster, G.M. (2004). Petroleum waxes. Manual 37—fuels and lubricants handbook: Technology, properties, performance, and testing. *ASTM Manual Series*, 37, 525–556.
67. Maurer, H.W. (2009). Starch in the paper industry. In: BeMiller, J., Whistler, R. (Eds.), *Starch: Chemistry and Technology* (pp. 657–713). Academic Press.
68. Maurer, H.W., Kearney, R.L. (1998). Opportunities and challenges for starch in the paper industry. *Starch-Stärke*, 50, 396–402.
69. Michelman, J.S., Capella, D.M. (1991). Repulpability of coated corrugated paperboard, *Tappi Journal*, 74, 79–82.
70. Morgan, J.L., McNamara, J.T., Zimmer, J. (2014). Mechanism of activation of bacterial cellulose synthase by cyclic di-GMP. *Nature Structural & Molecular Biology*, 21, 489–496.
71. Mujtaba, M., Lipponen, J., Ojanen, M., Puttonen, S., Vaittinen, H. (2022). Trends and challenges in the development of bio-based barrier coating materials for paper/cardboard food packaging; a review. *Science of the Total Environment*, 851, 158328.
72. Nazhad, M.M., Sridach, W., Retulainen, E., Kuusipalo, J., Parkpian, P. (2006). Biodegradation potential of some barrier-coated boards in different soil environments. *Journal of Applied Polymer Science*, 100, 3193–3202.
73. Nechita, P., Roman, M. (2020). Review on polysaccharides used in coatings for food packaging papers. *Coatings*, 10, 566.
74. Ni, S., Zhang, H., Godwin, P.M., Dai, H., Xiao, H. (2018). ZnO nanoparticles enhanced hydrophobicity for starch film and paper. *Materials Letters*, 230, 207–210.
75. Ni, S., Zhang, H., Dai, H., Xiao, H. (2018). Starch-based flexible coating for food packaging paper with exceptional hydrophobicity and antimicrobial activity. *Polymers*, 10, 1260.
76. Otto, S., Strenger, M., Maier-Nöth, A., Schmid, M. (2021). Food packaging and sustainability—Consumer perception vs. correlated scientific facts: A review. *Journal of Cleaner Production*, 298, 126733.
77. Pinto, P.I., Magina, S., Fateixa, S., Pinto, P.C., Liebner, F., Evtuguin, D.V. (2022). Modification of paper surface by all-lignin coating formulations. *Materials*, 15, 7869.
78. Ragnarsson, M., Engström, G., Järnström, L. (2013). Coating: Porosity variations in coating layers–impact on backtrap mottle. *Nordic Pulp & Paper Research Journal*, 28, 257–263.
79. Rao, X., Barros, J. (2023). Modeling lignin biosynthesis: A pathway to renewable chemicals. *Trends in Plant Science*, 29, 546–559. doi: 10.1016/j.tplants.2023.09.011
80. Rastogi V.K, Samyn P. (2015). Bio-based coatings for paper applications. *Coatings*, 5:887–930.

81. Reis, A.B., Yoshida, C.M., Reis, A.P.C., Franco, T.T. (2011). Application of chitosan emulsion as a coating on Kraft paper. *Polymer International*, 60, 963–969.
82. Runte, S., Putz, H.J., Bussini, D., Limongi, L., Elegir, G. (2015). Recyclability criteria for paper-based packaging products. *Cellulose Chemistry and Technology*, 49, 8.
83. Ruwoldt, J., Blindheim, F.H., Chinga-Carrasco, G. (2023). Functional surfaces, films, and coatings with lignin–a critical review. *RSC Advances*, 13, 12529–12553.
84. Shankar, S., Rhim, J.W. (2018). Antimicrobial wrapping paper coated with a ternary blend of carbohydrates (alginate, carboxymethyl cellulose, carrageenan) and grapefruit seed extract. *Carbohydrate Polymers*, 196, 92–101.
85. Sharma, M., Aguado, R., Murtinho, D., Valente, A.J., De Sousa, A.P.M., Ferreira, P.J. (2020). A review on cationic starch and nanocellulose as paper coating components. *International Journal of Biological Macromolecules*, 162, 578–598.
86. Smook, G. (2016). Overview of pulp and paper technology. In: Smook, G. (Eds.), *Handbook for Pulp and Paper Technologists*. TAPPI Press: Atlanta, GA, USA.
87. Šumiga, B., Šumiga, B., Ravnjak, D., Boh Podgornik, B. (2019). Antimicrobial paper coatings containing microencapsulated *Cymbopogon citratus* oil. *Coatings*, 9, 470.
88. Sundar, N., Kumar, A., Pavithra, A., Ghosh, S. (2020). Studies on semi-crystalline poly lactic acid (PLA) as a hydrophobic coating material on kraft paper for imparting barrier properties in coated abrasive applications. *Progress in Organic Coatings*, 145, 105682.
89. Tang, X., Yan, X. (2017). Dip-coating for fibrous materials: Mechanism, methods and applications. *Journal of Sol-Gel Science and Technology*, 81, 378–404.
90. Tanpichai, S., Witayakran, S., Wootthikanokkhan, J., Srimarut, Y., Woraprayote, W., Malila, Y. (2020). Mechanical and antibacterial properties of the chitosan coated cellulose paper for packaging applications: Effects of molecular weight types and concentrations of chitosan. *International Journal of Biological Macromolecules*, 155, 1510–1519.
91. Technical Association of the Pulp and Paper Industry- TAPPI Test Method T 559 cm-02; Grease Resistance Test for Paper and Paperboard. Tappi Press: Atlanta, GA, USA, 2002.
92. Technical Association of the Pulp and Paper Industry—TAPPI Test Method 458 cm-14: Surface wettability of paper (angle of contact method). Atlanta: Tappi Press. GA, USA, 2014.
93. Tyagi, P., Salem, K.S., Hubbe, M.A., Pal, L. (2021). Advances in barrier coatings and film technologies for achieving sustainable packaging of food products–A review. *Trends in Food Science & Technology*, 115, 461–485.
94. UNE 2001—Spanish Association for Standardization—EN 13432: 2001 Packaging—Requirements for Packaging Recoverable Through Composting and Biodegradation—Test Scheme and Evaluation Criteria for the Final Acceptance of Packaging.
95. Vartiainen, J., Vähä-Nissi, M., Harlin, A. (2014). Biopolymer films and coatings in packaging applications—A review of recent developments. *Materials Sciences and Applications*, 5, 708–718.
96. Velázquez, M.E., Ferreiro, O.B., Menezes, D.B., Corrales-Ureña, Y., Vega-Baudrit, J. R., Rivaldi, J.D. (2022). Nanocellulose extracted from paraguayan residual agro-industrial biomass: Extraction process, physicochemical and morphological characterization. *Sustainability*, 14, 11386.

97. Vijayan, S.P., Aparna, S., Sahoo, S.K. (2023). Effect of beeswax on hydrophobicity, moisture resistance and transparency of UV curable linseed oil-based coating for compostable paper packaging. *Industrial Crops and Products*, 197, 116645.
98. Vikman, J. Vartiainen, I. Tsitko, P. (2015). Korhonen, biodegradability and compostability of nanofibrillar cellulose-based products, *Journal of Polymers and the Environment*, 23, 206–215.
99. Vrabič Brodnjak, U., Tihole, K. (2020). Chitosan solution containing zein and essential oil as bio-based coating on packaging paper. *Coatings*, 10, 497.
100. Warsiki, E., Aprilliani, F., Iskandar, A. (2019). The effects of the use of corrugated cardboards covered with ethylene absorbers on mango fruit quality after short-term storage (L.). *Journal of Horticultural Research*, 27, 65–70.
101. World Health Organization (2011). Codex Alimentarius: general standard for food additives.
102. Xie, H., Zhang, H., Liu, X., Tian, S., Liu, Y., Fu, S. (2021). Ag immobilized lignin-based PU coating: A promising candidate to promote the mechanical properties, thermal stability, and antibacterial property of paper packaging. *International Journal of Biological Macromolecules*, 189, 690–697.
103. Yan, X., Meng, F., Van, T.T., Wigati, L.P., Nkede, F.N., Hamayoon, W., Tanaka, F. (2024). Water barrier coating of chitosan/poly(vinyl alcohol)/trans-cinnamaldehyde regulated postharvest quality and reactive oxygen species metabolism of 'Frutica'tomato (*Solanum lycopersicum* L.). *Postharvest Biology and Technology*, 212, 112910.
104. Yaradoddi, J.S., Banapurmath, N.R., Ganachari, S.V., Soudagar, M.E.M., Mubarak, N.M., Hallad, S., Hugar, S., Fayaz, H. (2020). Biodegradable carboxymethyl cellulose based material for sustainable packaging application. *Scientific Reports*, 10, 21960.
105. Yoo, C.G., Meng, X., Pu, Y., Ragauskas, A.J. (2020). The critical role of lignin in lignocellulosic biomass conversion and recent pretreatment strategies: A comprehensive review. *Bioresource Technology*, 301, 122784.
106. Zakaria, S., Chin, H.C., Ahmad, W.H.W., Kaco, H., Soon, W.C., Chi, H.C. (2015). Mechanical and antibacterial properties of paper coated with chitosan. *Sains Malaysiana*, 44, 905–911.
107. Zhang, Z., Zhou, X., Wang, D., Fang, C., Zhang, W., Wang, C., Huang, Z. (2021). Lysozyme-based composite membranes and their potential application for active packaging. *Food Bioscience*, 43, 101078.
108. Zhu, R., Fu, X., Jin, S., Ma, R., He, Z., Zhang, D., Long, Z. (2023). Water and oil-resistant paper materials based on sodium alginate/hydroxypropyl methylcellulose/polyvinyl butyral/nano-silica with biodegradable and high barrier properties. *International Journal of Biological Macromolecules*, 225, 162–171.
109. Battisti, R., Fronza, N., Júnior, Á.V., da Silveira, S.M., Damas, M.S.P., Quadri, M.G.N. (2017). Gelatin-coated paper with antimicrobial and antioxidant effect for beef packaging. *Food Packaging and Shelf life*, 11, 115–124. http://dx.doi.org/10.1016/j.fpsl.2017.01.009
110. Roman, M., Nechita, P., Vasile, M.A., & Cantaragiu Ceoromila, A.M. (2023). Barrier and antimicrobial properties of coatings based on xylan derivatives and chitosan for food packaging papers. *Coatings*, 13 (10), 1761. https://doi.org/10.3390/coatings13101761

CHAPTER 5

Starch-Based Biopolymers as Sustainable Food Packaging Materials

MAGALI ORDÓÑEZ-GARCÍA and
JUAN CARLOS BUSTILLOS-RODRÍGUEZ*

Tecnológico Nacional de México, Instituto Tecnológico de Cd. Cuauhtémoc, Mexico

*Corresponding author

ABSTRACT

Nowadays, with the globalization of our food supply, technology and logistics allow us to consume food from different regions of the world. However, the need for effective and sustainable food packaging materials is becoming increasingly important. To fulfill its main function, there is a variety of packaging materials that maintain and protect the properties and characteristics of the food. Currently, most of these materials are conventionally manufactured from plastic, glass, paper, cardboard, and metal. Unfortunately, most of these materials are not biodegradable, take a long time to integrate into the environment, and contribute to contamination, as they are mostly made from nonrenewable sources, particularly oil-based materials. Recently, to reduce the use of nonbiodegradable packaging, global legislation toward the food industry has been laid down, and eco-friendly materials are being taken as alternatives while ensuring the quality and safety of food. In this regard, biopolymers are a wide variety of natural polymers (e.g., polysaccharides, proteins, and lipids) derived directly from biomass and generally obtained from renewable sources. Starch, cellulose, chitin, lignin, and polylactic acid are among the most utilized biopolymers. These biopolymers have a wide range of industrial applications, including

Sustainable Food Packaging Materials and Technology. Juan Manuel Tirado-Gallegos, Juan Daniel Rivaldi Chávez, and José Juan Buenrostro-Figueroa (Eds)

DOI: 10.1201/9781779641434-5

the development and composition of food packaging material. Therefore, this chapter delves into the properties of starch-based biopolymers, their advantages over traditional packaging materials, and the latest advancements in their application.

5.1 INTRODUCTION

Packaging is known as a protective layer outside the food and tends to protect it. Various commercial polymers have been studied that contribute substantially to the formation of packaging films and are used in many industries, including food, electronic devices, medical, and pharmaceutical fields [1, 2]. Polymers are macromolecules that are widely used in several industrial applications and can be classified into synthetic and natural according to their source. Synthetic polymers are made up of petroleum-based resources; hence, they are limited, are not biodegradable (their decomposition period ranges from hundreds to sometimes thousands of years), have a short lifespan, and contribute to global pollution through waste generated, presenting a very serious environmental threat, including the terrestrial and aquatic ecosystems [3, 4]. Because of the mass production of huge quantities and the excessive usage of synthetic polymers (plastics), there exists a serious concern about the high toxicity and nonbiodegradability [5]. By 2050, it is expected that the annual output of global plastic will amount to 1800 million tons [6]. The manufacturing process of petroleum-based plastics is an energy-exhaustive technique and causes the emanation of a huge amount of greenhouse gases, such as carbon dioxide (CO_2) [7]. The CO_2 concentration and global temperature in 2019 were over 400 ppm and had risen by 1°C, respectively, which is a severe environmental issue [6]. Being necessary innovative sustainable solutions to attend waste management and contribute to environmental sustainability. On the other hand, natural polymers called biopolymers that possess plastic-like properties are obtained from biosources and are renewable, helpful toward the reduction of carbon footprint due to their origin from organic resources [8].

In recent decades, industry-generated packaging, including that used by the food sector, has caused great concern [9]. The packaging industry consumes the highest plastic volume and has become the primary source of the accumulation of plastic waste [10]. Currently, considerable interest in the application of biopolymers made from eco-friendly and renewable sources, using diverse polysaccharides, proteins, and lipids. Exist some categories of biopolymers, extracted from natural raw materials, such as cellulose, starch, protein, and marine prokaryotes; produced by chemical synthesis; and produced by microorganisms [11]. These biopolymers have been employed in the conformation of food packaging materials due to their biodegradability and biocompatibility, and nontoxic properties [12]. However, although biopolymers offer an environmentally friendly alternative, their industrial applications are limited [13]. The chemical composition of these biopolymers has a significant impact on the options that are available to manage plastic waste responsibly [5].

Nowadays, it is crucial to save packaged food from spoilage; hence, preplanned strategies to prevent food wastage and degradation are necessary [1]. One of the essential characteristics that a food packaging material must have is to be cost-effective; satisfy industry requirements and consumers (convenience and communication); reduce food waste; protect it from external influences (biological, chemical, and physical); and avoid affecting its smell, taste, and nutritional content conserve [2, 14]. For which, to know the viability of a material in food packaging applications, it is essential to investigate its properties [15]. Likewise, the presence of antimicrobials in packaging materials could be an alternative to enhance their efficiency [16]. Antimicrobial packaging, a promising version of active packaging, has received attention as a potential pathogen intervention strategy for various foods [17]. Based on the properties mentioned above, starch is one of the most widely used sources of natural materials as an alternative to the application of nonbiodegradable compounds in food packaging. Given that among all the biopolymers, starch stands out because of its abundance and low cost; the frequent use of starch-based products in the food industry is due to their characteristics such as edibility, biodegradability, ability to produce nontoxic films, and abundance and is indicated as one of the rising renewable resources [18]. The main starch sources are wheat, corn, potatoes, cassava, and rice. The most important difference between types of starch is the proportion of amylose and amylopectin that they have [19].

However, in the case of native starch, it presents poor mechanical properties, and it cannot be used directly as a plastic material; hence, to overcome these drawbacks, this starch can be converted into a thermoplastic material through the addition of a plasticizer; besides, starch should be modified completely or partially and transformed into amylose/amylopectin semicrystalline matrix under high pressure and temperature [20]. Starches present varied morphologies, structures, and properties depending on their extraction source, agronomic conditions, and genetic diversity from which the starch is extracted, and these factors influence the physical properties and consequently their usefulness, including food packaging applications [21, 22].

5.2 COMBINATIONS WITH STARCH AND OTHER BIOPOLYMERS

The use of starch-based food packaging is due to its advantages, including its contribution to reducing fossil content, lack of toxins, the origin being plant sources (renewable resources), biodegradability and biocompatibility, low cost, and abundance [18]. Nowadays, starch-based bioplastics contribute about 66% of the global bioplastics market [23]. However, starch presents some drawbacks that limit its applications and uses such as water sensitivity and poor mechanical properties. Hence, several studies have focused on improving the functionality of starch in water presence [22]. Reinforcing starch with natural fibers or other biopolymers has been proven to improve its drawbacks [24]. In this sense, efforts have been made to improve the properties of starch-based materials, such as starch modification, matrix reinforcement, and blending with other biopolymers, before their acceptance as an ideal food packaging material [10, 25].

5.2.1 CELLULOSE

The natural most abundant and available polymer is cellulose, which is a polysaccharide with high molecular conformed by linear chains of β-1,4 linked D-glucose units, that help in the maintenance of the cell structure of organisms and some microorganisms, which can be a possible substitute for petroleum-based materials because of its properties such as low cost, availability, renewability, lightweight, nontoxicity, sustainability, and biodegradability, and acquired GRAS standard by the Food and Drug Administration (FDA) can be used as reinforcing fillers in biopolymers based on starch in food packaging [4, 26]. The incorporation of natural fibers into starch-based materials reduces their water uptake [22]. Cellulose fibers are useful materials with many properties; for example, they are decomposable and renewable and can be used in combination with different polymer composites.

Cellulose also provides barrier properties to the packaging material, but one of the major functions of cellulose is that they are used as a carrier for antimicrobial agents and antioxidants; moreover, it serves as a vehicle for the incorporation of numerous food additives, and it is used for the formation of multilayer packaging material along with other polymer films such as starch; hence, cellulose and its derivatives are an excellent packaging material for the preservation of various food items [1]. Starch has been mainly reinforced by filler materials such as cellulosic materials, due to a strong tendency to obtain natural and biodegradable materials. Due to the presence of active hydroxyl groups (–OH) in the cellulose molecular framework, it is possible to introduce several chemical modifications; the cellulose's basic structure can be improved by reducing its fiber size to the micrometric/nanometric scale [27]. For example, the addition of small amounts of cellulose nanocrystals (CNCs) to extruded native starch forms strong hydrogen–hydroxyl bonds, resulting in nanocomposites with improved properties such as higher tensile strength, lower water absorption, water vapor permeability (WVP), and good antimicrobial activities [28]. In the same way, the elaboration of starch films formulated with cellulose nanofibers (CNFs) to form biocomposites has been shown to enhance the functional properties, which makes it an ideal component as a raw material for food packaging [29]. Likewise, the incorporation of CNF into starch increases the tensile strength and decreases the WVP [10].

In one study conducted by Otenda et al. [30], the incorporation of the nanofibrils with a nanocomposite hibiscus starch-based improved the mechanical properties of the films since it imparted good flexibility against starch films, ensuring that they are used in food packaging material, given that they can prolong food shelf life during transportation, storage, and handling. Balakrishnan et al. [31] studied the impact of filler reinforcement on the potato starch–cellulose films and showed that the water contact angle increases with the presence of CNF nanocomposites, and the WVP decreases; also, several enhanced properties were observed for the final product quality, including that the potato starch–cellulose nanocomposite films were found to be UV resistant. On the other hand, Fazeli et al. [32] prepared thermoplastic starch (TPS)-CNF films and observed that CNF-loaded TPS films improved tensile strength and elastic modulus, the water vapor transmission rate and oxygen transmission rate, and water vapor barrier properties were improved since water vapor transmission rate decreased by approximately 86%; the oxygen permeability was considerably reduced by 94% compared with TPS film. Bionanocomposite films of cassava starch added with a mixture of sucrose, inverted sugar, and CNCs were done by da Silva et al. [33] and observed that the presence of low concentrations of CNCs improved their properties such as the maximum tensile strength and elastic modulus since the cellulose nanofillers contribute to the stability properties of the nanocomposites; hence, they may be used for the elaboration of biodegradable packaging and prolong the shelf life of food products.

5.2.2 CHITOSAN

Chitosan is a polysaccharide composed of β-1,4 linked D-glucosamine and *N*-acetyl-D-glucosamine; it is synthesized by the deacetylation of chitin, is the second most abundant polysaccharide in nature, and it is found in the crustacean exoskeleton; hence, it can be obtained from fisheries waste, giving added value to these high-volume and low-cost residues; other sources are arthropods and cell walls of fungi, and like its derivatives, it is biodegradable, bioactive, biocompatible, and has adequate water vapor barrier, is low oxygen permeable, and excellent film-forming ability and presents antimicrobial properties [34–36]. In this sense, the biological activity of chitosan depends on some factors such as the degree of deacetylation, derivatization, and molecular weight, and chitosan antimicrobial properties could be enhanced by decreasing its molecular size; the use of low molecular weight chitosan as antimicrobial agent of starch packaging is an encouraging strategy not only to increase its antimicrobial effect but also to favor its diffusion ability [16]. High molecular weight chitosan has high polymerization degree chains, which favors polymer–polymer interactions, leading to strong matrices and decreasing to decrease the molecular weight [15]. In the same way, chitosan shows antimicrobial activity against diverse microorganisms such as foodborne bacteria, fungi, and yeast. This property has been proposed for using chitosan as a reference polymer to make food packaging and improve food safety [15].

The interaction of starch/chitosan is a good alternative to improve biodegradable matrices' properties focused on food packaging, due to the compatibility between both polysaccharides [37]. The hydrophobic behavior of other polymers such as chitosan protects starch from moisture uptake and helps to improve the mechanical characteristics; in the same way, some mechanical and hydrophobicity properties of chitosan could be enhanced by blending with starch, generating a synergism between both biopolymers [10]. The films formulated through chitosan and starch are recommended for food packaging because a mixture of these polymers can enhance its mechanical properties and may be favorable in terms of the resistance of biodegradable materials, due to the presence of chitosan in the film starch, which causes a decrease in water vapor transmission values [22].

The formulated starch–chitosan films could be used for the development of bags for packaging, particularly, these films could be used for the design of packaging for organic products, since consumers of this type of food value the biodegradability of the material very positively [37]. A study conducted by Mutmainna et al. [38] found that the biodegradation test shows that the bioplastic samples (starch–chitosan) lost 80% of the weight in 3–14 days when stockpiled on the ground. Durango et al. [39] developed an edible antimicrobial film based on yam starch (*Dioscorea alata*) and chitosan and evaluated its antimicrobial activity on *Salmonella enteritidis*; they observed that the chitosan-treated films caused a reduction of one log cycle of microorganisms, besides films showed good flexibility. Ren et al. [17] conducted a study where corn starch and chitosan films were developed; they showed that the incorporation of chitosan enhanced some properties, such as solubility, tensile strength, and elongation at break, besides decreased Young's modulus and WVP. These results showed that these biodegradable films presented the potential to be applied as food packaging. In the same way, Shapi'i et al. [40] evaluated antimicrobial properties of starch–chitosan nanoparticles films *via in vitro* and *in vivo*; and they founded a inhibitory zone of 15% and 20% for all the tested bacteria, including *Bacillus cereus and Staphylococcus* aureus; the *in vivo* study revealed that the starch–chitosan nanoparticles films via *in vitro* and *in vivo*, the *in vivo* study revealed that the starch–chitosan nanoparticles film (15% w/w) was more effective in avoiding the microbial presence compared to neat starch film, thus showed the potential of the films with possible uses such as food packaging.

5.2.3 LIGNIN

Lignin is a polymer that acts as a structural material that contributes to the structure and strength of the plant cell walls; lignin has been used as a macromolecular compound by selective functionalization, improving its compatibility and performance in composite, copolymer materials, adhesives, and resins [41]. Lignin is the second most abundant polymer after cellulose; it possesses powerful antioxidant activity due to the complex phenolic structure, which gives the biopolymer an additional functional characteristic [42]. Lignin comes from several sources, mainly residues from agricultural waste and woody biomass, and presents some functional groups, such as carboxyl, methoxy groups, and phenolic hydroxyl, which facilitate its interaction with different polymers, including starch [43]. This hydrophobic biopolymer presents antimicrobial, UV light absorption, and antioxidant properties and can be considered as a mechanical reinforcement for polymer matrices with compatible food packaging materials such as biopolymers [44]. Lignin appreciation has generated diverse subproducts, including fuels, chemicals, and materials, and its products have shown promise in reducing environmental effects, making pulp and paper more profitable based on an eco-friendly economy [42]. The antiradical and antioxidant characteristics of lignin make it suitable to be used in film plastic as a component of food packaging; besides, antimicrobial properties in lignin also justify its utilization as a protector of products from microorganisms or environmental disruption [43].

Different kinds of fractionated or modified lignins [kraft lignins (KLs)] have been investigated in association with TPS for the preparation of biodegradable films [45]. Some properties of starch can be improved with the incorporation of natural fillers or reinforcements; for example, the addition of KL increases the thermal and mechanical properties of this polymer [46]. Spiridon et al. [47] studied the effects of the addition of lignin on the mechanical, thermal, and water sorption properties of starch–lignin composite and observed that the films starch–lignin presented higher tensile strength and were more rigid comparatively to those without lignin. In one of the studies conducted by Bhat et al. [48], results showed that sago palm (*Metroxylon sagu*) starch films produced by the incorporation of isolated lignin from oil palm black liquor waste (from empty fruit bunch) improved selected thermomechanical and barrier properties with a decrease in WVP. In a study conducted by Espinoza Acosta et al. [49], alcohol-soluble lignin (ASL) was used to prepare starch–ASL films, and ASL promoted a significant decrease of the tensile strength and elastic modulus. Likewise, films presented resistance to thermal degradation; in addition, the starch–ASL films exhibited antioxidant activity, which increased with increasing ASL content. Blend starch–lignin results in an alternative that is low cost and can be used to make biodegradable composite materials, which may have wide applications in food packaging [47].

5.2.4 POLYLACTIC ACID

Polylactic acid (PLA) is derived from lactic acid, which in turn is the product of a fermentative process of vegetal sources renewable and biodegradable sources (such as corn, sugar beet, tapioca roots, and potato starch); it shows similar behavior compared to polymers such as polystyrene and polyethylene terephthalate in mechanical, thermal, and barrier properties, PLA and starch are studied and evaluated as replacements for petroleum polymers based on their functional properties and cost [50, 51]. In addition, it has good properties, such as a high melting point, high tensile modulus, and a high degree of transparency, making PLA a high-performance polymer attractive for multiple industrial uses, including food packaging [52]. The properties of PLA that make it work excellently as food and beverage packaging are that it has a high transparency with excellent tensile strength and wearability, very stiff, glossy, and clear [23]. However, it is well known that the starch–PLA composite films normally possess several drawbacks (rigidity, fragility, and poor processability) caused by phase separation, which makes them unable to simultaneously satisfy the essential requirements of flexibility, stretchability, and durability; due to this, hydrophobic modification on the starch surface is a strategy to solve the existing problems between starch and PLA due to incompatibility [53].

PLA is highly used in the food packaging industry, with an estimated annual production of 140,000 tons, and its mass production employs several well-established polymer manufacturing techniques [2]. Besides, since lactides are formed by the agricultural by-products' microbial fermentation (mainly the carbohydrate-rich substances) allows the use of some agroindustrial waste [6]. PLA's characteristics have been converted into a substitute for traditional plastics for packaging usage. PLA has the potential to exceed or blend with other biodegradable polymers to use in food packaging, this polymer being recognized as safe (GRAS) and approved by the FDA of the United States, and this substance results in optimal use in food and beverage packaging [23, 52, 54]. In this sense,

PLA is advised as a harmless food packaging material because the migration of lactic acid from PLA packaging containers to food is negligible [2].

PLA–starch films can be used as packaging due to their biodegradability and easy use because PLA addition reduces and improves biodegradation properties [10]. The degradation half-life is dependent on the PLAs' stereochemistry, with an average time between 6 and 24 months [6]. Likewise, it is known that PLA is blended with starch to decrease its cost [22]. In this way, Yusoff et al. [23] suggested mixing PLA with tapioca starch as a bioplastic material to be used as a packaging material, as it is safe for living beings, low cost, and environmentally friendly. In a study conducted by Mangaraj et al. [7], the biodegradable film fabricated by PLA and cassava starch was appropriate for packaging food products. The outcomes of the work indicated that the bio-based film was handy for packaging applications and may replace the use of nonbiodegradable materials for the packaging of fresh produce. In the same way, Trinh et al. [55] elaborated films comprised of starch and PLA; this combination provided a 1300% improvement in moisture barrier compared to starch films and an improvement in oxygen barrier properties against PLA films. Zúniga et al. [56] showed that the addition of modified avocado starch resulted in an incredible increase in the elongation at break of PLA approximately 730% improvement; likewise, this blend displayed remarkably higher UV-blocking values around 70–84% for UVB and 40–51% for UVA wavelengths; compared to those of pure PLA, the water vapor and oxygen barrier were dramatically improved, indicated a permeability reduction of 54.8% and 97.5%, respectively. Blending PLA with starch contributes to lowering the price, decreasing the glass transition temperature, and increasing several properties in films used in food packaging, such as biodegradability, crystallinity, and transparency [13].

5.3 CONCLUSION

Even though biopolymer-based films are not the ultimate replacement for petroleum-based plastics, it is necessary to replace as many polymers as possible in food packaging through the development of cheap, functional, and eco-friendly materials such as biopolymer-based films. Achieving this goal requires considerable efforts from both the scientific and industrial spheres. Various studies have investigated the possible use of biopolymers for food packaging. According to the results, biopolymer blends based on starch for food packaging are very functional, promising, and have great potential to minimize the amount of synthetic residues generated and discarded in the environment. Further studies are necessary to evaluate all properties of these starch-based films as food packaging for further commercial exploitation and use.

KEYWORDS

- **starch-based**
- **eco-friendly**
- **packaging**
- **preservation**
- **value-added**

REFERENCES

1. Liu, Y., Ahmed, S., Sameen, D.E., Wang, Y., Lu, R., Dai, J., Li, S., Qin, W. (2021). A review of cellulose and its derivatives in biopolymer-based for food packaging application. *Trends in Food Science & Technology*, 112, 532–546.
2. Swetha, T.A., Bora, A., Mohanrasu, K., Balaji, P., Raja, R., Ponnuchamy, K., Muthusamy, G., Arun, A. (2023). A comprehensive review on polylactic acid (PLA)–Synthesis, processing and application in food packaging. *International Journal of Biological Macromolecules*, 234, 123715.
3. Matheus, J.R.V., Dalsasso, R.R., Rebelatto, E.A., Andrade, K.S., Andrade, L.M.d., Andrade, C.J.d., Monteiro, A.R., Fai, A.E.C. (2023). Biopolymers as green-based food packaging materials: A focus on modified and unmodified starch-based films. *Comprehensive Reviews in Food Science and Food Safety*, 22, 1148–1183.
4. Kumari, S.V.G., Pakshirajan, K., Pugazhenthi, G. (2022). Recent advances and future prospects of cellulose, starch, chitosan, polylactic acid and polyhydroxyalkanoates for sustainable food packaging applications. *International Journal of Biological Macromolecules*, 221, 163–182.
5. Mukherjee, C., Varghese, D., Krishna, J., Boominathan, T., Rakeshkumar, R., Dineshkumar, S., Rao, C.B., Sivaramakrishna, A. (2023). Recent advances in biodegradable polymers—Properties, applications and future prospects. *European Polymer Journal*, 192, 112068.
6. Taib, N.-A.A.B., Rahman, M.R., Huda, D., Kuok, K.K., Hamdan, S., Bakri, M.K.B., Julaihi, M.R.M.B., Khan, A. (2023). A review on poly lactic acid (PLA) as a biodegradable polymer. *Polymer Bulletin*, 80, 1179–1213.

7. Mangaraj, S., Thakur, R.R., Yadav, A. (2022). Development and characterization of PLA and Cassava starch-based novel biodegradable film used for food packaging application. *Journal of Food Processing and Preservation*, 46, e16314.
8. Agarwal, S., Singhal, S., Godiya, C.B., Kumar, S. (2023). Prospects and applications of starch based biopolymers. *International Journal of Environmental Analytical Chemistry*, 103, 6907–6926.
9. Sundqvist-Andberg, H., Åkerman, M. (2021). Sustainability governance and contested plastic food packaging–an integrative review. *Journal of Cleaner Production*, 306, 127111.
10. Bangar, S.P., Whiteside, W.S., Ashogbon, A.O., Kumar, M. (2021). Recent advances in thermoplastic starches for food packaging: A review. *Food Packaging and Shelf life*, 30, 100743.
11. Cha, D.S., Chinnan, M.S. (2004). Biopolymer-based antimicrobial packaging: A review. *Critical Reviews in Food Science and Nutrition*, 44, 223–237.
12. Bose, I., Nousheen, Roy, S., Yaduvanshi, P., Sharma, S., Chandel, V., Biswas, D. (2023). Unveiling the potential of marine biopolymers: Sources, classification, and diverse food applications. *Materials*, 16, 4840.
13. Malek, N., Faizuwan, M., Khusaimi, Z., Bonnia, N., Rusop, M., Asli, N. (2021). Preparation and characterization of biodegradable polylactic acid (PLA) film for food packaging application: A Review. *Journal of Physics: Conference Series*, 1892, 012037.
14. Khan, B., Bilal Khan Niazi, M., Samin, G., Jahan, Z. (2017). Thermoplastic starch: A possible biodegradable food packaging material—A review. *Journal of Food Process Engineering*, 40, e12447.
15. Cazón, P., Vázquez, M. (2019). Applications of chitosan as food packaging materials. In: Crini, G., Lichtfouse, E. (Eds.), *Sustainable Agriculture Reviews* (pp. 81–123). Springer: Cham.
16. Castillo, L.A., Farenzena, S., Pintos, E., Rodríguez, M.S., Villar, M.A., García, M.A., López, O.V. (2017). Active films based on thermoplastic corn starch and chitosan oligomer for food packaging applications. *Food Packaging and Shelf life*, 14, 128–136.
17. Ren, L., Yan, X., Zhou, J., Tong, J., Su, X. (2017). Influence of chitosan concentration on mechanical and barrier properties of corn starch/chitosan films. *International Journal of Biological Macromolecules*, 105, 1636–1643.
18. Onyeaka, H., Obileke, K., Makaka, G., Nwokolo, N. (2022). Current research and applications of starch-based biodegradable films for food packaging. *Polymers*, 14, 1126.
19. Nasiri, A., Wearing, J., Dubé, M.A. (2020). Using lignin to modify starch-based adhesive performance. *ChemEngineering*, 4, 3.
20. Müller, C.M., Laurindo, J.B., Yamashita, F. (2012). Composites of thermoplastic starch and nanoclays produced by extrusion and thermopressing. *Carbohydrate Polymers*, 89, 504–510.
21. Kringel, D.H., Dias, A.R.G., Zavareze, E.d.R., Gandra, E.A. (2020). Fruit wastes as promising sources of starch: Extraction, properties, and applications. *Starch-Stärke*, 72, 1900200.

22. García-Guzmán, L., Cabrera-Barjas, G., Soria-Hernández, C.G., Castaño, J., Guadarrama-Lezama, A.Y., Rodríguez Llamazares, S. (2022). Progress in starch-based materials for food packaging applications. *Polysaccharides*, 3, 136–177.
23. Yusoff, N.H., Pal, K., Narayanan, T., de Souza, F.G. (2021). Recent trends on bioplastics synthesis and characterizations: Polylactic acid (PLA) incorporated with tapioca starch for packaging applications. *Journal of Molecular Structure*, 1232, 129954.
24. Nazrin, A., Sapuan, S., Zuhri, M. (2020). Mechanical, physical and thermal properties of sugar palm nanocellulose reinforced thermoplastic starch (TPS)/poly(lactic acid) (PLA) blend bionanocomposites. *Polymers*, 12, 2216.
25. Jayarathna, S., Andersson, M., Andersson, R. (2022). Recent advances in starch-based blends and composites for bioplastics applications. *Polymers*, 14, 4557.
26. Gutiérrez, T.J., Alvarez, V.A. (2017). Cellulosic materials as natural fillers in starch-containing matrix-based films: A review. *Polymer Bulletin*, 74, 2401–2430.
27. Sahnoun, N., Abdelaziz, A., Tarchoun, A.F., Boukeciat, H., Mezroua, A., Trache, D. (2022). Nitrostarch as a promising insensitive energetic biopolymer: Synthesis, characterization, and thermal decomposition kinetics. *Industrial Crops and Products,* 189, 115774.
28. Muñoz-Gimena, P.F., Oliver-Cuenca, V., Peponi, L., López, D. (2023). A review on reinforcements and additives in starch-based composites for food packaging. *Polymers*, 15, 2972.
29. Othman, S.H., Wane, B.M., Nordin, N., Noor Hasnan, N.Z., A.Talib, R., Karyadi, J.N.W. (2021). Physical, mechanical, and water vapor barrier properties of starch/cellulose nanofiber/thymol bionanocomposite films. *Polymers*, 13, 4060.
30. Otenda, B., Kareru, P., Madivoli, E., Salim, A., Gichuki, J., Wanakai, S. (2022). Starch-hibiscus-cellulose nanofibrils composite films as a model antimicrobial food packaging material. *Journal of Natural Fibers*, 19, 12371–12384.
31. Balakrishnan, P., Gopi, S., MS, S., Thomas, S. (2018). UV resistant transparent bionanocomposite films based on potato starch/cellulose for sustainable packaging. *Starch-Stärke*, 70, 1700139.
32. Fazeli, M., Keley, M., Biazar, E. (2018). Preparation and characterization of starch-based composite films reinforced by cellulose nanofibers. *International Journal of Biological Macromolecules*, 116, 272–280.
33. da Silva, J.B., Pereira, F.V., Druzian, J.I. (2012). Cassava starch-based films plasticized with sucrose and inverted sugar and reinforced with cellulose nanocrystals. *Journal of Food Science*, 77, N14–N19.
34. Flórez, M., Guerra-Rodríguez, E., Cazón, P., Vázquez, M. (2022). Chitosan for food packaging: Recent advances in active and intelligent films. *Food Hydrocolloids*, 124, 107328.
35. Ghasemzadeh, H., Afraz, S., Moradi, M., Hassanpour, S. (2021). Antimicrobial chitosan-agarose full polysaccharide silver nanocomposite films. *International Journal of Biological Macromolecules*, 179, 532–541.
36. Jha, P. (2020). Effect of plasticizer and antimicrobial agents on functional properties of bionanocomposite films based on corn starch-chitosan for food packaging applications. *International Journal of Biological Macromolecules*, 160, 571–582.

37. Bof, M.J., Locaso, D.E., García, M.A. (2021). Corn starch-chitosan proportion affects biodegradable film performance for food packaging purposes. *Starch-Stärke*, 73, 2000104.
38. Mutmainna, I., Tahir, D., Lobo Gareso, P., Ilyas, S. (2019). Synthesis composite starch-chitosan as biodegradable plastic for food packaging. *Journal of Physics: Conference Series*, 1317, 012053.
39. Durango, A., Soares, N., Benevides, S., Teixeira, J., Carvalho, M., Wobeto, C., Andrade, N. (2006). Development and evaluation of an edible antimicrobial film based on yam starch and chitosan. *Packaging Technology and Science*, 19, 55–59.
40. Shapi'i, R.A., Othman, S.H., Nordin, N., Basha, R.K., Naim, M.N. (2020). Antimicrobial properties of starch films incorporated with chitosan nanoparticles: In vitro and in vivo evaluation. *Carbohydrate Polymers*, 230, 115602.
41. Torres, L.A.Z., Woiciechowski, A.L., de Andrade Tanobe, V.O., Karp, S.G., Lorenci, L.C.G., Faulds, C., Soccol, C.R. (2020). Lignin as a potential source of high-added value compounds: A review. *Journal of Cleaner Production*, 263, 121499.
42. Anushikha., Gaikwad, K.K. (2023). Lignin as a UV blocking, antioxidant, and antimicrobial agent for food packaging applications. *Biomass Conversion and Biorefinery*, 14, 16755–16767.
43. Agustiany, E.A., Rasyidur Ridho, M., Rahmi DN, M., Madyaratri, E.W., Falah, F., Lubis, M.A.R., Solihat, N.N., Syamani, F.A., Karungamye, P., Sohail, A. (2022). Recent developments in lignin modification and its application in lignin-based green composites: A review. *Polymer Composites,* 43, 4848–4865.
44. Quilez-Molina, A.I., Chandra Paul, U., Merino, D., Athanassiou, A. (2022). Composites of thermoplastic starch and lignin-rich agricultural waste for the packaging of fatty foods. *ACS Sustainable Chemistry & Engineering,* 10, 15402–15413.
45. Çalgeris, İ., Çakmakçı, E., Ogan, A., Kahraman, M.V., Kayaman-Apohan, N. (2012). Preparation and drug release properties of lignin–starch biodegradable films. *Starch-Stärke*, 64, 399–407.
46. de Freitas, A.d.S., Rodrigues, J.S., Maciel, C.C., Pires, A.A., Lemes, A.P., Ferreira, M., Botaro, V.R. (2021). Improvements in thermal and mechanical properties of composites based on thermoplastic starch and Kraft Lignin. *International Journal of Biological Macromolecules*, 184, 863–873.
47. Spiridon, I., Teaca, C.-A., Bodirlau, R. (2011). Preparation and characterization of adipic acid-modified starch microparticles/plasticized starch composite films reinforced by lignin. *Journal of Materials Science*, 46, 3241–3251.
48. Bhat, R., Abdullah, N., Din, R.H., Tay, G.-S. (2013). Producing novel sago starch based food packaging films by incorporating lignin isolated from oil palm black liquor waste. *Journal of Food Engineering*, 119, 707–713.
49. Espinoza Acosta, J.L., Torres Chávez, P.I., Ramírez-Wong, B., Bello-Pérez, L.A., Vega Ríos, A., Carvajal Millán, E., Plascencia Jatomea, M., Ledesma Osuna, A.I. (2015). Mechanical, thermal, and antioxidant properties of composite films prepared from durum wheat starch and lignin. *Starch-Stärke*, 67, 502–511.
50. Muller, J., González-Martínez, C., Chiralt, A. (2017). Combination of poly (lactic) acid and starch for biodegradable food packaging. *Materials*, 10, 952.

51. Siracusa, V., Blanco, I., Romani, S., Tylewicz, U., Rocculi, P., Rosa, M.D. (2012). Poly (lactic acid)-modified films for food packaging application: Physical, mechanical, and barrier behavior. *Journal of Applied Polymer Science*, 125, E390–E401.
52. Martinez Villadiego, K., Arias Tapia, M.J., Useche, J., Escobar Macías, D. (2022). Thermoplastic starch (TPS)/polylactic acid (PLA) blending methodologies: A review. *Journal of Polymers and the Environment*, 30, 75–91.
53. Zhou, X., Cheng, R., Wang, B., Zeng, J., Xu, J., Li, J., Kang, L., Cheng, Z., Gao, W., Chen, K. (2021). Biodegradable sandwich-architectured films derived from pea starch and polylactic acid with enhanced shelf life for fruit preservation. *Carbohydrate Polymers*, 251, 117117.
54. Gürler, N., Paşa, S., Temel, H. (2021). Silane doped biodegradable starch-PLA bilayer films for food packaging applications: Mechanical, thermal, barrier and biodegradability properties. *Journal of the Taiwan Institute of Chemical Engineers*, 123, 261–271.
55. Trinh, B.M., Chang, C.C., Mekonnen, T.H. (2021). Facile fabrication of thermoplastic starch/poly (lactic acid) multilayer films with superior gas and moisture barrier properties. *Polymer*, 223, 123679.
56. Zúniga, M.M.G., Ding, R., Oh, E., Nguyen, T.B., Trung, T.T., Nam, J.-D., Suhr, J. (2024). Avocado seed starch utilized in eco-friendly, UV-blocking, and high-barrier polylactic acid (PLA) biocomposites for active food packaging applications. *International Journal of Biological Macromolecules*, 265, 130837.

CHAPTER 6

Polybutylene Succinate (PBS): Sustainable Applications and Environmental Benefits in Food Preservation

JOSÉ ROBERTO VEGA-BAUDRIT,[1,2*] YENDRY CORRALES,[1] DIEGO BATISTA,[1] SERGIO A. PANIAGUA,[1] SUSANA MESÉN,[1] ANDREA ARAYA,[1] REBECA CORRALES,[1] LUIS CASTILLO HENRÍQUEZ,[1,3] CLAUDIA CHAVES VILLAREAL,[1] JUAN MIGUEL ZÚÑIGA-UMAÑA,[1] and MARY LOPRETTI[4]

[1]*LANOTEC CeNAT CONARE, San José, Costa Rica*

[2]*Escuela de Química, Universidad Nacional de Costa Rica, Heredia, Costa Rica*

[3]*Facultad de Farmacia, Universidad de Costa Rica, San Pedro, Costa Rica*

[4]*UDELAR, Montevideo, Uruguay*

*Corresponding author

ABSTRACT

The field of biodegradable plastics is expanding rapidly, with numerous research groups actively seeking innovative uses for these environmentally friendly materials. One area that has garnered significant attention is the development of renewable resources for bio-based polymers and the latter synthesis of degradable polymers. Polybutylene succinate (PBS) has appeared as a noteworthy candidate. PBS is easily synthesized from biological resources and has relatively low production costs compared to

Sustainable Food Packaging Materials and Technology. Juan Manuel Tirado-Gallegos, Juan Daniel Rivaldi Chávez, and José Juan Buenrostro-Figueroa (Eds)

DOI: 10.1201/9781779641434-6

other biodegradable polymers like polylactic acid (PLA). As a potential alternative to polyolefins and nondegradable polymers, PBS, an aliphatic polyester, displays physical properties that make it an appealing choice for biodegradable materials. With a melting temperature ranging from 105 to 115°C, PBS can be processed efficiently, reducing environmental impact. Its tensile properties are also commendable compared to other biodegradable polymers. Moreover, PBS strongly resists oil, grease, and odors, making it a promising candidate for biodegradable packaging and agricultural applications. Ongoing research on BPS has revealed promising prospects for its implementation across various industries. Derived from sustainable sources, PBS has inherent qualities that position it as a frontrunner in pursuing sustainable materials. Its ease of synthesis from biological resources further enhances its appeal and contributes to a cost-efficient production process. These advantages distinguish PBS from its alternatives, paving the way for widespread adoption.

6.1 INTRODUCTION

Polymers are lightweight, durable, and rigid materials, characteristics that are fueling high demand for polymer-based materials today. Although beneficial, most commodity polymers are not degradable, and the disposal of polymers has become an environmental problem. Landfill disposal wastes valuable resources and brings no benefits to nature, while burning produces CO_2 gas, contributing to global warming. Slow decomposition and low recycling rate of polymers have resulted in the Earth's and its ecosystem's environmental pollution. Various nations have endeavored to bring these pollution problems under control, led to research on developing degradable and eco-friendly materials to replace the current polymer materials [1–8].

The need for degradable plastics has increased, and many developed and developing nations have been working on environmentally friendly plastic materials. These materials are needed in applications such as agricultural mulch films, rope, and bags, disposed of after usage, and durable materials unsuitable for the environment. Materials used in food packaging are another widespread example of one-time-use materials, and switching to degradable materials could reduce plastic waste worldwide. This increasing demand and necessity for degradable polymers has resulted in the search for alternative materials to the current polymers [9–10].

Various environmentally friendly plastics have been developed, and one of the materials that contributes significantly to the environment is biodegradable polyesters and their copolymers. Polybutylene succinate (PBS) belongs to the biodegradable polyester family. Biodegradable polymers have been designed to counter current environmental problems caused by other nondegradable polymer materials. PBS has been one of the materials thoroughly investigated over the last few years and is said to be a noteworthy material for replacing current polymers. This chapter shall assess the properties and progression of this material from earlier studies [11–15].

6.1.1 BACKGROUND

Traditional nondegradable polymers or plastics are environmentally resistant due to their high molecular weight and synthetic structure. It is estimated that in the year 2000, the production of plastics was 100 million tons per year and might rise in the following years. Most of the plastic waste is disposed of in landfills, causing soil and water contamination, or left in the environment and becoming plastic waste known as an eyesore, threatening wildlife, and taking years to degrade. Plastic waste is hard to degrade or biodegrade due to its structure. The interest in bio-based and biodegradable polymers comes from global issues like increasing waste disposal costs and the ever-decreasing landfill space. Depletion of fossil fuel resources and the environmental problems caused by the emission of carbon dioxide have also lent a hand in driving the interest in biodegradable polymers. PBS is a biodegradable semicrystalline polymer that has recently gained attention due to its long-term potential in substituting current nondegradable polymers in various applications (Figure 6.1). PBS shows favorable mechanical properties and is processable, much like polypropylene. Due to this, it has been considered a potential general-purpose biodegradable polymer [11, 16–20].

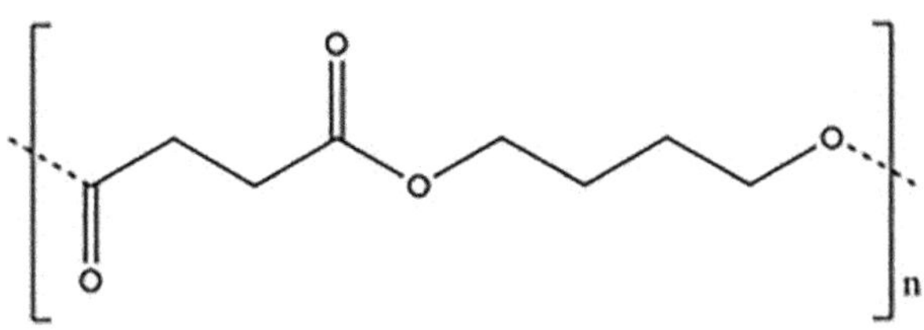

FIGURE 6.1 Chemical structure of PBS.

PBS can be synthesized from bio-based succinic acid and 1,4-butanediol (BDO) through polycondensation or bio-based methods such as bacterial fermentation. Bacterial fermentation of succinic acid has potential via genetically modified organisms and immobilized enzymes. PBS can vary in crystallinity, which affects its thermal and mechanical properties. High crystallinity can lead to high stiffness, viscoelasticity, and size changes due to absorbing water. Low crystallinity leads to ductile properties. PBS is also reported to have poor resistance to greases and oils. Due to this, it is often blended with other polymers or additives during synthesis to achieve different properties [21–24].

Reliable and efficient methods for the large-scale production of PBS have been developed over time: during synthesis, several critical components are carefully combined and subjected to various parameters such as temperature, pressure, and catalysts to manipulate the molecular structure of PBS, resulting in specific properties and characteristics [11, 21]. The synthesis of PBS is significant in various industries, such as automotive, aerospace, electronics, agricultural, packaging, disposable, textile, and healthcare (Figure 6.2) due to its high strength, flexibility, thermal stability, and processability through injection molding [17, 22, 25].

FIGURE 6.2 PBS is important in various industries, such as automotive, aerospace, electronics, and healthcare.

PBS degradation in soil or compost occurs by random chain scission. It has been reported that water absorption into the amorphous region of PBS causes hydrolysis of ester bonds, and this reaction tries gradually, effectively showing that the degradation rate of PBS can be controlled by changing the degree of crystallinity and molecular weight. Tertiary-butyl alcohol and carbon dioxide are the final products of chain scission, and these substances are expected to have minimal effect on the natural environment [26–28].

Organic pigments can be mixed in with PBS, which, in addition to its excellent weatherability, transparency, and gloss, makes it attractive for industry printability. These properties indicate that PBS may be a suitable biodegradable polyester; a rapid increase in its production and development has been seen in recent years [21, 22, 29–31].

6.2 CHEMICAL STRUCTURE, PROPERTIES, AND COMPOSITION

6.2.1 MOLECULAR FORMULA

The widespread use of butylene glycol produces highly crystalline PBS with a high melting temperature (~115°C), close to that of polyethylene (PE). By changing the butylene isomer, various grades of PBS can be produced, which have different properties and are suited for specific applications. This unique feature of PBS is essential in tailoring the polymer for other uses [21, 32, 33].

The general molecular formula of succinate polyesters is HO(CO)R_1(COOR_2) OR, where R_1 and R_2 are aliphatic groups and various isomers are possible. PBS is synthesized from BDO (also known as butylene glycol) and succinic acid, which undergoes a condensation reaction with the elimination of water to form oligomers, which can proceed with further trans-esterifications to produce the polymer. Butylene glycol ($HO(CH_2)_4OH$) is a four-carbon atom chain with three possible isomers. Given that isomer mixtures of butylene glycol are affordable and readily available, the molecular structure of PBS has not been well defined, and each isomer of butylene glycol may produce different molecular weights during polymerization, and hence the polymer properties [18, 25, 34, 35].

6.2.2 POLYMERIZATION REACTION

The primary catalyst for the polycondensation of PBS is titanium(IV) 2-ethylhexoxide and tin(II) 2-ethylhexanoate. The former catalyst has a higher catalytic activity and produces a high molecular weight PBS. Yamane et al. examined the effects of the main reaction variables, including temperature, reaction time, and the amount of catalyst, on the molecular weight of PBS using titanium(IV) 2-ethylhexoxide as a polymerization catalyst [17, 18, 21, 22]. The molecular weight of the PBS increased with catalyst concentration and reaction time, but decreased with higher temperatures. Flores et al. [40] further clarified the correlation between the molecular weight of PBS and its physical properties. Syndiotactic high molecular weight (1.0×10^6) PBS was synthesized using titanium(IV) 2-ethylhexoxide at 145°C. High molecular weight PBS has superior impact strength and heat resistance, while low molecular weight PBS has higher crystallinity, melting temperature, and biodegradability [17, 21, 22, 29, 37, 38].

6.2.3 PHYSICAL AND MECHANICAL PROPERTIES

The properties of recycled PBS are comparable to those of fresh material, and recycling and recovery do not require separate systems. The pyrolysis of PBS occurs at a relatively low temperature and can produce oil, gas, and char in a high-weight yield [39–45]. PBS has excellent dielectric properties. It offers better electrical resistance and lower dielectric loss than polystyrene and keeps most mechanical properties despite high humidity or immersion in water. PBS has good weatherability and resistance to both prolonged heat and gamma radiation. Its ultraviolet resistance is also good, around that of PE [46].

6.3 SYNTHESIS METHODS

The following sections focus on the production of PBS through direct polycondensation (esterification), its reaction mechanisms, and how the method compares to PBS production through trans-esterification and ring-opening polymerization (ROP). A brief mention of the monomers is presented first. Regarding BDO, its industrial production is typically the Reppe process, which consists of an acetylene reaction with two equivalents of formaldehyde to form butyne-1,4-diol, which is then hydrogenated. The Davy process is also employed, consisting of dimethyl maleate's hydrogenation from maleic anhydride and methanol. While BDO is included in the US Department of Justice Special Surveillance List (US Drug Enforcement Administration, 2023), it is not a controlled substance or listed chemical under the Controlled Substances Act (CSA).

Current succinic acid production still involves the petrochemical Roche–Wacker process; however, succinic acid can also be produced by microbial fermentation of a renewable substrate. This method is more favorable regarding sustainability and the production of a biodegradable material. Still, its cost-effectiveness and energy savings compared to the established methods are yet to be proven [21, 47].

6.3.1 DIRECT ESTERIFICATION

In a typical esterification process, succinic acid and BDO are heated in the presence of a catalyst and solvent. Water and low-molecular-weight by-products are removed from the PBS until the desired molecular weight is reached. Direct polycondensation can be achieved using various catalysts and reaction conditions [33, 39]. Isotactic PBS can be achieved using condensation polymerization; cadmium catalysts are effective in attaining isotactic polymer, though concerns over toxicity have shifted interests to the use of rare earth metal catalysts, which are not only effective for isotactic polymerization but also the syndiotactic polymerization of succinic acid with BDO to yield various stereoisomeric PBS and its copolymers [48–50].

6.3.2 TRANSESTERIFICATION

The oligomers from direct esterification are trans-esterified under vacuum to form a high molar mass polymer, with the use of appropriate catalysts such as titanium, zirconium, tin, or germanium organo-compounds [51, 52]. Metal oxides can also be used in the presence of hydroxy acids.

6.3.3 RING-OPENING POLYMERIZATION

This method is the most direct and does not involve an intermediate such as a prepolymer. The mechanism for ROP indicates that initially, the monomer undergoes a nucleophilic attack from a chain initiator's –OH or $–NH_2$ group. It forms a covalent bond between the monomer and the chain initiator, and later, a molecule of water or alcohol is eliminated. After this, only the monomer found at the chain terminal can undergo addition to another monomer, and this continues until a high molecular weight polymer has been formed [52, 53].

β-Butyrolactone, the central building unit of PBS, can be polymerized in the presence of a catalyst to yield high molecular weight PBS. Various lactones have been polymerized to form PBS and its copolymers; successfully synthesized a random copolymer of PBS with PE succinate using ε-caprolactone as a comonomer. While the polymerization of lactones is the most effective method to synthesize PBS, the lack of stereospecificity in lactone monomers can generate a head-to-tail and head-to-head (with bridging oxygen atom) configuration, resulting in a stereoisomeric mixture of the polymer. High isotacticity is essential for obtaining high melting points and good crystallinity in PBS [49, 54–56] (Scheme 6.1).

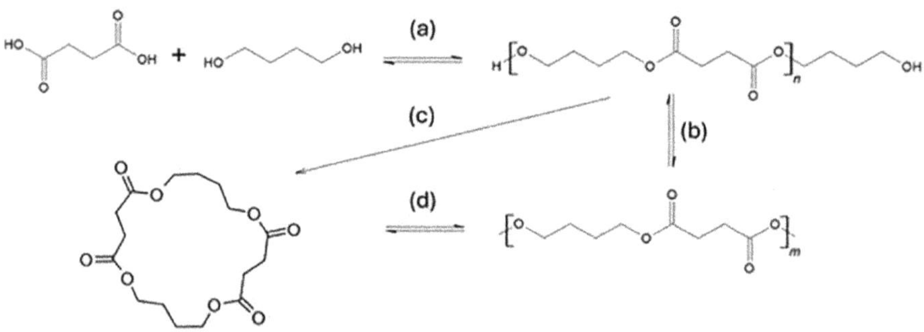

SCHEME 6.1 PBS synthesis via (a) direct esterification (typical condition: 225°C, 2 bars), (b) trans-esterification (typical condition: catalyst, 230°C, vacuum), (c) oligomer depolymerization (typical condition: catalyst, 290°C, vacuum), and (d) ring-opening polymerization (typical condition: catalyst, 115°C).

Source: Adapted with permission from Refs. [51, 57].

6.4 BIODEGRADATION MECHANISMS

Various methods have been researched to enhance the degradation rate of biodegradable polymers in the interest of shortening the recycling process. Many such methods involve the design of copolymers or liquid crystal polymers, grafting, or adding fillers such as starch, cellulose, or protein. However, the relatively slow degradation of PBS may be advantageous when a long product lifetime is desired. Changes in the degradation rate would give PBS a broad range of applications in biodegradable products for which the time of use before disposal varies [18, 25, 37, 58].

By incorporating bio-based fillers derived from biofuel coproducts, the biodegradation rate of PBS can be accelerated. Integrating isosorbide—a carbohydrate-based compound—into PBS copolyesters enhances thermal properties and improves biodegradability through hydrolytic and enzymatic degradation.

Incorporating materials like organo-montmorillonite and compatibilizers such as maleic anhydride-grafted PBS changes the degradation pace and the mechanical characteristics after biodegradation. Surface treatments like photografting with hydrophilic or hydrophobic membranes offer control over the material's biodegradability and reusability.

Under thermophilic anaerobic conditions, the degradation of PBS is significantly enhanced, helped by hydrolytic enzymes in the sludge. Combining PBS with polycaprolactone (PCL) in composite materials improves biodegradability, particularly in noncrystalline regions, while overall crystallinity increases after degradation. The biodegradation of PBS can be effectively modulated using various strategies, such as employing bio-based fillers, copolymerization with hydrophilic compounds, interactions with soil microorganisms, and forming composites or nanocomposites. These enhancements elevate the environmental viability of PBS, marking it as a preferable alternative to traditional plastics [25, 29, 59–63].

6.4.1 MICROBIAL DEGRADATION

Microbial degradation is the most critical first step in the chain of events that lead to polymer breakdown. It is the process in which microorganisms such as bacteria, fungi, and algae break down polymers into simpler compounds. The changes in the chemical structure of polymers are due to the enzymatic action of microorganisms. These changes include cleavage of the polymer chains, biotransformation, and mineralization. While some polymers may biodegrade to compounds that the organism can use as a carbon and energy source, many polymers are utilized as a carbon and energy source without changing the chemical structure of the starting material [64–67].

One of the reasons that many conventional plastics are resistant to microbial attack is their hydrophobic and inert nature. It can be attributed to the fact that many polymers are derived from petroleum-based products and, therefore, have no preestablished mechanism for microbial degradation. The general insolubility of many polymers in water and organic solvents makes it difficult for microorganism colonization. In the case of plastics, physical and chemical factors such as UV radiation and oxygenation can lead to chain scission and surface cracking. While this may create more sites for the microorganisms to colonize, it usually leads to fragmentation, releasing smaller polymer chains into the environment [27, 65, 68, 69]. The controlled degradation of PBS to lower molecular weight compounds, which are more accessible to microorganisms to consume, can be achieved by enzyme-catalyzed processes such as hydrolysis, alcoholysis, and ester exchange. Host–guest chemistry studies have even been conducted to create pseudo "enzyme reactors." These involved encapsulating an enzyme in a cage of PBS copolymers. Through varying the copolymer, the size of the holes in the cage could be adjusted to enable the entry of only specific molecules. A study has reported the encapsulation of lipase in a poly(butylene oxide) cage that catalyzed the hydrolysis of PBS into oligomers. Step-growth aliphatic polyesters such as PBS are commonly hydrolyzed using an acid or base-catalyzed trans-esterification. The chain scission occurs by a back-and-forth exchange of the alcohol and acid groups between the catalyst and the polymer.

6.4.2 ENZYMATIC DEGRADATION

Enzymatic degradation occurs when a polymer is subjected to a change in its chemical structure due to the reaction with enzymes in living organisms. The enzymes catalyze the reaction and can be specific to the type of polymer or have a range of activity. The organism often uses the breakdown products as a source of energy. Enzymes are specific in their action and usually only act on a few substrates. The choice of enzymes to be used by an organism may be due to the process of coevolution with a polymer [70]. Enzymatic degradation has been studied on various aliphatic polyesters via *in vitro* and *in vivo* tests using enzymes isolated from microorganisms known to degrade these polymers. These enzymes are usually esterases due to the nature of the linkages in these polymers or lipases [71]. Ester bonds are the predominant linkages in polyesters, and the degradation process is a simple hydrolysis reaction of these bonds to diols and dicarboxylic acids. High molecular-weight polymers are broken down into low molecular-weight polymers, oligomers, and monomers [18, 25, 58, 72–74].

A study on the degradation of PBS and polyhydroxybutyrate found that the polymers could function as inducers or substrates to produce extracellular esterases. The polymers can induce the enzymes in an organism, which is a separate process from an inducer being transported into a cell or one enzyme acting on a specific substrate [75–78].

6.4.3 ENVIRONMENTAL FACTORS AFFECTING BIODEGRADATION

Moisture content significantly affects a polymer's biodegradation. In general, water is needed for biodegradation processes to occur, as enzymes are secreted from microbial cells in a water-soluble form, and biodegradation occurs extracellularly. However, PBS and other biodegradable polymers can be sensitive to change in the biodegradation rate at different moisture levels. High moisture levels can cause the polymer to swell and result in abiotic chain scission, thus weakening the polymer. While there may be a first increase in the biodegradation rate due to the weakened state of the polymer, the overall effect is less biodegradation of the polymer. Elevated levels of moisture have also been shown to lead to anaerobic conditions, which can result in the formation of methane gas but retard the biodegradation of the polymer as the methane formation inhibits microbial growth [79–83].

pH affects biodegradation due to its effect on the microorganisms responsible for it. Different biodegradation processes are optimal at various pH levels, and when the pH deviates too much from the optimum, biodegradation will be significantly slowed. Looking at PBS, and, in most cases, biodegradable polymers, the optimum pH is near neutral since the microorganisms that break down the polymer are optimized to digest at near-neutral pH, and acidic and alkaline conditions have been shown to slow the biodegradation of the polymers significantly [58, 84, 85].

Phosri et al. [86] compared the biodegradability of PBS with that of PLA and low-density polyethylene (LDPE) floated on the seawater surface and when submerged in the sand (under the seawater). The floating PBS on seawater took 22 weeks to degrade and needed 26–30 weeks under the sand, while PLA did not degrade fully but cracked into small pieces in 56 and 66 weeks, respectively. These degradations occurred due to both hydrolysis and bioactive consumption. No degradation was observed on LDPE in up to 100 weeks. The microplastics from PBS degradation had disappeared after 100 weeks, which was not the case for PLA microplastics.

6.5 APPLICATIONS

This section mainly explores the potential use of PBS in food preservation and its applications as a packaging material and in coatings and films compared to conventional packaging materials and methods (Figure 6.3).

FIGURE 6.3 Polybutylene succinate (PBS) and its copolymers are aliphatic polyesters, showing increasing interest as a biodegradable alternative.

6.5.1 PBS PACKAGING MATERIAL IN FOOD PRESERVATION

PBS has drawn broad attention in food packaging and drug delivery due to its biodegradability in soil and compost and its nontoxicity. PBS can produce various food packaging materials with different shapes and forms through common plastic processing methods such as film blowing, injection molding, and sheet thermoforming. PBS has been processed as a film wrapper for vegetables or fruits and a bag for vegetables, fruit, meat, and bread to maintain the food's quality and freshness for longer. Unpreserved food decays during storage through the action of food spoilage organisms such as microorganisms, enzymes, and oxidation. Coatings and films made from biodegradable polymers can be used as a means to improve the shelf life of food products by providing a protective layer to prevent mechanical damage or by incorporating active agents to provide antimicrobial and antioxidant effects [21, 37, 87].

6.5.1.1 PACKAGING MATERIALS

PBS can maintain an atmosphere within the package, delay the respiration rate, which is a good indicator of the food's freshness, reduce the weight loss of food, keep the original color and flavor, and extend the shelf life of the food. A suitable gas barrier property of the film achieves all these. PBS exhibits high compatibility when blended with other polymers and additives. This provides an excellent opportunity to modify the properties of PBS film to suit various food packaging applications. For instance, blending with PE and adding a slip agent can produce a low-cost and low-galling resistance film suitable for accessible open-packaging applications such as a bag for salad or potato chips. PBS is blended with ethylene vinyl alcohol copolymer (EVOH) to enhance the film's gas barrier property. Substantial progress has been made in producing PBS/EVOH multilayer barrier films [88–94].

6.5.1.2 COATINGS AND FILMS

Two types of coatings can be used on food products: edible and nonedible. Edible coatings comprise lipids, proteins, or polysaccharides and generally provide a barrier against moisture and oxygen. Nonedible coatings are designed to offer excellent mechanical protection and can be used in a broader range of food products. PBS is potentially a biodegradable food coating due to its thermoplastic processability and film properties. PBS has a lower gas permeability for O_2 and CO_2 than conventional plastic materials such as PEs, but higher permeability to water [95]. A study by Yamada et al. (2002) showed that the oxygen scavenging rate of a PBS film was significantly improved by adding carbon particles. The film added 10 wt% of channel-type carbon fiber and carbon black, demonstrating an oxygen scavenging rate of 14 mL/m^2 h. It was shown that the film maintained a low oxygen concentration inside a package of meat or fish while stored in aerobic conditions. This could be a particularly effective method for the preservation of red meat products where an oxygen-free packaging environment is highly desirable [44, 83, 93, 96–99].

6.5.1.3 CONTROLLED RELEASE SYSTEMS

Antimicrobial agent allyl isothiocyanate (AITC) is an effective antimicrobial agent against a broad array of microorganisms, including food poisoning bacteria; however, it has caused the breakdown of films when contacting the plasticizer and also leaves a pungent odor due to the residual amount of AITC that is not released from the film. Therefore, a film must be able to change the duration of release by altering the concentration of AITC in the different wash-out fractions. This would benefit active packaging to prevent food oxidative and bacterial spoilage. A proposed film was based on a blend of poly(butylene adipate-*co*-terephthalate) (PBAT) and PBS at a 7:3 weight ratio. There was a correlation between the increase of AITC concentration and surface antimicrobial activity. *In vitro* testing showed that the films were fully effective in inhibiting the growth of *Escherichia coli* and *Rhodotorula glutinis* [100–103] (Figure 6.4).

FIGURE 6.4 Packing materials for food.

6.5.1.4 ACTIVE PACKAGING SOLUTIONS

Active packaging is a new concept in the packaging field and is still in the exploration stage. Active packaging is designed to extend shelf life and maintain food quality by interacting selectively with the food. Various substances can be incorporated into the package to interact with the food and indicate contamination or the time–temperature history of the package. Another example is the development of indicator films to check the package, which is sterility. Active packaging can be divided into two types: sachet and release packaging. In the sachet, active components are enclosed in a separate part of the package. In contrast, with release packaging, the active components are incorporated into the packaging material and released into the package's headspace [104–107].

PBS has established a robust platform as an active/bright packaging material. This material has the potential to store and release substances on demand due to its excellent gas permeability. This material can store and release the preservatives at the most appropriate time and into the selected part of the packed food. An example of this would be using PBS in composite resin-type packaging to extend the shelf life of cheese. In this packaging, PBS will release a preservative by detecting the change in pH of the cheese. Preservatives can be stored and released by utilizing the difference in solubility of the substance in various food products or by sealing them in a microcapsule. Storage and release can be regulated through the concentration and molecular weight of the substance. Another benefit would be prolonging food durability with a deficient concentration of preservatives compared to packaging that releases into the entire package. This will reduce the taint from the preservative and is a very effective way to preserve food, such as the modified atmosphere packaging of meat. Step one would store the antimicrobial gas in PBS and release it onto the wrapped meat. Since PBS has high gas barrier properties, storing gases for an extended period will be possible [31, 108].

6.5.2 OTHER APPLICATIONS

6.5.2.1 AGRICULTURAL FILMS

Agricultural films primarily aim to change and improve a seedbed's soil environment and protect the crops by covering the films. These applications have the potential benefits to contribute to sustainable crop production. Developments and modifications of biodegradable PBS products are needed to reach these goals. There are many types of agricultural films: mulch, row cover, and greenhouse covering films. Mulch films are used to better the soil environment of a seedbed. High-quality mulch films can improve soil temperature and moisture and keep a clean environment around the crops [17, 109–111].

Usually, the row-parallel laid films are winded once the left and right edges are cut. PBS or PBS blend is suitable for this application, and the biodegradable mulch films will reduce the soil pollution caused by the aged films left on the field. Row cover films protect the crop from damage from insects and wind. Nonwoven fabric is used as it is easy to handle, but the film type has cost benefits. PBS resin is suitable for making thin and nonwoven fabrics and can be the raw material for nonwoven fabrics. It is also possible to make films for easy installation. Greenhouse covering films are used to protect the crops from rain and wind.

In most cases, it is used as a long-life type of film, but the fact that the films need to be replaced due to deterioration and damage means they are consumable. The degradable greenhouse covering films will avoid the pollution problem. However, compared to the other types of agricultural films, PBS-based greenhouse-covering films are costly, so developments in cost reductions for film manufacturing are needed [112–115].

6.5.2.2 MEDICAL DEVICES

The medical device industry is an expansive and continually expanding sector characterized by the ever-increasing volume of disposable and single-use devices. As technology advances and healthcare demands rise, more innovative products are being introduced to address various medical needs. This trend has contributed to the significant growth and development of the industry, with a steady influx of novel devices effectively catering to the dynamic healthcare landscape. With an emphasis on convenience, safety, and efficiency, the industry has seen a surge in the production and utilization of disposable and single-use devices, further solidifying their pivotal role in modern healthcare practices.

A novel development has been conducted on producing PBS sutures and fasteners for joining tissue. Compared to the current market standard of polyglycolic acid sutures, PBS sutures have been shown to have much slower degradation rates and better maintenance of tensile strength in

PBS, while having an equivalent or slightly better performance in PBSA (succinate-*co*-adipate copolymer). In an *in vivo* study using rabbits, PBS sutures caused less tissue reaction and faster healing after skin closure than polyglycolic acid (PGA). It is an important finding, as the rate of healing in tissue joining is crucial and can directly affect the mechanical support provided by the suture during the early stages of wound healing. Surgeons have also noted that the passage of PBS sutures into the tissue is easier during use [116, 117].

The effect of environmental stress cracking, a widespread problem in plastics that leads to a reduction in mechanical properties and failure in the presence of an active agent has been studied in PBS and PBSA. A mechanical test was carried out in tensile bars to simulate an *in vivo* setting, and it showed that PC resisted cracking in PBS. At the same time, there was less sensitivity in PBSA but still an improvement over other materials [21, 87].

6.6 PERFORMANCE AND LIMITATIONS

PBS has good mechanical properties regarding stiffness, hardness, and strength. It has been found to have double the stiffness and strength compared to PE and similar strength to polypropylene. This makes it a suitable candidate for short-lived, low-load applications such as packaging. PBS is often touted as having performances like polyolefins and is directly replaceable at comparable costs. PBS had double the elongation at the polybutylene (PB) break but the same tensile strength and modulus. PBS has a lower glass transition temperature, typically well below room temperature. This means that PBS is often in a glassy state at store temperatures and is more impact-brittle. It can limit its applications, and some studies have tried to modify the PBS, such as copolymerizing it with other butylene esters to alter the T_g [29, 34, 118, 119].

The above mechanical properties suggest that PBS may require improvement for some applications via copolymerization or blending, especially if it is to replace specific existing polymers in more demanding applications. PBS is sensitive to the processing conditions. For example, it has been shown that increased mold temperature and lower injection pressure increase the crystallinity and, therefore, stiffness and strength of the PBS. On the other hand, increasing the temperature or shear rate during processing when using PBS can cause severe thermal degradation. It will reduce molecular weight and mechanical performances during the processing, so the most notable effects will be near the surface of the formed article [12, 46, 120, 121].

6.6.1 MECHANICAL STRENGTH

The mechanical strength of PBS has been characterized by its modulus, tensile strength, elongation at break, and impact strength. PBS possesses high stiffness, as indicated by its high modulus compared to other currently used biodegradable polymers. The modulus of PBS can range from 1500 to 2400 MPa, which can be altered by copolymerization with other polymers. The high modulus and excellent tensile strength ranging from 50 to 80 MPa means PBS can replace certain petroleum-based thermoplastics. A reduction in elongation at break and impact strength of PBS compared with other aliphatic polyesters may limit the range of applications for PBS in its current form. Softer materials such as PE have an elongation at break of around 200–700%, compared with only 20–40% for PBS, and this can reduce toughness and ductility for specific applications. PBS has been produced in grades to offer a rubbery polymer with higher impact strength, which could potentially expand the range of applications for PBS in the future [25, 44, 122, 123].

6.6.2 THERMAL STABILITY

The full crystalline melting temperature of the PBS stored at 25°C is around 114°C, and the melting behavior of spherulites of PBS is seen at about 125°C. Thus, a significant thermal degradation test is the weight loss of PBS stored in air (static air oven tests). PBS can be easily crystallized from the melt and exhibits an appreciable crystallinity ranging from 30% to 50%. PBS with a low molecular weight of <50,000 is soluble in chloroform and has a crystallinity of around 30%. The sample heats up to 75°C with no apparent melting taking place. After an 8-h oven test at 100°C or 110°C, there will be a rapid weight loss and significant changes in the physical properties, as shown by the test piece becoming opaque and rubbery due to partial melting [16, 88, 124, 125].

PBS has quite an excellent oxygen index of 47%, and there is little change in this index and tensile strength after aging in air at 100°C for four months. It indicates that PBS has good long-term heat stability at around 100°C, which would be a significant criterion for its applicability to conventional polymers [21].

6.7 COMPARATIVE ANALYSIS WITH OTHER BIODEGRADABLE POLYMERS

Table 6.1 shows the physical and chemical properties of PBS, the comparable PLA, and other commonly studied biodegradable plastics.

TABLE 6.1 Comparative Table of the Physical and Chemical Properties Between Polybutylene Succinate (PBS) and Other Biodegradable Plastics [21, 126–134]

Property	PBS	PLA	PHA	PBAT
Chemical structure	Polyester of succinic acid and butanediol	Polyester of lactic acid	Polyesters of hydroxyalkanoic acids	Copolyester of butylene adipate and terephthalic acid
Biodegradability	Good, in both aerobic and anaerobic conditions	Good, primarily in industrial composting conditions	Excellent biodegrades in various environments, including marine	Good, primarily in soil and composting facilities
Melting point (°C)	90–120	150–170	80–180, depending on the copolymer composition	100–120
Glass transition temperature (T_g) (°C)	−40 to −10	55–65	−10 to 10	−30 to 0
Tensile strength (MPa)	25–45	50–70	20–40	20–30
Elongation at break (%)	150–400	3–10	100–1000, depending on composition	300–600
Density (g/cm³)	1.25–1.27	1.24–1.26	1.2–1.3	1.25–1.30
Water absorption	Moderate, depending on crystallinity	High	Low to moderate	Moderate
Resistance to oils and greases	Good	Moderate	Variable, generally good	Good
Applications	Packaging, agriculture, biodegradable medical devices	Packaging, disposable utensils, medical devices	Packaging, medical applications, agricultural films	Flexible packaging, agricultural films, compost bags

6.7.1 POLYLACTIC ACID

PLA has shown potential for short-term, single-use applications and long-term implants. Currently, all commercial PLAs are produced from the high-value D,L-lactide. Due to the nature of lactic acid formation, PLA can be made with very low toxicity and from renewable resources. It means that it is a highly sustainable and environmentally friendly material. Its biodegradability has been extensively evaluated both *in vivo* and *in vitro* and is one of the primary properties that make PLA an attractive alternative to synthetic polymers. *In vivo* tests have shown that the degradation time of PLA can be tailored depending on the intended application, from a few weeks to several years. It is achieved by varying the molecular weight and crystallinity of the polymer. High-crystalline, high-molecular-weight PLA can be used for applications such as bone fixation devices, where an extended degradation time is beneficial. Although the nature of *in vivo* testing is somewhat controversial, one must recognize the potential benefits of a biodegradable polymer in the medical industry.

As discussed earlier, PLA has shown potential for short-term, single-use applications and long-term implants. It opens many doors to new applications for PLA as technology in the medical industry continues to advance [135–140]. PLA can be used in medical sutures, where making a suture of a particular lifetime is possible. Initially, PCL was used for this, but now it is possible to get the same behavior from PGA sutures. Changing the crystallinity or making block copolymers with polycarbonates can alter the degradation rate of PLA to obtain the desired lifetime. PLA and PCL block copolymers can also be used for specific applications. Block copolymers of equivalent lactic acid and ε-caprolactone have been shown to degrade to nontoxic products *in vitro* tests and are possible candidates for biodegradable implant materials [134, 141–143].

6.7.2 POLYHYDROXYALKANOATES

Compared to PLA, PBS is considered a less optimal candidate for synthetic modification of PHA biosynthesis because it is insoluble and possesses a highly crystalline structure. However, PHB biosynthesis in recombinant *E. coli* has been significantly improved by culturing the bacteria in PBS media, and this has led to the highest productivities of

PHB reported to date, with dry cell weights of 40 g/L, cell mass PHB content of 80%, and overall PHB productivity of 2.9 g/L h. It has been achieved by the use of a constitutive promoter and the removal of β-oxidation pathways using a thiolase II mutant, and valuable information was gained on the genetic manipulations required to overcome the feedback inhibitions of CoA-esters on the vital biosynthetic enzymes in *E. coli.* Other *Pseudomonas* species and *Ralstonia eutropha* have been cultured in PBS media to enhance PHB productivity. *Pseudomonas resinovorans* has been shown to produce mcl-PHA using PBS as a sole carbon source. This cultivation strategy of using PBS as a growth medium for PHA accumulation has been termed the "matrix polymer strategy" and is likely to be very useful for producing various PHAs using a wide range of substrates. Thus, PBS may produce PHB or other PHAs cost-effectively using recombinant organisms. PBS would also be a good candidate for biodegradable packaging using a PHA coating [131, 132, 144–146].

6.7.3 POLYCAPROLACTONE

PCL is another semicrystalline and biodegradable polymer synthesized from ROP between caprolactone monomers. PCL is one of the first discovered biodegradable aliphatic polyesters that has found many applications primarily due to its low melting point and low glass transition temperature, which make it very easy to process. PCL is also very flexible and has a longer degradation time of about 1–2 years compared to other aliphatic polyesters. This longer bio-degradation time allows PCL to be a better candidate for applications such as drug-delivery systems and scaffolds for tissue engineering, where they need a material that will maintain its structure for a while and release drugs or degrade in the body later [74, 147–149]. PCL is nontoxic and nonimmunogenic. When implanted in the body, only a mild inflammatory reaction is elicited, and it is completely absorbed without any significant tissue response. The aliphatic ester linkages in PCL also contribute to the degraded products' excellent biocompatibility and low toxicity. It contrasts with aromatic polymers, whose acidic degradation products can cause an inflammatory reaction [150, 151].

6.8 FUTURE DIRECTIONS AND CHALLENGES

6.8.1 IMPROVED SYNTHESIS TECHNIQUES

The formulation of PBS is of utmost importance in determining the polymer's physical and chemical properties. Using tin catalysts, a highly heat-resistant polymer grade of PBS can be produced through ROP of succinic anhydride and BDO. The reaction between the carboxylic acid group and the alcohol, with the aid of an acid catalyst, has also been reported to yield a high molecular weight polymer suitable for thermoforming and casting applications. Various methods have been investigated to enhance the polymer's reaction rate and molecular weight. These methods include microwave assistance, ultrasonication, supercritical CO_2, and DMC solvents. However, the reverse polycondensation of succinate esters with butanediol has proven to be an energy-efficient approach to producing high molecular weight PBS with minimal downstream processing. Despite advancements in achieving high molecular weight polymers, PBS still undergoes degradation during the polymerization process. The thermomechanical degradation of the polymer can result in a decrease in molecular weight and chain scission, which can adversely affect the biodegradation of the polymer. This degradation can be controlled using isothermal polymerization techniques and exploring alternative polymerization methods. Investigating different monomer sources for PBS synthesis could lead to a narrower distribution of molecular weight and improved flexibility of the polymer. A better understanding of the role of catalysts and reaction mechanisms will also be essential for future research in achieving controlled PBS polymerization [18, 25, 152].

6.8.2 INDUSTRIAL-SCALE PRODUCTION

A significant technological challenge in the industrial production of thermoplastic aliphatic polyesters and aliphatic–aromatic copolyesters is their high production costs. These costs are mainly due to the complexity of the manufacturing process and the extended reaction times required. Comparative studies indicate that the production costs of aliphatic–aromatic copolyesters are approximately 20% higher than those of conventional commodity plastics such as polypropylene and PE [153]. To render these biodegradable plastics competitive, it is crucial to employ strategies aimed at cost reduction. Several approaches have been proposed to achieve this:

- Developing methods to accelerate the synthesis reaction significantly reduces the production costs of biodegradable plastics.
- Improve the biodegradability of plastics to ensure they can be processed alongside other organic waste after use. This strategy also contributes to the conservation of fossil resources, which is particularly important as the disposal of fossil-based commodity plastics by incineration or landfills faces increasing restrictions.
- To expand the production volume, which is intrinsically linked to market size. A larger production volume can substantially decrease the unit cost of production, making these sustainable alternatives more economically viable [154–156].

These initiatives are essential for the broader adoption of biodegradable plastics, aligning environmental sustainability with economic feasibility.

6.9 CONCLUSION

PBS is the biodegradable aliphatic polyester that most closely resembles the desired properties of general-purpose polyolefins. It has garnered considerable interest and attention due to its immense potential in addressing the numerous ecological challenges of using and disposing of nonbiodegradable petroleum-based polymers. PBS boasts an exceptional balance of mechanical and thermal properties and commendable degradability. PBS can be transformed into films, bags, or containers for food and cosmetic products, including disposable items like tableware or medical supplies. In the agricultural sector, PBS can be useful for creating mulching films or controlled-release materials for pesticides and fertilizers. In other areas, material recovery and recycling postuse is a challenge. In the medical field, PBS could serve as a biodegradable system for drug encapsulation and is also being explored for use in implants.

KEYWORDS

- **polybutylene succinate**
- **biopolymer**
- **biodegradability**
- **sustainability**
- **food packaging**
- **barrier properties**
- **shelf life**
- **circular economy**

REFERENCES

1. Agarwal, S. (2020). Biodegradable polymers: Present opportunities and challenges in providing a microplastic-free environment. *Macromolecular Chemistry and Physics*, 221, 2000017.
2. Amulya, K., Katakojwala, R., Ramakrishna, S., Mohan, S.V. (2021). Low carbon biodegradable polymer matrices for sustainable future. *Composites Part C: Open Access,* 4, 100111.
3. Law, K.L. Narayan, R. (2022). Reducing environmental plastic pollution by designing polymer materials for managed end-of-life. *Nature Reviews Materials*, 7, 104–116.
4. Manfra, L., Marengo, V., Libralato, G., Costantini, M., De Falco, F., Cocca, M. (2021). Biodegradable polymers: A real opportunity to solve marine plastic pollution? *Journal of Hazardous Materials*, 416, 125763.
5. Millican, J.M., Agarwal, S. (2021). Plastic pollution: A material problem? *Macromolecules*, 54, 4455–4469.
6. Moshood, T.D., Nawanir, G., Mahmud, F., Mohamad, F., Ahmad, M.H., AbdulGhani, A. (2022). Sustainability of biodegradable plastics: New problem or solution to solve the global plastic pollution?. *Current Research in Green and Sustainable Chemistry*, 5, 100273.
7. Sazali, N., Ibrahim, H., Jamaludin, A.S., Mohamed, M.A., Salleh, W.N.W., Abidin, M.N.Z. (2020). A short review on polymeric materials concerning degradable polymers. *IOP Conference Series: Materials Science and Engineering,* 788, 012047.
8. Wang, C., Liu, Y., Chen, W.Q., Zhu, B., Qu, S., Xu, M. (2021). Critical review of global plastics stock and flow data. *Journal of Industrial Ecology*, 25, 1300–1317.
9. Filiciotto, L., Rothenberg, G. (2021). Biodegradable plastics: Standards, policies, and impacts. *ChemSusChem*, 14, 56–72.

10. Shen, M., Song, B., Zeng, G., Zhang, Y., Huang, W., Wen, X., Tang, W. (2020). Are biodegradable plastics a promising solution to solve the global plastic pollution? *Environmental Pollution*, 263, 114469.
11. Nakayama, K., Masuda, T., Cao, A., Vega-Baudrit, J., Pereira, R. (2003). Orientation effect in poly(butylene succinate) fibers. *Polymers and Polymer Composites*, 11, 51–56. doi: 10.1177/096739110301100106.
12. Nanni, A., Ricci, A., Versari, A., Messori, M. (2020). Wine derived additives as poly(butylene succinate) (PBS) natural stabilizers for different degradative environments. *Polymer Degradation and Stability*, 182, 109381.
13. Satti, S.M., Shah, A.A. (2020). Polyester-based biodegradable plastics: an approach towards sustainable development. *Letters in Applied Microbiology*, 70, 413–430.
14. Zheng, C., Zhu, G., Shi, Y., Liu, L.Z., Ren, M., Zhang, W., Han, L. (2021). Crystallization, structures, and properties of biodegradable poly(butylene succinate-co-butylene terephthalate) with a symmetric composition. *Materials Chemistry and Physics*, 260, 124183.
15. Zhou, S.J., Wang, H.M., Xiong, S.J., Sun, J.M., Wang, Y.Y., Yu, S., Sun, Z., Wen, J.L., Yuan, T.Q. (2021). Technical lignin valorization in biodegradable polyester-based plastics (BPPs). *ACS Sustainable Chemistry & Engineering*, 9, 12017–12042.
16. de Matos-Costa, A.R., Crocitti, A., Hecker de Carvalho, L., Carroccio, S.C., Cerruti, P., Santagata, G. (2020). Properties of biodegradable films based on poly(butylene succinate) (PBS) and poly(butylene adipate-co-terephthalate) (PBAT) blends. *Polymers (Basel),* 12, 2317.
17. Rafiqah, S.A., Khalina, A., Harmaen, A.S., Tawakkal, I.A., Zaman, K., Asim, M., Nurrazi, MN., Lee, C.H. (2021). A review of properties and application of bio-based poly(butylene succinate). *Polymers (Basel),* 13, 1436.
18. Savitha, K.S., Paghadar, B.R., Kumar, M.S., Jagadish, R.L. (2022). Polybutylene succinate, a potential bio-degradable polymer: Synthesis, copolymerization, and bio-degradation. *Polymer Chemistry*, 13, 3562–3612.
19. Vega-Baudrit, J., Pereira, R., Nakayama, K., Masuda, T., Cao, A. (2004). Structure and properties of biodegradable polymers: Rolling effect in poly(butylene succinate) sheets. *Polymers and Polymer Composites,* 12, 63–73. doi: 10.1177/096739110401200106.
20. Zhao, X., Zhang, D., Yu, S., Zhou, H., Peng, S. (2021). Recent advances in compatibility and toughness of poly(lactic acid)/poly(butylene succinate) blends. *e-Polymers*, 21, 793–810.
21. Barletta, M., Aversa, C., Ayyoob, M., Gisario, A., Hamad, K., Mehrpouya, M., Vahabi, H. (2022). Poly (butylene succinate) (PBS): Materials, processing, and industrial applications. *Progress in Polymer Science*, 132, 101579.
22. Platnieks, O., Gaidukovs, S., Thakur, V.K., Barkane, A., Beluns, S. (2021). Bio-based poly(butylene succinate): Recent progress, challenges and future opportunities. *European Polymer Journal*, 161, 110855.
23. Thakur, S., Chaudhary, J., Singh, P., Alsanie, W.F., Grammatikos, S.A., Thakur, V.K. (2022). Synthesis of bio-based monomers and polymers using microbes for a sustainable bioeconomy. *Bioresource Technology*, 344, 126156.
24. Zhu, Y., Yang, J., Mei, F., Li, X., Zhao, C. (2022). Bio-based 1, 4-butanediol and tetrahydrofuran synthesis: Perspective. *Green Chemistry*, 24, 6450–6466.

25. Peñas, M.I., Pérez-Camargo, R.A., Hernández, R., Müller, A.J. (2022). A review on current strategies for the modulation of thermomechanical, barrier, and biodegradation properties of poly(butylene succinate) (PBS) and its random copolymer. *Polymers (Basel)*, 14, 1025.
26. Liu, B., Guan, T., Wu, G., Fu, Y., Weng, Y. (2022). Biodegradation behavior of degradable mulch with poly(butylene adipate-co-terephthalate) (PBAT) and poly(butylene succinate) (PBS) in simulation marine environment. *Polymers,* 14, 1515.
27. Liu, L., Xu, M., Ye, Y., Zhang, B. (2022). On the degradation of (micro) plastics: Degradation methods, influencing factors, environmental impacts. *Science of The Total Environment*, 806, 151312.
28. Palai, B., Mohanty, S., Nayak, S.K. (2021). A comparison on biodegradation behaviour of polylactic acid (PLA) based blown films by incorporating thermoplasticized starch (TPS) and poly(butylene succinate-co-adipate) (PBSA) biopolymer in Soil. *Journal of Polymers and the Environment*, 29, 2772–2788.
29. Aliotta, L., Seggiani, M., Lazzeri, A., Gigante, V., Cinelli, P. (2022). A brief review of poly(butylene succinate) (PBS) and its main copolymers: Synthesis, blends, composites, biodegradability, and applications. *Polymers (Basel),* 14, 844.
30. Gucker, F.E., Sayer, C., de Oliveira, D., Hermes de Araújo, P.H., Oechsler, B.F. (2023). Current status and perspectives on the green synthesis of succinic polyesters for value-added applications. *Macromolecular Reaction Engineering*, 17, 2200061.
31. Zhang, F., Zhang, M., Chen, Y., Ouyang, J., Wang, Y., Yang, H., Luo, X., Zhang, D., Lu, Y., Yu, H., Wang, Y. (2021). Antimicrobial, anti-biofilm properties of three naturally occurring antimicrobial peptides against spoilage bacteria, and their synergistic effect with chemical preservatives in food storage. *Food Control*, 123, 107729.
32. Borges, G.R., Aboelkheir, M.G., de Souza Junior, F.G., Waldhelm, K.C., Kuster, R.M. (2020). Poly (butylene succinate) and derivative copolymer filled with *Dendranthema grandiflora* biolarvicide extract. *Environmental Science and Pollution Research*, 27, 23575–23585.
33. Zarei, M., El Fray, M. (2021). Synthesis of hydrophilic poly(butylene succinate-butylene dilinoleate) (PBS-DLS) copolymers containing poly(ethylene glycol) (PEG) of variable molecular weights. *Polymers (Basel),* 13, 3177.
34. Di Lorenzo, M.L. (2021). Poly(l-lactic acid)/poly(butylene succinate) biobased biodegradable blends. *Polymer Reviews*, 61, 457–492.
35. Im, D., Gavande, V., Lee, H., Iwata, T., Lee, W.K. (2021). Compatibility and hydrolytic behaviors of polylactide isomer/poly(butylene succinate) mixtures by the Langmuir technique. *Polymer Degradation and Stability*, 186, 109517.
36. Flores, E.D., Funabashi, M., Kunioka, M. (2009). Mechanical properties and biomass carbon ratios of poly(butylene succinate) composites filled with starch and cellulose filler using furfural as plasticizer. *Journal of Applied Polymer Science*, 112, 3410–3417. doi: 10.1002/app.29777.
37. Rajgond, V., Mohite, A., More, N., More, A. (2023). Biodegradable polyester-polybutylene succinate (PBS): A review. *Polymer Bulletin*, 81, 5703–5752.
38. Sokołowska, M., Nowak-Grzebyta, J., Stachowska, E., El Fray, M. (2022). Enzymatic catalysis in favor of blocky structure and higher crystallinity of poly(butylene

succinate)-co-(dilinoleic succinate) (PBS-DLS) copolymers of variable segmental composition. *Materials*, 15, 1132.

39. Guckert, F.E., Sayer, C., de Oliveira, D., de Araújo, P.H.H., Oechsler, B.F. (2022). Synthesis of polybutylene succinate via lipase-catalyzed transesterification: Enzyme stability, reuse and PBS properties in bulk polycondensations. *European Polymer Journal,* 179, 111573.
40. Kumar, R., Sadeghi, K., Jang, J., Seo, J. (2023). Mechanical, chemical, and bio-recycling of biodegradable plastic: A review. *Science of the Total Environment*, 882, 163446.
41. Mtibe, A., Motloung, M.P., Bandyopadhyay, J., Ray, S.S. (2021). Synthetic biopolymers and their composites: Advantages and limitations—An overview. *Macromolecular Rapid Communications*, 42, 2100130.
42. Reichert, C.L., Bugnicourt, E., Coltelli, M.B., Cinelli, P., Lazzeri, A., Canesi, I., Braca, F., Martínez, B.M., Alonso, R., Agostinis, L., Verstichel, S., Six, L, Mets, S.D., Gómez, E.C., Ißbrücker, C., Geerinck, R., Nettleton, D.F., Campos, I., Sauter, E., Pieczyk, P., Schmid, M. (2020). Bio-based packaging: Materials, modifications, industrial applications and sustainability. *Polymers*, 12, 1558.
43. Sheldon, R.A., Norton, M. (2020). Green chemistry and the plastic pollution challenge: Towards a circular economy. *Green Chemistry*, 22, 6310–6322.
44. Zhang, J., Hirschberg, V., Rodrigue, D. (2022). Mechanical fatigue of biodegradable polymers: A study on polylactic acid (PLA), polybutylene succinate (PBS) and polybutylene adipate terephthalate (PBAT). *International Journal of Fatigue*, 159, 106798.
45. Zhao, X., Cornish, K., Vodovotz, Y. (2020). Narrowing the gap for bioplastic use in food packaging: An update. *Environmental Science & Technology*, 54, 4712–4732.
46. Hallstein, J., Gomoll, A., Lieske, A., Büsse, T., Balko, J., Brüll, R., Malz, F., Metzsch-Zilligen, E., Pfaendner, R., Zehm, D. (2021). Unraveling the cause for the unusual processing behavior of commercial partially bio-based poly(butylene succinates) and their stabilization. *Journal of Applied Polymer Science*, 138, 50669.
47. Murmu, S.B., Nayak, L.K., Lakhmanan, A., Baite, H., Nayak, D., & Huirem, B. (2024). Constraints on industrial-scale application of lignocellulosic fiber and biomass in food packaging. *Iranian Polymer Journal*, 33, 855–875.
48. Huang, W., Zhang, G., Zhou, Y., Joziasse, C.A., Wang, R., du Sart, G.G., Chen, P. (2024). Superior heat-resistant polylactide/poly(butylene succinate) blend fibers via in-situ reactive compatibilization. *Polymers for Advanced Technologies*, 35, e6345.
49. Rennert, M., Hiller, B.T. (2023). Influence of coffee variety and processing on the properties of parchments as functional bioadditives for biobased poly(butylene succinate) composites. *Polymers (Basel),* 15, 2985.
50. Zhang, W., Sun, W.H. (2024). Progressive Metal Complex Catalysts for Ring-Opening Polymerization of Cyclic Esters. In Synthesis and Applications in Chemistry and Materials: Volume 14: Biomass and Waste Valorisation, Functional Materials, Energy Conversion and Supercritical Systems 217–313.
51. Jacquel, N., Freyermouth, F., Fenouillot, F., Rousseau, A., Pascault, J.P., Fuertes, P., Saint-Loup, R. (2011). Synthesis and properties of poly(butylene succinate): Efficiency of different transesterification catalysts. *Journal of Polymer Science Part A: Polymer Chemistry*, 49, 5301–5312. doi: 10.1002/pola.25009.

52. Plengsangsri, T. (2022). Control of a polybutylene succinate (PBS) polymerization in a semi-batch reactor by neural network model-based predictive controller. *Chulalongkorn University Theses and Dissertations (Chula ETD)*. 5774.
53. Gkountela, C. (2024). Development of environmentally friendly polymerization processes.
54. Al-Shok, L., Haddleton, D.M., Adams, F. (2022). Progress in catalytic ring-opening polymerization of biobased lactones. In: *Synthetic Biodegradable and Biobased Polymers: Industrial Aspects and Technical Products* (pp. 197–267). Springer International Publishing: Cham.
55. de Ilarduya, A.M. Guerra, S.M. (2020). Ring-opening polymerization of macrocyclic oligoesters derived from renewable sources. *Polymer Chemistry*, 11, 4850–4860.
56. Phan, H., Kortsen, K., Englezou, G., Couturaud, B., Nedoma, A.J., Pearce, A.K., Taresco, V. (2020). Functional initiators for the ring-opening polymerization of polyesters and polycarbonates: An overview. *Journal of Polymer Science*, 58, 1911–1923.
57. Labruyère, C., Talon, O., Berezina, N., Khousakoun, E., Jérôme, C. (2014). Synthesis of poly(butylene succinate) through oligomerization–cyclization–ROP route. *RSC Advances*, 4, 38643–38648. doi: 10.1039/C4RA03373F.
58. Liu, T.Y., Huang, D., Xu, P.Y., Lu, B., Wang, G.X., Zhen, Z.C., Ji, J. (2022). Biobased seawater-degradable poly(butylene succinate-l-lactide) copolyesters: exploration of degradation performance and degradation mechanism in natural seawater. *ACS Sustainable Chemistry & Engineering*, 10, 3191–3202.
59. Anstey, A., Muniyasamy, S., Reddy, M., Misra, M., Mohanty, A. (2014). Processability and biodegradability evaluation of composites from poly(butylene succinate) (PBS) bioplastic and biofuel co-products from Ontario. *Journal of Polymers and the Environment*, 22, 209–218. doi: 10.1007/s10924-013-0633-8.
60. Mizuno, S., Maeda, T., Kanemura, C., Hotta, A. (2015). Biodegradability, reprocessability, and mechanical properties of polybutylene succinate (PBS) photografted by hydrophilic or hydrophobic membranes. *Polymer Degradation and Stability*, 117, 58–65. doi: 10.1016/J.POLYMDEGRADSTAB.2015.03.015.
61. Phua, Y., Lau, N., Sudesh, K., Chow, W., Ishak, Z. (2012). Biodegradability studies of poly(butylene succinate)/organo-montmorillonite nanocomposites under controlled compost soil conditions: Effects of clay loading and compatibilizer. *Polymer Degradation and Stability*, 97, 1345. doi: 10.1016/J.POLYMDEGRADSTAB.2012.05.024.
62. Qi, J., Wu, J., Chen, J., Huaping, W. (2019). An investigation of the thermal and (bio) degradability of PBS copolyesters based on isosorbide. *Polymer Degradation and Stability*, 160, 229–241. doi: 10.1016/J.POLYMDEGRADSTAB.2018.12.031.
63. Tanunchai, B., Kalkhof, S., Guliyev, V., Wahdan, S., Krstic, D., Schädler, M., Geissler, A., Glaser, B., Buscot, F., Blagodatskaya, E., Noll, M., Purahong, W. (2022). Nitrogen-fixing bacteria facilitate microbial biodegradation of a bio-based and biodegradable plastic in soils under ambient and future climatic conditions. *Environmental Science Processes & Impacts,* 24, 233–241. doi: 10.1039/d1em00426c.
64. Delangiz, N., Aliyar, S., Pashapoor, N., Nobaharan, K., Lajayer, B.A., Rodríguez-Couto, S. (2022). Can polymer-degrading microorganisms solve the bottleneck of plastics' environmental challenges? *Chemosphere*, 294, 133709.

65. Elahi, A., Bukhari, D.A., Shamim, S., Rehman, A. (2021). Plastics degradation by microbes: A sustainable approach. *Journal of King Saud University-Science*, 33, 101538.
66. Jaiswal, S., Sharma, B., Shukla, P. (2020). Integrated approaches in microbial degradation of plastics. *Environmental Technology & Innovation,* 17, 100567.
67. Oliveira, J., Belchior, A., da Silva, V.D., Rotter, A., Petrovski, Ž., Almeida, P.L., Lourenço, N.D., Gaudêncio, S.P. (2020). Marine environmental plastic pollution: Mitigation by microorganism degradation and recycling valorization. *Frontiers in Marine Science*, 7, 567126.
68. Ali, S.S., Elsamahy, T., Koutra, E., Kornaros, M., El-Sheekh, M., Abdelkarim, E.A., Zhu, D., Sun, J. (2021). Degradation of conventional plastic wastes in the environment: A review on current status of knowledge and future perspectives of disposal. *Science of The Total Environment*, 771, 144719.
69. Inderthal, H., Tai, S.L., Harrison, S.T.L. (2021). Non-hydrolyzable plastics–an interdisciplinary look at plastic bio-oxidation. *Trends in Biotechnology,* 39, 12–23.
70. Brunner, P.C., Torriani, S.F.F., Croll, D., Stukenbrock, E.H., McDonald, B.A. (2013). Coevolution and life cycle specialization of plant cell wall degrading enzymes in a hemibiotrophic pathogen. *Molecular Biology and Evolution*, 30, 1337–1347. doi: 10.1093/molbev/mst041.
71. Lee, C.W., Akashi, M., Kimura, Y. et al. (2017). Synthesis and enzymatic degradability of an aliphatic/aromatic block copolyester: Poly(butylene succinate)-multi-poly(butylene terephthalate). *Macromolecular Research*, 25, 54–62. doi: 10.1007/s13233-017-5011-2.
72. Dutra, L., Pinto, M.C., Lima, R.C., Franco, M., Viana, M., Cipolatti, E.P., Manoel, E.A., Freire, D.M.G., Pinto, J.C. (2021). Preparation of polymer microparticles through non-aqueous suspension polycondensations: Part VI—Analyses of chemical and enzymatic degradation of poly(butylene succinate) (PBS). *Journal of Polymers and the Environment*, 30, 1893–1907.
73. Penas, M.I., Criado-Gonzalez, M., de Ilarduya, A.M., Flores, A., Raquez, J.M., Mincheva, R., Müller, A.J., Hernández, R. (2023). Tunable enzymatic biodegradation of poly(butylene succinate): Biobased coatings and self-degradable films. *Polymer Degradation and Stability,* 211, 110341.
74. Rosato, A., Romano, A., Totaro, G., Celli, A., Fava, F., Zanaroli, G., Sisti, L. (2022). Enzymatic degradation of the most common aliphatic bio-polyesters and evaluation of the mechanisms involved: An extended study. *Polymers*, 14, 1850.
75. Bher, A., Mayekar, P.C., Auras, R.A., Schvezov, C.E. (2022). Biodegradation of biodegradable polymers in mesophilic aerobic environments. *International Journal of Molecular Sciences*, 23, 12165.
76. Gricajeva, A., Nadda, A.K., Gudiukaite, R. (2022). Insights into polyester plastic biodegradation by carboxyl ester hydrolases. *Journal of Chemical Technology & Biotechnology*, 97, 359–380.
77. Urbanek, A.K., Mirończuk, A.M., García-Martín, A., Saborido, A., de la Mata, I., Arroyo, M. (2020). Biochemical properties and biotechnological applications of microbial enzymes involved in the degradation of polyester-type plastics. *Biochimica et Biophysica Acta (BBA)-Proteins and Proteomics*, 1868, 140315.

78. Zadjelovic, V., Chhun, A., Quareshy, M., Silvano, E., Hernandez-Fernaud, J.R., Aguilo-Ferretjans, M.M., Bosch, R., Dorador, C., Gibson, M.I., Christie-Oleza, J.A. (2020). Beyond oil degradation: Enzymatic potential of Alcanivorax to degrade natural and synthetic polyesters. *Environmental Microbiology*, 22, 1356–1369.
79. Bher, A., Cho, Y., Auras, R. (2023). Boosting degradation of biodegradable polymers. *Macromolecular Rapid Communications*, 44, 2200769.
80. Bu, Y., Kim, H.K., Lee, J.S., Jang, H.G., Jeong, J.H., Chun, S.W., Sharma, M., Kim, B.S. (2023). Improvement of biodegradable polymer film properties by incorporating functionalized few-layer graphene. *Food Packaging and Shelf life*, 40, 101205.
81. Glaskova-Kuzmina, T., Starkova, O., Gaidukovs, S., Platnieks, O., Gaidukova, G. (2021). Durability of biodegradable polymer nanocomposites. *Polymers*, 13, 3375.
82. Vikhareva, I.N., Buylova, E.A., Yarmuhametova, G.U., Aminova, G.K., Mazitova, A.K. (2021). An overview of the main trends in the creation of biodegradable polymer materials. *Journal of Chemistry*, 2021, 1–15.
83. Wu, F., Misra, M., Mohanty, A.K. (2021). Challenges and new opportunities on barrier performance of biodegradable polymers for sustainable packaging. *Progress in Polymer Science*, 117, 101395.
84. La Fuente, C.I.A., Maniglia, B.C., Tadini, C.C. (2023). Biodegradable polymers: A review about biodegradation and its implications and applications. *Packaging Technology and Science*, 36, 81–95.
85. Nabeoka, R., Suzuki, H., Akasaka, Y., Ando, N., Yoshida, T. (2021). Evaluating the ready biodegradability of biodegradable plastics. *Environmental Toxicology and Chemistry,* 40, 2443–2449.
86. Phosri, S., Kunjiek, T., Mukkhakang, C., Suebthep, S., Sinsup, W., Phornsirigarn, S., Charoeythornkhajhornchai, P. (2022). Biodegradability of bioplastic blown film in a marine environment. *Frontiers in Marine Science*, 9, 917397.
87. Mochane, M.J., Magagula, S.I., Sefadi, J.S., Mokhena, T.C. (2021). A review on green composites based on natural fiber-reinforced polybutylene succinate (PBS). *Polymers (Basel),* 13, 1200.
88. Aziman, N., Kian, L.K., Jawaid, M., Sanny, M., Alamery, S. (2021). Morphological, structural, thermal, permeability, and antimicrobial activity of PBS and PBS/TPS films incorporated with biomaster-silver for food packaging application. *Polymers (Basel)*, 13, 391.
89. Bumbudsanpharoke, N., Wongphan, P., Promhuad, K., Leelaphiwat, P., Harnkarnsujarit, N. (2022). Morphology and permeability of bio-based poly(butylene adipate-co-terephthalate) (PBAT), poly(butylene succinate) (PBS) and linear low-density polyethylene (LLDPE) blend films control shelf life of packaged bread. *Food Control*, 132, 108541.
90. Messin, T., Marais, S., Follain, N., Guinault, A., Gaucher, V., Delpouve, N., Sollogoub, C. (2020). Biodegradable PLA/PBS multinanolayer membrane with enhanced barrier performances. *Journal of Membrane Science*, 598, 117777.
91. Mohamad, N., Mazlan, M.M., Tawakkal, I.S.M.A., Talib, R.A., Kian, L.K., Jawaid, M. (2022). Characterization of active polybutylene succinate films filled essential oils for food packaging application. *Journal of Polymers and the Environment*, 30, 585–596.

92. Moe, N.C., Winotapun, C., Hararak, B., Wanmolee, W., Suwanamornlert, P., Leelaphiwat, P., Boonruang, K., Chinsirikul, W., Chonhenchob, V. (2023). Application of lignin nanoparticles in polybutylene succinate based antifungal packaging for extending the shelf life of bread. *Food Packaging and Shelf life*, 39, 101127.
93. Nazrin, A., Sapuan, S.M., Zuhri, M.Y.M., Ilyas, R.A., Syafiq, R.S.F.K.S., Sherwani, S.F.K. (2020). Nanocellulose-reinforced thermoplastic starch (TPS), polylactic acid (PLA), and polybutylene succinate (PBS) for food packaging applications. *Frontiers in Chemistry*, 8, 213.
94. Phothisarattana, D., Wongphan, P., Promhuad, K., Promsorn, J., & Harnkarnsujarit, N. (2021). Biodegradable poly(butylene adipate-co-terephthalate) and thermoplastic starch-blended TiO2 nanocomposite blown films as functional active packaging of fresh fruit. *Polymers (Basel),* 13, 4192.
95. Łopusiewicz, Ł., Zdanowicz, M., Macieja, S., Kowalczyk, K., Bartkowiak, A. (2021). Development and characterization of bioactive poly(butylene-succinate) films modified with quercetin for food packaging applications. *Polymers*, 13, 1798.
96. Aversa, C., Barletta, M., Puopolo, M., Vesco, S. (2020). Cast extrusion of low gas permeability bioplastic sheets in PLA/PBS and PLA/PHB binary blends. *Polymer-Plastics Technology and Materials*, 59, 231–240.
97. Cabrini, A., Esfahani, A.G., Petraconi, A., Lavorgna, M., De Nardo, L., Buonocore, G.G., Andrade, R.J.E., Cerruti, P. (2023). Ultrasonic spray deposition of PEGDE-crosslinked chitosan/graphene oxide coatings for enhancing gas barrier properties of polybutylene succinate films. *Progress in Organic Coatings*, 183, 107760.
98. Govindan, S., Ramos, M., Al-Jumaily, A.M. (2023). A review of biodegradable polymer blends and polymer composite for flexible food packaging application. *Materials Science Forum*, 1094, 51–60.
99. Shaikh, S., Yaqoob, M., Aggarwal, P. (2021). An overview of biodegradable packaging in food industry. *Current Research in Food Science*, 4, 503–520.
100. Basbasan, A.J., Hararak, B., Winotapun, C., Wanmolee, W., Chinsirikul, W., Leelaphiwat, P., Chonhenchob, V., Boonruang, K. (2023). Lignin nanoparticles for enhancing physicochemical and antimicrobial properties of polybutylene succinate/ thymol composite film for active packaging. *Polymers (Basel),* 15, 989.
101. Hernández-García, E., Vargas, M., González-Martínez, C., Chiralt, A. (2021). Biodegradable antimicrobial films for food packaging: Effect of antimicrobials on degradation. *Foods*, 10, 1256.
102. Montero, Y., Souza, A.G., Oliveira, É.R., dos Santos Rosa, D. (2021). Nanocellulose functionalized with cinnamon essential oil: A potential application in active biodegradable packaging for strawberry. *Sustainable Materials and Technologies*, 29, e00289.
103. Yusof, N.L., Mutalib, N.A.A., Nazatul, U.K., Nadrah, A.H., Aziman, N., Fouad, H., Sain, M. (2021). Efficacy of biopolymer/starch-based antimicrobial packaging for chicken breast fillets. *Foods,* 10, 2379.
104. Deshmukh, R.K., Gaikwad, K.K. (2024). Natural antimicrobial and antioxidant compounds for active food packaging applications. *Biomass Conversion and Biorefinery*, 14, 4419–4440.

105. Firouz, M.S., Mohi-Alden, K., Omid, M. (2021). A critical review on intelligent and active packaging in the food industry: Research and development. *Food Research International*, 141, 110113.
106. Huang, J., Hu, Z., Hu, L., Li, G., Yao, Q., Hu, Y. (2021). Pectin-based active packaging: A critical review on preparation, physical properties and novel application in food preservation. *Trends in Food Science & Technology*, 118, 167–178.
107. Nian, L., Wang, M., Sun, X., Zeng, Y., Xie, Y., Cheng, S., Cao, C. (2024). Biodegradable active packaging: Components, preparation, and applications in the preservation of postharvest perishable fruits and vegetables. *Critical Reviews in Food Science and Nutrition*, 64, 2304–2339.
108. Petchwattana, N., Naknaen, P., Cha-Aim, K., Suksri, C., Sanetuntikul, J. (2021). Controlled release antimicrobial sachet prepared from poly(butylene succinate)/geraniol and ethylene vinyl alcohol coated paper for bread shelf life extension application. *International Journal of Biological Macromolecules*, 189, 251–261.
109. Akhir, M.A.M., Mustapha, M. (2022). Formulation of biodegradable plastic mulch film for agriculture crop protection: A review. *Polymer Reviews*, 62, 890–918.
110. Serrano-Ruiz, H., Martin-Closas, L., Pelacho, A.M. (2021). Biodegradable plastic mulches: Impact on the agricultural biotic environment. *Science of The Total Environment*, 750, 141228.
111. Song, J., Dou, Y., Niu, Y., He, N. (2021). Properties of HA/PBS biodegradable film and evaluation of its influence on the growth of vegetables. *Polymer Testing*, 95, 107137.
112. Mtibe, A., Muniyasamy, S., Mokhena, T.C., Ojijo, O.O.V., John, M. (2023). Recent insight into the biomedical applications of polybutylene succinate and polybutylene succinate-based materials. *Express Polymer Letters*, 17, 2–28.
113. Pakolpakçil, A. (2023). An experimental design approach to examine the influencing parameters of poly (butylene succinate) (PBS) nanofibrous nonwoven by solution blow spinning. *Tekstil ve Konfeksiyon*. doi: 10.32710/tekstilvekonfeksiyon.1215378
114. Sonseca, A., Sahay, R., Stepien, K., Bukala, J., Wcislek, A., McClain, A., El Fray, M. (2020). Architectured helically coiled scaffolds from elastomeric poly(butylene succinate) (PBS) copolyester via wet electrospinning. *Materials Science and Engineering: C*, 108, 110505.
115. Tangnorawich, B., Magmee, A., Roungpaisan, N., Toommee, S., Parcharoen, Y., Pechyen, C. (2023). Effect of polybutylene succinate additive in polylactic acid blend fibers via a melt-blown process. *Molecules*, 28, 7215.
116. Luan, Z., Liu, S., Wang, W., Xu, K., Ye, S., Dan, R., Zhang, H., Shu, Z., Wang, T., Fan, C., Xing, M., Yang, S. (2022). Aligned nanofibrous collagen membranes from fish swim bladder as a tough and acid-resistant suture for pH-regulated stomach perforation and tendon rupture. *Biomaterials Research*, 26, 60.
117. Wu, F., Zhou, Y., Pei, W., Jiang, Y., Yan, X., Wu, H. (2022). Biosynthesis of Poly-(3-hydroxybutyrate) under the control of an anaerobically induced promoter by recombinant Escherichia coli from sucrose. *Molecules*, 27, 294.
118. Myalenko, D., Fedotova, O. (2023). Physical, mechanical, and structural properties of the polylactide and polybutylene adipate terephthalate (PBAT)-based biodegradable polymer during compost storage. *Polymers*, 15, 1619.

119. Taleb, K., Saidi-Besbes, S., Pillin, I., Grohens, Y. (2022). Biodegradable poly(butylene succinate) nanocomposites based on dimeric surfactant organomodified clays with enhanced water vapor barrier and mechanical properties. *ACS Omega*, 7, 43254–43264.
120. Chen, P., Zhao, L., Gao, X., Xu, Z., Liu, Z., Hu, D. (2021). Engineering of polybutylene succinate with long-chain branching toward high foamability and degradation. *Polymer Degradation and Stability*, 194, 109745.
121. Li, Y., Zhang, Z., Wang, W., Gong, P., Yang, Q., Park, C.B., Li, G. (2022). Ultra-fast degradable PBAT/PBS foams of high performance in compression and thermal insulation made from environment-friendly supercritical foaming. *The Journal of Supercritical Fluids*, 181, 105512.
122. Post, W., Kuijpers, L.J., Zijlstra, M., Van Der Zee, M., Molenveld, K. (2021). Effect of mineral fillers on the mechanical properties of commercially available biodegradable polymers. *Polymers (Basel),* 13, 394.
123. Wu, F., Misra, M., Mohanty, A.K. (2020). Tailoring the toughness of sustainable polymer blends from biodegradable plastics via morphology transition observed by atomic force microscopy. *Polymer Degradation and Stability*, 173, 109066.
124. Feng, C.S., Chen, Y., Shao, J., Li, G., Hou, H.Q. (2020). The crystallization and melting behaviors of PDLA-b-PBS-b-PDLA triblock copolymers. *Chinese Journal of Polymer Science*, 38, 298–310.
125. Guidotti, G., Soccio, M., Gazzano, M., Bloise, N., Bruni, G., Aluigi, A., Visai, L., Munari, A., Lotti, N. (2020). Biocompatible PBS-based copolymer for soft tissue engineering: Introduction of disulfide bonds as winning tool to tune the final properties. *Polymer Degradation and Stability*, 182, 109403.
126. Ali, F., Mohan, R. (2009). Thermal, mechanical, and rheological properties of biodegradable polybutylene succinate/carbon nanotubes nanocomposites. *Polymer Composites, 31,* 1309–1314. doi: 10.1002/PC.20913.
127. Behtaj, S., Karamali, F., Masaeli, E., Anissimov, Y.G., Rybachuk, M. (2021). Electrospun PGS/PCL, PLLA/PCL, PLGA/PCL and pure PCL scaffolds for retinal progenitor cell cultivation. *Biochemical Engineering Journal*, 166, 107846.
128. Kanemura, C., Nakashima, S., Hotta, A. (2012). Mechanical properties and chemical structures of biodegradable poly(butylene-succinate) for material reprocessing. *Polymer Degradation and Stability*, 97, 972–980. doi: 10.1016/J.POLYMDEGRADSTAB.2012.03.015.
129. Kim, H., Yang, H., Kim, H. (2005). Biodegradability and mechanical properties of agro-flour-filled polybutylene succinate biocomposites. *Journal of Applied Polymer Science*, 97, 1513–1521. doi: 10.1002/APP.21905.
130. Kim, S., Cho, J., Cho, D., Jung, H., Kim, B., Bhatia, S., Park, S., Park, K., Yang, Y. (2022). Acceleration of polybutylene succinate biodegradation by *Terribacillus* sp. JY49 isolated from a marine environment. *Polymers (Basel)*, 14, 3978. doi: 10.3390/polym14193978.
131. Ma, H., Zhao, Y., Huang, W., Zhang, L., Wu, F., Ye, J., Chen, G.Q. (2020). Rational flux-tuning of Halomonas bluephagenesis for co-production of bioplastic PHB and ectoine. *Nature communications*, 11, 3313.
132. Napathorn, S.C., Visetkoop, S., Pinyakong, O., Okano, K., Honda, K. (2021). Polyhydroxybutyrate (PHB) production using an arabinose-inducible expression

system in comparison with cold shock inducible expression system in Escherichia coli. *Frontiers in Bioengineering and Biotechnology*, 9, 661096.
133. Shen, K. (2023). Synthesis, properties and applications of poly(butylene succinate). *MATEC Web of Conferences*, 386, 5. doi: 10.1051/matecconf/202338601005.
134. Vlachopoulos, A., Karlioti, G., Balla, E., Daniilidis, V., Kalamas, T., Stefanidou, M., Bikiaris, N.D., Christodoulou, E., Koumentakou, I., Karavas, E., Bikiaris, D.N. (2022). Poly (lactic acid)-based microparticles for drug delivery applications: An overview of recent advances. *Pharmaceutics*, 14, 359.
135. Arif, Z.U., Khalid, M.Y., Noroozi, R., Sadeghianmaryan, A., Jalalvand, M., Hossain, M. (2022). Recent advances in 3D-printed polylactide and polycaprolactone-based biomaterials for tissue engineering applications. *International Journal of Biological Macromolecules*, 218, 930–968.
136. DeStefano, V., Khan, S., Tabada, A. (2020). Applications of PLA in modern medicine. *Engineered Regeneration*, 1, 76–87.
137. Ebrahimi, F., Ramezani Dana, H. (2022). Polylactic acid (PLA) polymers: From properties to biomedical applications. *International Journal of Polymeric Materials and Polymeric Biomaterials*, 71, 1117–1130.
138. Liu, S., Qin, S., He, M., Zhou, D., Qin, Q., Wang, H. (2020). Current applications of poly(lactic acid) composites in tissue engineering and drug delivery. *Composites Part B: Engineering*, 199, 108238.
139. Teixeira, S., Eblagon, K.M., Miranda, F., R. Pereira, M.F., Figueiredo, J.L. (2021). Towards controlled degradation of poly(lactic) acid in technical applications. *C*, 7, 42.
140. Zhang, Q., Song, M., Xu, Y., Wang, W., Wang, Z., Zhang, L. (2021). Bio-based polyesters: Recent progress and future prospects. *Progress in Polymer Science*, 120, 101430.
141. Arun, Y., Ghosh, R., Domb, A.J. (2021). Biodegradable hydrophobic injectable polymers for drug delivery and regenerative medicine. *Advanced Functional Materials*, 31, 2010284.
142. Karava, V., Siamidi, A., Vlachou, M., Christodoulou, E., Bikiaris, N.D., Zamboulis, A., Kostoglou, M., Gounari, E., Barmpalexis, P. (2021). Poly (L-lactic acid)-co-poly(butylene adipate) new block copolymers for the preparation of drug-loaded long-acting injectable microparticles. *Pharmaceutics*, 13, 930.
143. Li, G., Zhao, M., Xu, F., Yang, B., Li, X., Meng, X., Teng, L., Sun, F., Li, Y. (2020). Synthesis and biological application of polylactic acid. *Molecules*, 25, 5023.
144. Wang, J., Ma, W., Fang, Y., Zhang, H., Liang, H., Li, Y., Wang, X. (2020). Truncating the structure of lipopolysaccharide in Escherichia coli can effectively improve poly-3-hydroxybutyrate production. *ACS Synthetic Biology*, 9, 1201–1215.
145. Qiao, J., Tan, X., Ren, H., Wu, Z., Hu, X., Wang, X. (2021). Construction of an Escherichia coli strain lacking fimbriae by deleting 64 genes and its application for efficient production of poly(3-hydroxybutyrate) and L-threonine. *Applied and Environmental Microbiology*, 87, e00381–e00321.
146. Wu, H., Guo, T., Zhou, F., Bu, J., Yang, S., Dai, Z., Teng, C., Ouyang, H., Wei, W. (2022). Surface coating prolongs the degradation and maintains the mechanical strength of surgical suture in vivo. *Colloids and Surfaces B: Biointerfaces*, 209, 112214.

147. Atanase, L.I., Salhi, S., Cucoveica, O., Ponjavic, M., Nikodinovic-Runic, J., Delaite, C. (2022). Biodegradability assessment of polyester copolymers based on poly(ethylene adipate) and poly(ε-caprolactone). *Polymers*, 14, 3736.
148. De Falco, F., Avolio, R., Errico, M.E., Di Pace, E., Avella, M., Cocca, M., Gentile, G. (2021). Comparison of biodegradable polyesters degradation behavior in sand. *Journal of Hazardous Materials*, 416, 126231.
149. Xu, Y., Zhou, F., Zhou, D., Mo, J., Hu, H., Lin, L., Wu, J., Yu, M., Zhang, M., Chen, H. (2020). Degradation behaviors of biodegradable aliphatic polyesters and polycarbonates. *Journal of Biobased Materials and Bioenergy,* 14, 155–168.
150. Fakhrali, A., Nasari, M., Poursharifi, N., Semnani, D., Salehi, H., Ghane, M., Mohammadi, S. (2021). Biocompatible graphene-embedded PCL/PGS-based nanofibrous scaffolds: A potential application for cardiac tissue regeneration. *Journal of Applied Polymer Science*, 138, 51177.
151. Yang, L., Li, X., Wang, D., Mu, S., Lv, W., Hao, Y., Lu, X., Zhang, G., Nan, W., Chen, H., Xie, L., Zhang, Y., Dong, Y., Zhang, Q., Zhao, L. (2020). Improved mechanical properties by modifying fibrin scaffold with PCL and its biocompatibility evaluation. *Journal of Biomaterials Science, Polymer Edition*, 31, 658–678.
152. Tolga, S., Kabasci, S., Duhme, M. (2020). Progress of disintegration of polylactide (PLA)/poly(butylene succinate) (PBS) blends containing talc and chalk inorganic fillers under industrial composting conditions. *Polymers (Basel)*, 13, 10.
153. Paek, K.H., Im, S.G. (2022). Biodegradable aromatic–aliphatic copolyesters derived from bis(2-hydroxyethyl) terephthalate for sustainable flexible packaging applications. *ACS Applied Polymer Materials*, 4, 5298–5307.
154. Abid, N., Khan, A.M., Shujait, S., Chaudhary, K., Ikram, M., Imran, M., Haider, J., Khan, M., Khan, Q., Maqbool, M. (2022). Synthesis of nanomaterials using various top-down and bottom-up approaches, influencing factors, advantages, and disadvantages: A review. *Advances in Colloid and Interface Science*, 300, 102597.
155. Sui, J., Yan, J., Liu, D., Wang, K., Luo, G. (2020). Continuous synthesis of nanocrystals via flow chemistry technology. *Small*, 16, 1902828.
156. Yuan, L., Qi, M.Y., Tang, Z.R., Xu, Y.J. (2021). Coupling strategy for CO2 valorization integrated with organic synthesis by heterogeneous photocatalysis. *Angewandte Chemie*, 60, 21150–21172.

CHAPTER 7

Polyethylene Furanoate: A Promising Biopolymer for Sustainable Food-Packaging Applications

JOSÉ ROBERTO VEGA-BAUDRIT,[1,2*] YENDRY CORRALES,[1] DIEGO BATISTA,[1] SERGIO A. PANIAGUA,[1] SUSANA MESÉN,[1] MAYNOR ALBERTO VARGAS-VARGAS,[3] GABRIELA MONTES DE OCA-VÁSQUEZ,[4] and MARY LOPRETTI[5]

[1]*LANOTEC CeNAT CONARE, San José, Costa Rica*

[2]*Escuela de Química, Universidad Nacional de Costa Rica, Heredia, Costa Rica*

[3]*Universidad Técnica Nacional, Alajuela, Costa Rica*

[4]*Centro de Estudios sobre Desarrollo Sostenible, Universidad Técnica Nacional, Alajuela, Costa Rica*

[5]*UDELAR, Montevideo, Uruguay*

*Corresponding author

ABSTRACT

Polyethylene furanoate (PEF) is an emerging biobased polyester with unique properties that make it a promising candidate for sustainable food-packaging applications. This chapter provides a comprehensive overview of PEF, focusing on its synthesis, material properties, and advantages over conventional petroleum-based plastics. The high barrier properties of PEF against oxygen and carbon dioxide, and its good mechanical strength and biodegradability, offer significant benefits for extending

Sustainable Food Packaging Materials and Technology. Juan Manuel Tirado-Gallegos, Juan Daniel Rivaldi Chávez, and José Juan Buenrostro-Figueroa (Eds)

DOI: 10.1201/9781779641434-7

food shelf life, reducing food waste, and promoting a circular economy. Recent research advancements in developing PEF-based food-packaging materials, including multilayer films, coatings, and nanocomposites, are explored. The chapter also discusses the challenges and opportunities associated with the large-scale production and commercialization of PEF for food-packaging applications, such as cost competitiveness, regulatory compliance, and consumer acceptance. Overall, this chapter highlights the potential of PEF as a sustainable alternative for food packaging and provides a valuable resource for researchers, industry stakeholders, and policymakers interested in this emerging material.

7.1 INTRODUCTION

Poly(ethylene 2,5-furan dicarboxylate), shown in Figure 7.1, emerges as a biobased polyester capable of supplanting petroleum-derived polyesters like polyethylene terephthalate (PET), attributing its potential to renewable origins but similar mechanical properties. This polymer is crafted from monomers sourced from renewable materials, typically sugars, and the recognition received from the USA National Renewable Energy Lab and the EU H2020 program highlights its potential impact [1, 2].

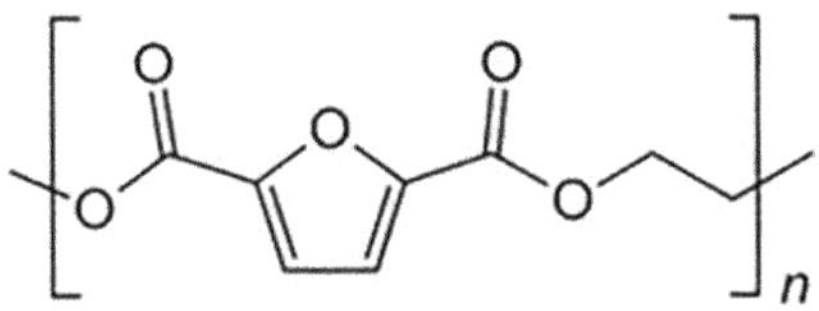

FIGURE 7.1 Poly(ethylene 2,5-furan dicarboxylate).

The synthesis process of PEF, if using high-yield ring-opening polymerization (ROP), yields polymers with molecular weights ranging from 50,000 to 60,000 g/mol. Its superior gas barrier properties are credited to the furan ring's higher dipole moment, a consequence of oxygen's greater electronegativity compared to sulfur in the thiophene ring, enhancing its utility [3, 4]. The ROP technique proves efficient in converting cyclic ethylene and butylene 2,5-furandicarboxylate oligoesters into PEF and poly(butylene 2,5-furan dicarboxylate) (PBF), maintaining consistent thermal properties [5].

In contrast to PET, PEF displays distinct crystallization kinetics, characterized by a heightened activation energy for chain diffusion, influencing its crystallization rate. As a polymer derived from C6 sugars, PEF presents a more sustainable alternative, offering advanced barrier properties, greener synthesis methods, and viable recycling and biodegradation pathways [6]. The properties of PEF are further modulated by varying catalysts and polycondensation techniques, where optimal conditions yield properties on par with PET. The copper-catalyzed oxidative methyl-esterification process transforms 5-hydroxymethylfurfural (HMF) into furan-2,5-dimethylcarboxylate (FDMC), facilitating the production of PEF, underscoring its potential as a biopolymer rival to fossil-based PET [3].

PEF is distinguished by its ability to undergo strain-induced crystallization under specific conditions, which is pivotal for processing applications such as bottle-blowing and film extrusion. This alipharomatic polyester shows thermal stability up to 325°C and a peak degradation rate at 438°C, showing advantages over PET and polyethylene naphthalate [7]. Compared to these other polymers, PEF shows a higher equilibrium melting point and divergent crystallization behavior, evidenced by increased nucleation density and reduced spherulite size. Furthermore, its crystallization kinetics reveal a higher activation energy for chain diffusion relative to PET, with the crystallization rate significantly influenced by the catalyst type [7]. This thermoplastic polymer can form an organized crystalline-like microstructure when uniaxially stretched. It shows strain hardening alongside strain rate and temperature sensitivities, marking it a viable PET substitute in stretching applications [8].

The polarity of the furan ring in PEF is instrumental in its outstanding gas barrier performance, surpassing that of polymers like PET. This attribute, alongside a higher dipole moment, underpins PEF's superior gas barrier characteristics. Focusing on PEF, research endeavors aim to develop environmentally friendly synthesis processes and solutions for end-of-life management, including recycling and biodegradation.

Emerging methods for synthesizing PEF precursors are pivotal for enhancing manufacturing efficiency, and sustainability of PEF manufacturing [3, 9], such as the oxidative methyl-esterification of HMF to FDMC, is discussed in more detail below.

PEF is a promising biobased polyester (Figure 7.2). It surpasses the performance of conventional petrochemical-based polyesters with its synthesis from renewable resources, strain crystallization capabilities, and superior barrier properties, thus rendering it suitable for a broad spectrum of applications. Ongoing research is dedicated to refining its production processes and enhancing its sustainability at the end of its lifecycle.

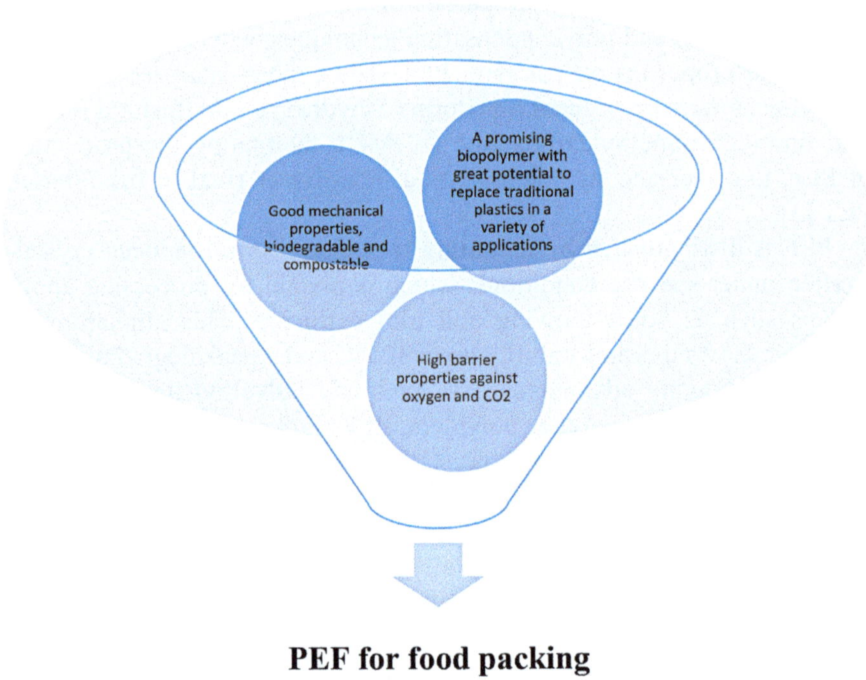

FIGURE 7.2 Properties of PEF for food-packaging applications.

7.2 PEF SYNTHESIS

As shown in Figure 7.3, PEF is produced through polycondensation or ROP using 2,5-furan dicarboxylic acid (FDCA) and ethylene glycol, where FDCA is sourced from renewable lignocellulosic biomass such as sugarcane, corn, or wood. The synthesis of FDCA from renewable sources, such as pectin-derived galactaric (mucic) acid, has been scaled up, showing the industrial feasibility of producing PEF-related monomers [10]. This process underscores the biobased nature of PEF, leveraging lignocellulosic biomass waste (LCBW) as a sustainable raw material.

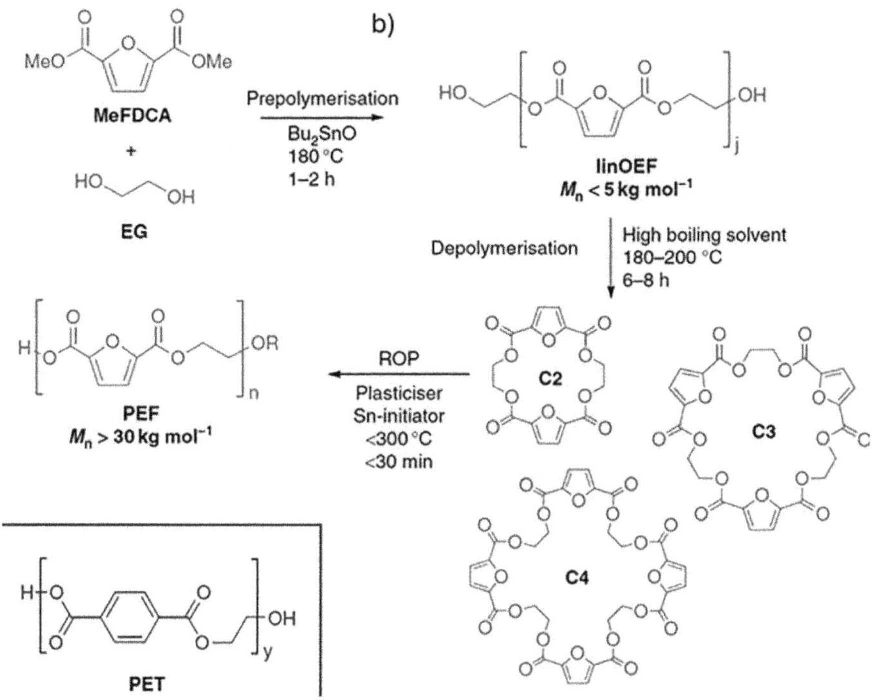

FIGURE 7.3 Synthesis of PEF via (a) polycondensation of FDCA and ethylene glycol, and (b) ROP.

Source: Reprinted from Rosemboom et al. [11]. https://creativecommons.org/licenses/by/4.0/

The schematic representation shown in Figure 7.4 outlines the bioplastics production pathways from LCBW, detailing the conversion processes involving key intermediates such as p-xylene, monoethylene glycol monomer, terephthalic acid, HMF, and FDCA, leading to the synthesis of PET and biobased PEF. This illustration effectively maps the transition from lignocellulosic biomass waste to valuable bioplastics, emphasizing the circular economy aspect of PEF production [12]. A practical and profitable synthetic route for PEF involves a direct esterification stage of FDCA using zinc acetate and aluminum acetylacetonate catalysts. This chapter shows a practical and profitable synthetic route for PEF using zinc acetate and aluminum acetylacetonate catalysts, offering environmental and food safety potential in packaging applications [13].

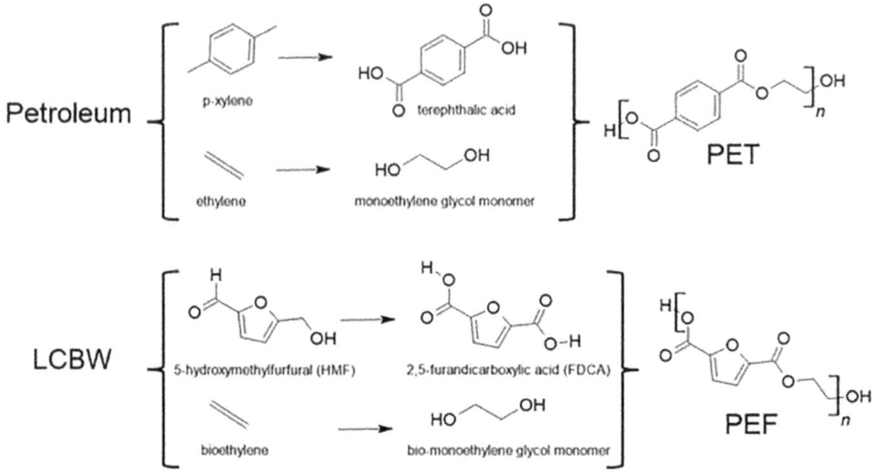

FIGURE 7.4 Comparison of PET synthesis with PEF bioplastic production from lignocellulosic biomass waste (LCBW).

These advancements highlight the shift toward more sustainable plastic production methods, where PEF stands out for its potential to reduce reliance on fossil fuels and enhance the environmental profiles of plastic products. The focus on using renewable resources to synthesize PEF contributes to reducing carbon footprint and shows the innovation in developing high-performance materials with reduced environmental impact.

The synthesis of PEF via ROP effectively produces polymers with molecular weights in the range of 50,000–60,000 g/mol. The ROP approach reveals that the polymerization rate for butylene cyclic oligofuranoates surpasses that of ethylene counterparts, with the thermal attributes of the resultant PEF aligning with those acquired through melt polycondensation methods, as detailed by Morales-Huerta et al. [5]. Such findings underscore the versatility and efficacy of ROP in synthesizing PEF with desirable molecular weights and thermal properties, which are essential for its application in various domains.

Further refinement in PEF's properties is achieved through solid-state polycondensation (SSP), a postpolymerization technique used to augment the molecular weight of PEF to achieve intrinsic viscosities of 1 dL/g. This threshold indicates the polymer's suitability for food-packaging applications, wherein higher molecular weights often correlate with better performance characteristics. The SSP method enhances the molecular weight and elevates the melting temperatures of PEF samples, suggesting their utility in high-performance applications. This advancement, as reported by Kasmi et al. [14], highlights the potential of SSP in improving the material properties of PEF, making it more competitive against traditional petrochemical-based polymers.

Additionally, the research conducted by Hwang et al. (2019) presents a novel approach for the high-yield production of furan dicarboxylate using methanol and ethylene glycol in concentrated solutions. This method significantly boosts the economic viability of biobased polyester production, achieving yields ranging from 80% to 95% for methyl furan-2,5-dicarboxylate and bis(2-hydroxyethyl). Such advancements not only optimize the synthesis process of PEF but also enhance its attractiveness for large-scale industrial applications, particularly in sectors keen on adopting sustainable and biobased materials.

These developments signify a positive trajectory toward commercializing PEF as a sustainable alternative to conventional petrochemical-derived polyesters. By leveraging innovative synthesis techniques and postpolymerization processes, the production of PEF becomes more efficient. It aligns with the growing demand for environmentally friendly and high-performance materials in various industries, including packaging, automotive, and textiles.

7.3 PEF PROPERTIES AND MODIFICATIONS

Table 7.1 shows some of the main properties of PEF. Its superior barrier properties distinguish it, and a lower gas diffusion rate than PET (see Table 7.2), a characteristic attributed to the reduced segmental motions and the polarity of the furan ring. Research into the enzymatic hydrolysis of PEF indicates its decomposability into monomers and oligomers, highlighting its recyclability. Enhancements in PEF's toughness have been achieved through the modification with biobased 1,5-pentanediol, simultaneously preserving its high tensile strength and oxygen barrier capabilities. Furthermore, the asymmetric nature of specific PEF monomers contributes to crystallization resistance, resulting in an amorphous polymer structure, broadening its application scope.

TABLE 7.1 Properties and Qualities of PEF

Properties and Qualities	References
PEF shows superior barrier properties and reduced diffusion of gases compared to PET, attributed to the suppressed segmental motions and furan ring polarity	[6] [33]
Enzymatic hydrolysis studies show that PEF can be broken down into monomers and oligomers, suggesting potential for recycling and surface functionalization	[47]
Modifying PEF with biobased 1,5-pentanediol can significantly improve its toughness while retaining high tensile strength and oxygen barrier properties	[57] [3] [11]
The asymmetric nature of specific PEF monomers leads to crystallization resistance and an amorphous polymer structure	[32]
PEF can be produced from lignocellulosic biomass waste, aligning with the principles of a circular bioeconomy and offering a sustainable alternative to PET	[50] [38]
Blending PEF with postconsumer PET-G can enhance thermal stability and mechanical properties, indicating the potential for creating environmentally friendly polymer blends	[35] [33]
PEF's compatibility with silver nanowires in conductive films shows its potential for use in flexible optoelectronic devices	[23]
Copolymerization of PEF with ethylene succinate can lead to a decrease in melting point and improved biodegradability	[28]
PEF/GZO films on biobased poly(ethylene furanoate) substrates show potential for high-antimicrobial food packaging due to their optimized optics, water vapor barrier, and antibacterial properties	[60] [37]
The flexibility and ductility of certain furanic polyesters, like PDDbF, are significantly higher than those of their monofuranic counterparts, making them suitable for applications that require extreme flexibility, such as flexible packaging	[60]
Incorporating additives, such as GZO, into PEF substrates can enhance properties such as visible light transmittance, water vapor barrier, and antibacterial activity, which are critical for food-packaging applications	[37]

TABLE 7.2 Comparison of PEF and PET

Property	PEF	PET	References
Biodegradability and recyclability	Typically biodegradable at 58°C (depends on copolymers and polyol used). Most PEF is chemically and mechanically recyclable	It is not biodegradable, but it is readily recycled mechanically and can be chemically recycled	[50] [43] [31] [25] [17] [18]

TABLE 7.2 Comparison

Property	PEF	PET	References
Barrier properties	A superior barrier to oxygen and carbon dioxide: amorphous PEF, the permeability for O_2 is 0.011 Barrer, and for CO_2, it is 0.026 Barrer	Amorphous PET is an excellent barrier to oxygen and carbon dioxide: its permeability to O_2 is 0.11 Barrer, and that to CO_2 is 0.49 Barrer	[57] [47] [25] [17]
Mechanical properties	High tensile strength and modulus (higher for ROP-derived PEF, up to 76 MPa/1.9 GPa, respectively)	High tensile strength and modulus (typically 55–75 MPa/>3 GPa, respectively)	[8] [29] [13] [56] [5] [16] [49]
Thermal properties	PEF synthesized by ROP has a lower melting point (~220°C) than PET but a similar glass transition temperature (~85°C).	Higher melting point (250–265°C) and similar glass transition temperature (~81°C).	[4] [55] [50] [43] [52] [1] [5] [56] [5] [38]
Cost	Currently, it is more expensive than PET. However, it should be similar to PET after scale-up.	The average cost is $1162/ton.	[19] [15] [44]
Sustainability	Made from renewable resources.	Terephthalate comes from nonrenewable resources, and diol can be from renewable sources.	[3] [25] [17]
Toxicology	PEF/GZO films on biobased poly(ethylene furanoate) substrates show potential for high-antimicrobial food packaging due to their optimized optics, water vapor barrier, and antibacterial properties.	Recent studies and analytical methodologies reveal the presence of PET, PBT, and PS oligomers in food-contact materials and food, which could have potential toxicological effects and challenge food safety.	[60] [26] [48]
Applications	Food packaging, beverage bottles, films, and other applications.	Food packaging, beverage bottles, fibers, electronics, automotive industries, and other applications.	[55] [41] [31] [5] [3] [20] [5] [18]

Blends of poly(ethylene terephthalate) glycol (PET-G) (a variant of PET modified with glycol) and biobased PEF, derived from postconsumer PET-G and biobased PEF, show environmentally favorable attributes. These include enhanced thermal stability and reduced water absorption, indicating an improvement over traditional materials and contributing to the environmental appeal of these bioplastics bioplastics [5, 6, 12, 15]. Mixing PEF with postconsumer PET-G has improved thermal stability and mechanical properties, paving the way for the development of environmentally friendly polymer blends [15, 16]. The material's compatibility with silver nanowires for conductive films extends its potential uses to flexible optoelectronics [17]. Moreover, copolymerization efforts with ethylene succinate have shown a reduction in melting points and enhanced biodegradability of PEF [12]. Innovations such as PEF/gallium-doped zinc oxide (GZO) films on biobased substrates reveal promising applications in antimicrobial food packaging, leveraging optimized optics, water vapor barriers, and antibacterial features [18, 19].

The reduction in gas diffusion within PEF, attributed to its amorphous phase, underscores a fundamental property needed for food packaging: its barrier qualities, which are not only better but also more environmentally friendly than the traditionally used PET and other conventional gas barrier polymers. PEF shows a significant reduction in oxygen permeability compared to PET, with an approximately 11-fold decrease at 35°C, attributed to reduced chain segment mobility due to hindered furan ring flipping. Carbon dioxide permeability in PEF is also substantially reduced by about 19 times compared to PET, despite an increase in solubility; this is offset by a 31-fold reduction in diffusivity, again likely due to restricted ring-flipping motions [20, 21]. Water diffusion in PEF is reduced by approximately five times over the entire water activity range compared to PET, with similar activation energies of diffusion for both PEF and PET, suggesting that the reduction in diffusion is not due to energetic barriers but possibly due to the physical structure of PEF [22].

The environmental advantages of substituting PET with PEF are substantial, as highlighted by Eerhart et al. [24], who suggest that adopting PEF in the bottle market could lead to considerable savings in nonrenewable energy use, estimated between 440 and 550 petajoules, and a reduction in greenhouse gas (GHG) emissions by approximately 20–35 million tonnes of CO_2 equivalent. These data underpin the broader environmental benefits of integrating PEF into industrial applications, emphasizing its role in promoting sustainability and reducing the ecological footprint of packaging materials [13]. Figure 7.5 shows the main PEF advantages.

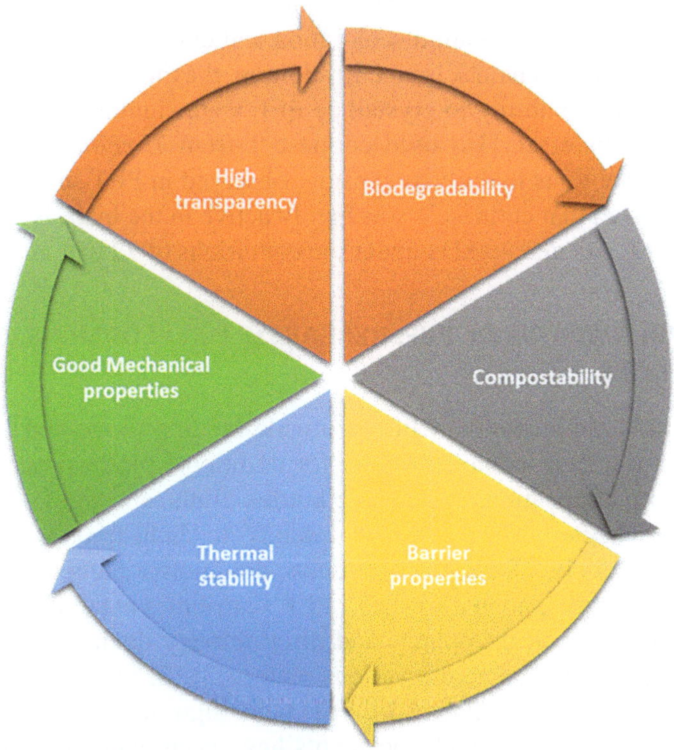

FIGURE 7.5 Main properties and qualities of PEF.

PEF stands out for its excellent barrier properties against oxygen and CO_2, positioning it as an ideal food and beverage packaging material. Alongside its barrier capabilities, PEF shows strong mechanical properties, such as tensile and flexural strength, and maintains high transparency. Notably, PEF is both biodegradable and compostable, allowing it to break down into biomass via microbial processes at the end of its life cycle, underscoring its environmental friendliness [4, 6, 23, 30].

Some studies have assessed the biodegradability and degradation of PEF. For instance, Weinberger et al. [48] studied the hydrolytic activity of cutinases from *Humicola insolens* and *Thermobifida cellulosilytica* on amorphous PEF films in different reaction media (potassium phosphate buffer [KPO] pH 8 and Tris–HCl pH 7) and at different temperatures (50 and 65°C). They found a 100% hydrolysis of the PEF films after 72 h of incubation with cutinase from *H. insolens* in KPO pH 8 at 65°C. Otherwise, Maaskant and van Es [49] determined that after 500 h of UV irradiation, PEF showed significant discoloration and evidence of cross-linking/chain extension and chain scission. In addition, they found that the thermal properties and the ability to crystallize PEF were significantly impacted. Similarly, Pellis et al. [16] used cutinase 1 from *T. cellulosilytica* as a biocatalyst due to its hydrolysis capacity observed in PET substrates. The authors showed that cutinase 1 was highly active in the hydrolysis of PEF powders of 6, 10, and 40 kDa under the conditions of 72 h at 50°C.

7.4 APPLICATIONS OF PEF: RESEARCH, DEVELOPMENT, AND CHALLENGES

The potential applications of PEF are wide ranging. Food and beverage packaging is a sustainable substitute for traditional plastics such as PET and PP, with applications in creating bottles, films, and rigid containers. Beyond packaging, PEF's utility extends to industrial applications, including producing films for electronics, automotive parts, and textiles. Additionally, in the biomedical field, PEF's attributes make it suitable for manufacturing medical implants and drug delivery devices, showing its versatility and potential across various sectors [3, 13, 31, 37].

Research and development of PEF have significantly intensified in recent years. Remarkable advancements have been made in various key areas, including synthesizing, processing, and characterizing PEF and its derivatives.

PEF and related furanose-based polyesters have emerged as promising candidates for sustainable applications in packaging and textiles due to their renewable origin and desirable mechanical properties. Blending furanoate-based polyesters with other biobased polymers, such as polylactic acid (PLA), enhances mechanical and gas barrier properties, rendering them ideal for food packaging. Moreover, these blends can be selectively enzymatically recycled [50].

Furanoate polyesters, such as poly(propylene 2,5-furanoate) (PPF) and PBF, or poly(propylene 2,5-thiophendicarboxylate (PPTF), can toughen PLA, improving its mechanical performance in bulk and fiber forms and expanding its use in packaging and textiles, as shown in Figure 7.6 [33, 51].

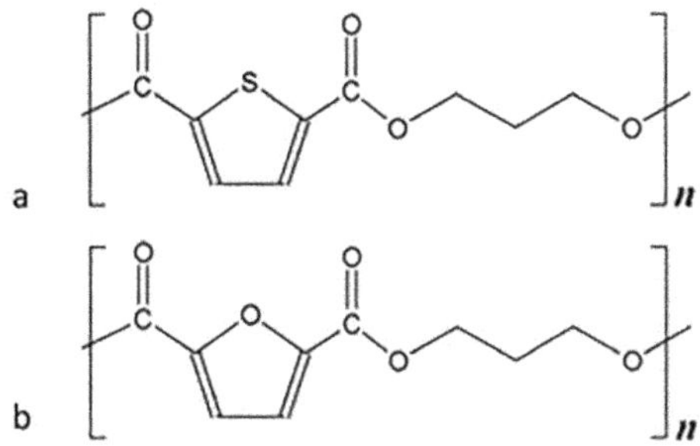

FIGURE 7.6 Chemical structures of furan-based polyesters: (a) poly(propylene 2,5-thiophenedicarboxylate) (PPTF) and (b) poly(propylene 2,5-furanoate) (PPF).

Novel biobased polyesters derived from furanic precursors show exceptional flexibility and thermal properties, making them competitive with traditional fossil-based plastics for flexible packaging applications [52]. Incorporating furan-dicarboxylate units into copolyesters such as poly(hexamethylene succinate-co-2,5-furandicarboxylate) improves their mechanical properties, such as tensile strength and elongation at break, indicating potential for high-performance materials [53]. Finally, fully biobased copolyesters combining furan dicarboxylate with other components like polycaprolactone (PCL) show enhanced thermal stability and flexibility, showing the potential for various industrial applications [54].

Despite its potential, PEF faces several challenges that must be addressed for its large-scale production and successful commercialization. Among these challenges are the following:

1. Cost: The production cost of PEF is currently higher than that of traditional polymers, which poses a significant barrier to its widespread adoption. Reducing these costs through technological advancements and economies of scale will make PEF a competitive alternative [6].
2. Availability of FDCA: The production of FDCA from sugars in lignocellulosic biomass, a key monomer for PEF, has yet to be developed commercially, although the technology is available [12]. It is also possible to source from lactose.
3. Recycling: Developing effective recycling processes for PEF is essential to ensure its fit within the circular economy model. This includes establishing collection systems and sorting technologies, and scaling up recycling methods to process PEF and efficiently reintegrate it into the product life cycle.

7.5 FINAL DISCUSSION ABOUT STUDYING PEF

PEF is a relatively new polymer that has attracted much attention from researchers and the industry. PEF shows a unique combination of properties, making it highly attractive as a potential replacement for well-established petrochemical polymers such as PET. First, PEF is a biodegradable material, which means it is made from renewable resources. It contrasts with petrochemical polymers, which are made from finite fossil resources. As the world seeks sustainable and environmentally friendly alternatives to traditional materials, there is a growing interest in developing and applying biopolymers such as PEF. Recent research has shown that modern life cycle assessments of PEF have revealed that PEF offers substantial environmental advantages over PET and other conventional plastics. However, for PEF to be used commercially and to achieve widespread success in the market, there needs to be more research in areas such as improving the synthetic routes involved in the production of PEF, investigating the most effective and efficient processing techniques for PEF, and expanding our scientific knowledge about the characterization of the properties and performance of PEF. This is why conducting a detailed analysis of PEF and its properties is essential to understanding this new polymer better and exploring potential industrial applications. Such knowledge will also guide and drive further research into PEF, which will positively impact the development of this exciting, sustainable biopolymer.

7.6 ENVIRONMENTAL AND SUSTAINABLE FACTORS OF PEF

Plastics are essential in society due to their low cost, lightweight nature, and valuable mechanical and thermal properties. However, they show limitations on agriculture, packaging, and transportation concerning plastic disposal and pollution associated with their widespread use [55].

PET is one of the most used polyester plastics worldwide, but it is challenging to hydrolyze in a natural environment for its disposal. PET advantages are associated with its lightweight, inexpensive, and durability, while the main disadvantage of PET is the single use of this material, which induces a significant environmental problem. Although PET plastics are a valuable resource in many aspects, the proliferation of plastic products promotes a negative ecological footprint [56]

The dependence on plastics derived from fossil feedstock promotes the development of biobased alternatives, such as PEF, with the potential to substitute PET [57].

PEF is an innovative bioplastic with superior characteristics to PET, particularly related to its mechanical and gas barrier properties. PEF can be 100% biobased, the monomers can be produced from biological resources, and it allows corecycling of at least up to 2%, which could accelerate its market induction. Its disposal after use is essential to study strategies that mitigate its environmental footprint [57].

Biobased plastic production has decreased the GHG emissions of fossil-based plastics [58], but it is end-of-life management dependent. Introducing biobased plastics needs early development of expertise, systems, and infrastructure to deal with challenges associated with waste disposal [57].

PEF production from biomass is promising but needs further research. PEF offers significant environmental advantages over traditional plastics. Its production from renewable biomass sources reduces reliance on fossil fuels and mitigates carbon emissions. Additionally, PEF is fully recyclable and biodegradable, minimizing plastic pollution and contributing to a more sustainable future [12].

7.7 CONCLUSIONS

While PEF shows significant potential due to its renewable origins and superior mechanical properties, its large-scale production and successful commercialization face several challenges. These challenges include the higher production cost of PEF compared to traditional polymers, the need for commercial-scale development of FDCA production from lignocellulosic biomass, and the current limitations in PEF's recyclability. Addressing these challenges is crucial for realizing PEFs' potential in sustainable applications.

Regarding advancements, PEF, and related furanoate-based polyesters are promising materials for sustainable packaging and textile applications. Blending these polyesters with other biobased polymers like PLA improves their mechanical and gas barrier properties, making them suitable for food packaging and introducing the possibility of selective enzymatic recycling. The scaled-up synthesis of FDCA esters from renewable sources like pectin-based galactaric acid shows the feasibility of producing PEF-related monomers on an industrial scale.

Further research into furanoate polyesters like PPF and PBF has shown their ability to enhance PLA's mechanical properties, expanding their application in packaging and textiles. Biobased polyesters derived from furanic precursors show remarkable flexibility and thermal properties, making them competitive alternatives to traditional fossil-based plastics, especially in flexible packaging applications. Incorporating furandicarboxylate units into copolyesters significantly improves mechanical properties, indicating their potential as high-performance materials. Developing fully plant-based sustainable copolyesters that combine furandicarboxylate with other components like PCL has enhanced thermal stability and flexibility, broadening their potential industrial applications.

Furanoate-based polyesters and their copolymers present a sustainable alternative to traditional plastics. They offer enhanced mechanical and thermal properties derived from renewable resources and hold promise for recyclability and industrial scalability. However, additional studies are still necessary to validate the polymer degradation processes through chemical or biological routes and determine the optimal conditions for the process to be efficient. These developments position them as attractive options for future material development in packaging food and textiles, emphasizing the importance of leveraging current obstacles to fully leverage their environmental and functional benefits.

KEYWORDS

- **polyethylene furanoate**
- **biopolymer**
- **biodegradability**
- **sustainability**
- **food packaging**
- **barrier properties**
- **shelf life**
- **circular economy**

REFERENCES

1. PEFerence. The project. Amsterdam, The Netherlands. https://peference.eu/the-project/.
2. Werpy, T., Petersen, G. (2004). Top value added chemicals from biomass: Volume I—Results of screening for potential candidates from sugars and synthesis gas. United States. doi: 10.2172/15008859.
3. Gupta, S., Vinu, A., Kantam, M. (2020). Copper-catalyzed oxidative methyl-esterification of 5-hydroxymethylfurfural using TBHP as an oxidizing and methylating reagent: A new approach for the synthesis of furan-2,5-dimethylcarboxylate. *Journal of Catalysis*, 389, 259–269. doi: 10.1016/j.jcat.2020.05.032.
4. Sun, L., Wang, J., Mahmud, S., Jiang, Y., Zhu, J., Liu, X. (2019). New insight into the mechanism for the excellent gas properties of poly(ethylene 2,5-furan dicarboxylate) (PEF): Role of furan ring's polarity. *European Polymer Journal*, 118, 642–650. doi: 10.1016/J.EURPOLYMJ.2019.06.033.
5. Morales-Huerta, J., Ilarduya, A., Muñoz-Guerra, S. (2016). Poly(alkylene 2,5-furandicarboxylate) (PEF and PBF) by ring-opening polymerization. *Polymer*, 87, 148–158. doi: 10.1016/J.POLYMER.2016.02.003.
6. Loos, K., Zhang, R., Pereira, I., Agostinho, B., Hu, H., Maniar, D., Sbirrazzuoli, N., Silvestre, A., Guigo, N., Sousa, A. (2020). A perspective on PEF synthesis, properties, and end-life. *Frontiers in Chemistry*, 8, 585. doi: 10.3389/fchem.2020.00585.
7. Berkel, J., Guigo, N., Kolstad, J., Sipos, L., Wang, B., Dam, M., Sbirrazzuoli, N. (2015). Isothermal crystallization kinetics of poly(ethylene 2,5-furandicarboxylate). *Macromolecular Materials and Engineering*, 300, 466–474. doi: 10.1002/MAME.201400376.
8. Menager, C., Guigo, N., Martino, L., Sbirrazzuoli, N., Visser, H., Boyer, S., Billon, N., Monge, G., Combeaud, C. (2018). Strain-induced crystallization in biobased Poly(ethylene 2,5 furandicarboxylate) (PEF); conditions for appearance and microstructure analysis. *Polymer*, 158, 364–371. doi: 10.1016/J.POLYMER.2018.10.054.
9. Liu, H., Ding, N., Wei, J., Tang, X., Zeng, X., Sun, Y., Lei, T., Fang, H., Li, T., Lin, L. (2020). CoCu-NC bimetallic catalyst with improved molecular oxygen activation ability and basicity: boosting the base-free oxidative esterification of 5-(hydroxymethyl)furfural. *ChemSusChem*, 13, 4151–4158. doi: 10.1002/cssc.202000537.
10. Zaidi, S., Bougarech, A., Abid, M., Abid, S., Silvestre, A., Sousa, A. (2023). Highly Flexible Poly(1,12-dodecylene 5,5′-isopropylidene-bis(ethyl 2-furoate)): A promising biobased polyester derived from a renewable cost-effective bisfuranic precursor and a long-chain aliphatic spacer. *Molecules*, 28, 4124. doi: 10.3390/molecules28104124.
11. Rosenboom, J., Hohl, D.K., Fleckenstein, P., Storti, G., Morbidelli, M. (2018). Bottle-grade polyethylene furanoate from ring-opening polymerization of cyclic oligomers. *Nature Communications*, 9, 2701. doi: 10.1038/s41467-018-05147-y.
12. Mendieta, C., González, G., Vallejos, M., Area, M. (2022). Bio-polyethylene furanoate (Bio-PEF) from lignocellulosic biomass adapted to the circular bioeconomy. *BioResources*, 17, 7313–7337. doi: 10.15376/biores.17.4.mendieta.

13. Banella, M., Bonucci, J., Vannini, M., Marchese, P., Lorenzetti, C., Celli, A. (2019). Insights into the synthesis of poly(ethylene 2,5-furandicarboxylate) from 2,5-furandicarboxylic acid: Steps toward environmental and food safety excellence in packaging applications. *Industrial & Engineering Chemistry Research*, 58, 8955–8962. doi: 10.1021/ACS.IECR.9B00661.
14. Kasmi, N., Papageorgiou, G., Achilias, D., Bikiaris, D. (2018). Solid-state polymerization of poly(ethylene furanoate) biobased polyester, II: An efficient and facile method to synthesize high molecular weight polyester appropriate for food packaging applications. *Polymers (Basel)*, 10, 471. doi: 10.3390/POLYM10050471.
15. Paszkiewicz, S., Irska, I., Piesowicz, E. (2020). Environmentally friendly polymer blends based on post-consumer glycol-modified poly(ethylene terephthalate) (PET-G) foils and poly(ethylene 2,5-furanoate) (PEF): Preparation and characterization. *Materials (Basel)*, 13, 2673. doi: 10.3390/ma13122673.
16. Pellis, A., Haernvall, K., Pichler, C.M., Ghazaryan, G., Breinbauer, R., Guebitz, G.M. (2016). Enzymatic hydrolysis of poly(ethylene furanoate). *Journal of Biotechnology*, 235, 47–53. doi: 10.1016/j.jbiotec.2016.02.006.
17. Lam, J., Shih, C., Lee, W., Chueh, C., Jang, G., Huang, C., Tung, S., Chen, W. (2018). Bio-based transparent conductive film consisting of polyethylene furanoate and silver nanowires for flexible optoelectronic devices. *Macromolecular Rapid Communications*, 39, e1800271. doi: 10.1002/marc.201800271.
18 Post, C., Maniar, D., Voet, V., Folkersma, R., Loos, K. (2023). Biobased 2,5-bis(hydroxymethyl)furan as a versatile building block for sustainable polymeric materials. *ACS Omega*, 8, 8991–9003. doi: 10.1021/acsomega.2c07629.
19. Zhu, C., Ye, D., Zhou, T., Cui, Y., Yin, J. (2023). High-antimicrobial gallium-doped zinc oxide thin films on bio-based poly(ethylene furanoate) substrates for food packaging application. *Membranes (Basel)*, 13, 239. doi: 10.3390/membranes13020239.
20. Burgess, S., Leisen, J., Kraftschik, B., Mubarak, C., Kriegel, R., Koros, W. (2014). Chain mobility, thermal, and mechanical properties of poly(ethylene furanoate) compared to poly(ethylene terephthalate). *Macromolecules*, 47, 1383–1391. doi: 10.1021/MA5000199.
21. Burgess, S., Kriegel, R., Koros, W. (2015). Carbon dioxide sorption and transport in amorphous poly(ethylene furanoate). *Macromolecules*, 48, 2184–2193. doi: 10.1021/ACS.MACROMOL.5B00333.
22. Edman, L., Ferry, A., Orädd, G. (2002). Analysis of diffusion in a solid polymer electrolyte in the context of a phase-separated system. *Physical Review. E, Statistical, Nonlinear, and Soft Matter Physics*, 65, 042803. doi: 10.1103/PHYSREVE.65.042803.
23. Xie, H., Wu, L., Li, B., Dubois, P. (2018). Modification of poly(ethylene 2,5-furan dicarboxylate) with biobased 1,5-pentanediol: significantly toughened copolyesters retaining high tensile strength and o2 barrier property. *Biomacromolecules,* 20 1, 353–364. doi: 10.1021/acs.biomac.8b01495.
24. Eerhart A.J.J., Faaij A.P.C., Patel M.K. (2012). Replacing fossil-based PET with biobased PEF: Process analysis, energy, and GHG balance. *Energy & Environmental Science*, 5, 6407. doi: 10.1039/C2EE02480B.
25. Nolasco, M., Araujo, C., Thiyagarajan, S., Rudić, S., Vaz, P., Silvestre, A., Ribeiro-Claro, P., Sousa, A. (2020). Asymmetric monomer, amorphous polymer?

structure-property relationships in 2,4-FDCA and 2,4-PEF. *Macromolecules*, 53, 1380–1387. doi: 10.1021/acs.macromol.9b02449.

26. Terzopoulou, Z., Tsanaktsis, V., Bikiaris, D., Exarhopoulos, S., Papageorgiou, D., Papageorgiou, G. (2016). Biobased poly(ethylene furanoate-co-ethylene succinate) copolyesters: Solid-state structure, melting point depression, and biodegradability. *RSC Advances*, 6, 84003–84015. doi: 10.1039/C6RA15994J.
27. Poulopoulou, N., Kasmi, N., Siampani, M., Terzopoulou, Z., Bikiaris, D., Achilias, D., Papageorgiou, D., Papageorgiou, G. (2019). Exploring next-generation engineering bioplastics: Poly(alkylene furanoate)/poly(alkylene terephthalate) (PAF/PAT) blends. *Polymers (Basel),* 11, 556. doi: 10.3390/polym11030556.
28. Soares, M., Dannecker, P., Vilela, C., Bastos, J., Meier, M., Sousa, A. (2017). Poly(1,20-eicosanediyl 2,5-furan dicarboxylate) is a biodegradable polyester from renewable resources. *European Polymer Journal*, 90, 301–311. doi: 10.1016/J.EURPOLYMJ.2017.03.023.
29. Nakajima, H., Dijkstra, P., Loos, K. (2017). The recent developments in biobased polymers toward general and engineering applications: Polymers that are upgraded from biodegradable polymers, analogous to petroleum-derived polymers, and newly developed. *Polymers (Basel)*, 9, 523. doi: 10.3390/polym9100523.
30. Hwang, K., Jeon, W., Lee, S., Kim, M., Park, Y. (2020). Sustainable bioplastics: Recent progress in the production of bio-building blocks for the bio-based next-generation polymer PEF. *Chemical Engineering Journal,* 390, 124636. doi: 10.1016/j.cej.2020.124636.
31. Jong, E., Visser, H., Dias, A., Harvey, C., Gruter, G. (2022). The road to bring FDCA and PEF to the market. *Polymers (Basel),* 14, 943. doi: 10.3390/polym14050943.
32. Cazón, P., Velázquez, G., Ramírez, J., Vázquez, M. (2017). Polysaccharide-based films and coatings for food packaging: A review. *Food Hydrocolloids*, 68, 136–148. doi: 10.1016/J.FOODHYD.2016.09.009.
33. Guidotti, G., Soccio, M., Lotti, N., Siracusa, V., Gazzano, M., Munari, A. (2019). New multi-block copolyester of 2,5-furandicarboxylic acid containing PEG-like sequences to form flexible and degradable films for sustainable packaging. *Polymer Degradation and Stability*, 169, 108963. doi: 10.1016/j.polymdegradstab.2019.108963.
34. Xie, H., Meng, H., Wu, L., Li, B., Dubois, P. (2019). In-situ synthesis, thermal and mechanical properties of biobased poly(ethylene 2,5-furandicarboxylate)/montmorillonite (PEF/MMT) nanocomposites. *European Polymer Journal*, 121, 109266. doi: 10.1016/j.eurpolymj.2019.109266.
35. Höhnemann, T., Steinmann, M., Schindler, S., Hoss, M., König, S., Ota, A., Dauner, M., Buchmeiser, M.R. (2021). Poly(ethylene furanoate) along its life cycle from a polycondensation approach to high-performance yarn and its recyclate. *Materials (Basel)*, 14, 1044. doi: 10.3390/ma14041044.
36. Takarada, W., Sugimoto, K., Nakajima, H., Visser, H., Gruter, G., Kikutani, T. (2021). Melt-spun fibers from bio-based polyester–fiber structure development in high-speed melt spinning of poly(ethylene 2,5-furandicarboxylate) (PEF). *Materials (Basel),* 14, 1172. doi: 10.3390/ma14051172.
37. Wu, J., Xie, H., Wu, L., Li, B., Dubois, P. (2016). DBU-catalyzed biobased poly(ethylene 2,5-furandicarboxylate) polyester with rapid melt crystallization:

Synthesis, crystallization kinetics, and melting behavior. *RSC Advances*, 6, 101578–101586. doi: 10.1039/C6RA21135F.

38. Wang, J., Liu, X., Jia, Z., Liu, Y., Sun, L., Zhu, J. (2017). Synthesis of bio-based poly(ethylene 2,5-furandicarboxylate) copolyesters: Higher glass transition temperature, better transparency, and good barrier properties. *Journal of Polymer Science Part A: Polymer Chemistry*, 55, 3298–3307. doi: 10.1002/POLA.28706.
39. Araujo, C., Nolasco, M., Ribeiro-Claro, P., Rudić, S., Silvestre, A., Vaz, P., Sousa, A. (2018). Inside PEF: Chain conformation and dynamics in crystalline and amorphous domains. *Macromolecules,* 51, 3515–3526. doi: 10.1021/ACS.MACROMOL.8B00192.
40. Bourdet, A., Esposito, A., Thiyagarajan, S., Delbreilh, L., Affouard, F., Knoop, R., Dargent, E. (2018). Molecular mobility in amorphous biobased poly(ethylene 2,5-furandicarboxylate) and poly(ethylene 2,4-furandicarboxylate). *Macromolecules*, 51, 1937–1945. doi: 10.1021/ACS.MACROMOL.8B00108.
41. Joshi, A., Alipourasiabi, N., Kim, Y., Coleman, M., Lawrence, J. (2018). Role of enhanced solubility in esterification of 2,5-furandicarboxylic acid with ethylene glycol at reduced temperatures: energy efficient synthesis of poly(ethylene 2,5-furandicarboxylate). *Reaction Chemistry and Engineering*, 3, 447–453. doi: 10.1039/C8RE00086G.
42. Gupta, V., Biswas, D., Roy, S. (2022). A comprehensive review of biodegradable polymer-based films and coatings and their food packaging applications. *Materials (Basel)*, 15, 5899. doi: 10.3390/ma15175899.
43. https://www.statista.com/statistics/1171088/price-polyethylene-terephthalate-forecast-globally/.
44. Lopes, J., Tsochatzis, E. (2023). Poly(ethylene terephthalate), poly(butylene terephthalate), and polystyrene oligomers: Occurrence and analysis in food contact materials and food. *Journal of Agricultural and Food Chemistry*, 71, 2244–2258. doi: 10.1021/acs.jafc.2c08558.
45. Svyntkivska, M., Makowski, T., Pawlowska, R., Kregiel, D., De Boer, E.L., Piorkowska, E. (2024). Cytotoxicity studies and antibacterial modification of poly(ethylene 2,5-furandicarboxylate) nonwoven. *Colloids and Surfaces B: Biointerfaces*, 233, 113609. doi: 10.1016/j.colsurfb.2023.113609.
46. Siracusa, V., Rocculi, P., Romani, S., Rosa, M. (2008). Biodegradable polymers for food packaging: A review. *Trends in Food Science and Technology*, 19, 634–643. doi: 10.1016/J.TIFS.2008.07.003.
47. Joshi, A., Lawrence, J., Coleman, M. (2019). Effect of biaxial orientation on microstructure and properties of renewable copolyesters of poly(ethylene terephthalate) with 2,5-furandicarboxylic acid for packaging application. *ACS Applied Polymer Materials*, 1, 1798–1810. doi: 10.1021/ACSAPM.9B00330.
48. Weinberger, S., Canadell, J., Quartinello, F., Yeniad, B., Arias, A., Pellis, A., Guebitz, G. M. (2017). Enzymatic degradation of poly(ethylene 2,5-furanoate) powders and amorphous Films. *Catalysts*, *7*, 318. doi: 10.3390/catal7110318.
49. Maaskant, E., van Es, D.S. (2021). Unexpected susceptibility of poly(ethylene furanoate) to UV irradiation: A warning light for furandicarboxylic acid? *ACS Macro Letters*, 10, 1616–1621. doi: 10.1021/acsmacrolett.1c00676.

50. Wang, C., Chen, M., Jiang, Z., Qiu, Z. (2023). Synthesis, thermal and mechanical properties of fully biobased poly(hexamethylene succinate-co-2,5-furandicarboxylate) copolyesters. *Polymers*, 15, 427. doi: 10.3390/polym15020427.
51. Siracusa, C., Quartinello, F., Soccio, M., Manfroni, M., Lotti, N., Dorigato, A., Guebitz, G., Pellis, A. (2023). On the selective enzymatic recycling of poly(pentamethylene 2,5-furanoate)/poly(lactic acid) blends and multiblock copolymers. *ACS Sustainable Chemistry & Engineering*, 11, 9751–9760. doi: 10.1021/acssuschemeng.3c01796.
52. Fredi, G., Zonta, E., Dussin, A., Bikiaris, D., Papageorgiou, G., Fambri, L., Dorigato, A. (2023). Toughening effect of 2,5-furandicaboxylate polyesters on polylactide-based renewable fibers. *Molecules*, 28, 4811. doi: 10.3390/molecules28124811.
53. van Strien, N., Niskanen, J., Berghuis, A., Pöhler, H., Rautiainen, S. (2024). Production of 2,5-furandicarboxylic acid methyl esters from pectin-based aldaric acid: From laboratory to bench scale. *ChemSusChem*, 17, e202300732. doi: 10.1002/cssc.202300732.
54. Paszkiewicz, S., Irska, I., Zubkiewicz, A., Walkowiak, K., Rozwadowski, Z., Dryzek, J., Linares, A., Nogales, A., Ezquerra, T. (2023). Supramolecular structure, relaxation behavior and free volume of bio-based poly(butylene 2,5-furandicarboxylate)-block-poly(caprolactone) copolyesters. *Soft Matter*, 19, 959–972. doi: 10.1039/d2sm01359b.
55. Koshti, R., Mehta, L.B., Samarth, N. (2018). Biological recycling of polyethylene terephthalate: A mini-review. *Journal of Polymers and the Environment*, 26, 3520–3529. doi: 10.1007/s10924-018-1214-7.
56. Soong, Y.V., Sobkowicz, M.J., Xie, D. (2022). Recent advances in biological recycling of polyethylene terephthalate (pet) plastic wastes. *Bioengineering*, 9, 98. doi: 10.3390/bioengineering9030098.
57. Silverwood, L., Mottoul, M., Dumont, M. (2024). A review of end-of-life pathways for poly(ethylene furanoate) and its derivatives. *Journal of Polymers and the Environment*, 32, 4130–4142. doi: 10.1007/s10924-024-03229-1.
58. Chen, G., Patel, M.K. (2011). Plastics derived from biological sources: Present and future: A technical and environmental review. *Chemical Reviews*, 112, 2082–2099. doi: 10.1021/cr200162d.

CHAPTER 8

Exploring Microbial Cellulose as a Biotechnological Alternative for Food Packaging

YURIDIA ORTIZ-RIVERA,[1*] JESÚS RUBÉN RODRÍGUEZ-NÚÑEZ,[2] and JOSÉ ALBERTO NÚÑEZ-GASTÉLUM[1]

[1]*Departamento de Ciencias Químico Biológicas, Instituto de Ciencias Biomédicas, Universidad Autónoma de Ciudad Juárez, Ciudad Juárez, Chihuahua, Mexico*

[2]*Programa de Biotecnología, División de Ciencias de la Salud e Ingenierías, Universidad de Guanajuato Campus Celaya-Salvatierra, Celaya, Guanajuato, Mexico*

*Corresponding author

ABSTRACT

There is a necessity to advance the technology of food packaging to meet the dual objectives of safeguarding food and increasing its shelf life while ensuring that such packaging is environmentally sustainable. Microbial cellulose (MC) is a polymer obtained by fermentation from fungi and bacteria that resembles plant cellulose. MC presents attractive physicochemical characteristics that make it a promising alternative to its plant-derived counterpart, including excellent mechanical properties, greater crystallinity, and affinity for water. Considering these particularities plus its innocuous properties, MC has the potential to be utilized in various industries, including the food industry. For this reason, the objective of this chapter is to analyze the prospect of cellulose from microbial origin as a component of food packaging. The analysis is approached based on

Sustainable Food Packaging Materials and Technology. Juan Manuel Tirado-Gallegos, Juan Daniel Rivaldi Chávez, and José Juan Buenrostro-Figueroa (Eds)

DOI: 10.1201/9781779641434-8

the structural properties of the biopolymer and the functionalities that derive from them. Then, it is followed by a description of the experimental and large-scale bioprocesses used to obtain MC. Finally, the application of MC as a food packaging material is presented, along with successful commercial cases.

8.1 INTRODUCTION

Food packaging is necessary to preserve the quality and safety of food during transportation and shelf life. Most of the packaging used in food is made of plastic; they are called single-use plastics because they are discarded immediately after use, causing their accumulation, which represents a serious environmental pollution problem [1]. The design of sustainable food packaging that preserves its integrity but is also environmentally friendly is necessary [1]. There are some options for macromolecules to be used in the development of sustainable packaging made from natural substances or molecules such as proteins, lipids, and polysaccharides [2]. Within these macromolecules, cellulose should be highlighted, which is the most abundant substance on the planet and has characteristics such as the ability to form films and transparency, which make it a viable option to be used in packaging [3]. Cellulose can be extracted from the cell wall of plants or from microbial sources through bioprocesses [4, 5]. The physicochemical characteristics of cellulose differ according to the source of production and extraction method, and these characteristics can be improved by adjusting the techniques or making mixtures with other substances [6, 7]. Cellulose of bacterial origin is the one that generates the best yields in its production, apart from the fact that it has fewer impurities compared to the cellulose that is extracted from plants. Cellulose has a wide application potential in the development of sustainable packaging; however, more research is still needed to make costs more accessible and properties more similar or superior to those of conventional plastics [8].

8.2 FOOD PACKAGING

The world population and the growing demand for food have caused food packaging to be produced in larger quantities [1]. Food packaging is important because it isolates the food from the environment, ensuring its protection against mechanical damage, odors, dust, gases, humidity, insects, and microorganisms [9]. In addition, food packaging must ensure its quality and safety. Packaging must preserve food during transportation and shelf life, which is a challenge due to the associated risks, such as contamination with pathogenic bacteria or chemical contaminants [1].

Most food packaging is based on petroleum-derived plastics, which are an important source of environmental pollution, making the development of environmentally friendly packaging essential [1]. The most used plastics for packaging are polyethylene, polyethylene terephthalate, polypropylene, polyvinyl chloride, and polystyrene [10]. Others such as glass and metal and even paper have also been used for food packaging [11]. Conventional plastics are used as food packaging because they have important advantages such as resistance, durability, and low cost; however, their main disadvantage is their low biodegradability and accumulation in the environment, especially because some plastics used in food are called plastic single-use only and discarded immediately. Considering plastic pollution, materials have been developed from waste that are easily degraded in the environment [2]. Therefore, sustainable packaging is necessary to reduce the accumulation of poorly degradable materials in the environment [1, 2]. Additionally, sustainable packaging can incorporate bioactive ingredients to improve food quality and safety. The main functions of active packaging are the absorption of oxygen, ethylene, moisture, and odor, and they can also incorporate antimicrobial activity, antioxidants, aromatic agents, and vitamins that ensure the quality and safety of food [2, 11].

Agricultural waste is viable for obtaining biomaterials such as cellulose that can be used for the development of food packaging; however, detailed studies of packaging made with biomaterials must be evaluated to know their life cycle and environmental impact, but they have the potential to improve the accumulation of plastic garbage [1]. Cellulose and its derivatives are good options for food packaging because it has an important filmogenic capacity, which results in the formation of biodegradable, nontoxic films with good mechanical, thermal, and barrier properties. However, it still has the disadvantage of its low solubility and indigestibility. It has the disadvantage of its low solubility and that it is indigestible, so its use in edible coatings continues to be a challenge [3].

The polymer matrices that can be used in the development of biomaterials for food packaging are carbohydrates such as cellulose, starch, or chitosan; proteins such as gelatin; and lipids such as beeswax [2, 3].

These biopolymers have been used in the development of coatings, films, and nanocomposites [12]. The use of biological materials such as chitosan, potato starch, carboxymethyl cellulose, corn starch, and gum Arabic facilitates packaging improvements because it is possible to incorporate antimicrobials and prevent the growth of foodborne pathogens [9]. In addition, packaging based on biomaterials can fertilize and condition the soil, which represents another important advantage for food production [10]. Smart packaging with electronic devices is also being developed to monitor food quality and safety and, thus, avoid the risk associated with the consumption of contaminated food and food waste [13]. However, research must still be carried out in this regard, especially to lower the costs of these packaging and achieve consumer acceptance [3]. Despite the multiple advantages, packaging based on biomaterials still lacks some characteristics of plastic packaging such as resistance, flexibility, and barrier properties, which is why research in this regard is completely necessary [10]. Efforts are focused on the development of combinations of these materials and additives that improve their properties [14].

8.3 CELLULOSE

Cellulose is a sustainable and economical biopolymer, for these reasons, which has been the subject of research for various applications. It is the most abundant biopolymer on the planet; for example, in plants, 40–45% of their biomass is cellulose. Cellulose fibers are obtained from the cell wall of plants. The degree of polymerization is determined by the source of cellulose or the plant from which it comes. In addition, cellulose has multiple hydroxyl groups that make it capable of forming hydrogen bonds and intra- and intermolecular interactions that make it suitable for modification and improvement. Cellulose is composed of glucose units with a high molecular weight linked by β-1,4-glycosidic bonds. It is composed of fibers with different morphologies depending on the source. Hydrogen bonds play an important role in determining the structural organization of cellulose [15, 16]. Cellulose comes from different natural sources such as plants; algae; fungi; marine animals; and even bacteria such as *Acetobacter xylinum*, *Agrobacterium tumefaciens*, *Rhizobium* spp., *Escherichia coli*, *Klebsiella pneumoniae*, *Salmonella typhimurium*, and *Sarcina ventriculi* [16]. The abundance and biodegradability make it an environmentally friendly material, with a unique, flexible chemical structure, with mechanical resistance and thermal stability that make it capable of being used for different purposes [17]. Even some celluloses may have antimicrobial properties such as those extracted from plants such as *Tridax procumbens*, *Catharanthus roseus*, and *Kigelia africana* [18]. The structure of the cellulose molecule, which depends on the conformation, configuration, and geometry, is responsible for its physical and chemical properties. This structure is determined by X-ray diffraction (XRD) techniques, electron microscopy, and other instrumental techniques [19]. Furthermore, these differences in the structure, development of derivatives, and improvements have made it possible for the properties to be better adapted according to the product to be obtained [20]. In recent years, due to recently developed instrumentation and techniques, the knowledge and potential application of cellulose in different materials have grown [16]. As the most abundant biomacromolecule, in addition to being extracted from agricultural waste, cellulose has been widely exploited by humans [21]. Cellulose has been used in diverse applications such as paper, food packaging, absorbent material, among others, due to the stable fibers of uniform sizes that form this biopolymer [22]. The innovative focus on developing sustainable products using cellulose fibers could replace the use of products based on plastic polymers on the market, which currently represent a serious pollutant for the planet [18].

8.4 SUPPLIES OF OBTAINING CELLULOSE AND PHYSICOCHEMICAL CHARACTERISTICS

The main source of obtaining cellulose is from solid waste of plant origin, since they are mainly made up of plant parts such as stems, roots, leaves, and seeds that remain as by-products of this agroindustrial productive sector. However, cellulose is in the form of fibers that also contain hemicellulose and lignin, which requires a purification process to obtain cellulose in pure form [4]. In this regard, Longaresi et al. [23] recovered α-cellulose from corn residue stalks with yields of up to 37%. The cellulose was purified by eliminating lignin and hemicellulose, obtaining cellulose microfibrils of 100–560 μm and 40.4–70.8 nm. In addition, the XRD patterns showed type I cellulose and type II cellulose, indicating that cellulose type II is produced by chemical treatments such as bleaching or purifying. In another recent study, Rahmi et al. [24] obtained cellulose from coconut waste and carried out purification with acid hydrolysis, eliminating lignin and hemicellulose. The FTIR results showed the elimination of lignin, by decreasing the band at 2843 and 2924 cm^{-1} corresponding to the methoxy group, while the XRD studies showed peaks at $2\theta = 10°$, 15.86°, and 19.7° related to amorphous structures due to impurities such as hemicellulose and lignin, which had to be eliminated with acid treatments. Similarly, the thermal stability of cellulose analyzed by TGA/DTGA showed an approximate degradation temperature of 350°C and indicated that unpurified cellulose increases up to 400°C due to the presence of hemicellulose and cellulose. In 2024, in a study carried out by Moghaddam et al. [4], they recovered cellulose microfibers from corn leaves left over from agriculture. Purification was performed with alkaline treatment with NaOH at 120°C. The FTIR results showed a decrease in the intensity of the peaks at 1732 and 1515 cm^{-1} related to the –C=O groups of hemicellulose and –C–O of the aromatic structure of lignin, respectively, demonstrating that the alkaline purification process reduces impurities but does not eliminate them completely. The crystallinity of cellulose was analyzed by XRD, showing crystalline and amorphous regions (semicrystalline character), while lignin and hemicellulose present only amorphous regions. Therefore, the crystallinity index of cellulose was high (65%) because chemical treatments (alkaline purification and bleaching) change the morphology of cellulose microfibers, decreasing compounds such as lignin, hemicellulose, and pectins and increasing the crystallinity index. This was demonstrated since the cellulose fibers presented a 65% crystallinity index, resulting in high values. The thermal properties evaluated by TGA/DTGA showed that the degradation temperature of cellulose, hemicellulose, and lignin begins at 200°C and ends at 400°C with a maximum mass loss peak at 353°C, resulting in the typical thermal degradation range of lignocellulosic materials. Table 8.1 shows sources of obtaining cellulose from recent research. In summary, the physicochemical characteristics of cellulose depend on the source of obtaining and the purification treatment used (alkaline, acid, and bleaching treatment). In this sense, Randis et al. [25] mentioned that the degradation temperature increases depending on the crystallinity; therefore, when purifying the cellulose, the crystalline region increases and with it the degradation temperature observed in the thermal analyzes by TGA/DTGA. Furthermore, it can be observed that purification processes significantly help to reduce noncellulosic material such as hemicelluloses, lignin, and pectins, and the FTIR method is ideal for the detection of these compounds.

TABLE 8.1 Sources of Cellulose and Physicochemical Characteristics

Source of Obtention	Physicochemical Characteristics	Author
Palm grass leaf fiber (*Curculigo capitulata*) (cellulose microfibers)	• Alkaline purification • Crystallinity index 84–85% by XRD • 2.4 nm cellulose microfibers • FTIR shows the presence of cellulose and lignin in microfibers	Kumar et al. [26]
Palm kernel meal (cellulose nanoparticles)	• Alkaline and acid purification • Crystallinity index 78.8% • XRD shows the presence of type I cellulose • FTIR shows the decrease in lignin and hemicellulose • TGA/DTGA shows a cellulose degradation temperature of 300–400°C	Chamathka et al. [27]
Jackfruit peel	• Alkaline treatment (NaOH 5%) • FTIR detected the presence of type I cellulose and the absence of hemicellulose and lignin • TGA/DTGA shows a cellulose degradation temperature of 200–300°C • XRD showed a crystallinity index of 30% in the cellulose obtained	Brahma and Ray [28]
Oil palm from biomass (cellulose microfibers)	• Alkaline treatment (NaOH 9%) • FTIR shows a decrease in hemicellulose and lignin • XRD shows the presence of type I cellulose and a crystallinity index of 71–76% • TGA/DTGA show a degradation temperature of 286–334°C in the treated cellulose microfibers	Randis et al. [25]
Leftover celery pulp (*Apium graveolens* sweet variety) (cellulose microfibers)	• Alkaline treatment with 2% NaOH • FTIR demonstrates the removal of lignin and hemicellulose • XRD shows the presence of type I (crystalline) cellulose with a crystallinity index of 45–46% • TGA/DTGA show that the degradation temperature of purified cellulose is 320–330°C	Abzan et al. [29]
Agrowaste of tamarind seed (*Tamarindus indica*)	• Alkaline, acid, and bleaching treatment • XRD showed a crystallinity index of 77.6% • TGA/DTGA had a degradation temperature higher than 230°C • FITIR showed the presence of α-cellulose and a decrease in the peaks corresponding to lignin and hemicellulose	Divakaran et al. [30]

8.5 BACTERIAL CELLULOSE AND PHYSICOCHEMICAL CHARACTERISTICS

Bacterial cellulose is an exopolymer synthesized mainly by bacteria of the genus *Acetobacteraceae*, highlighting the bacterium *A. xylinum* or *Gluconobacter xylinum* (renamed *Komagataeibacter xylinus*) as the main producer of bacterial cellulose, although *Acetobacter xylinoides*, *Acetobacter aceti*, and *Acetobacter pasteurianus* can also be found. In addition, other genera such as *Agrobacterium*, *Sarcina*, *Rhizobium*, *Pseudomonas*, *Alcaligenes*, *Achromobacter*, *Aerobacter*, *Azotobacter*, *Komagataeibacter*, *Dickeya*, and *Rhodobacter* have also been reported to have the capacity for extracellular cellulose biosynthesis [5]. Regarding the chemical composition, bacterial and plant cellulose are the same (repeating units of D-glucose, linked by β (1-4) bonds, not branched); however, bacterial cellulose has different structural compositions and physicochemical properties to the vegetal one [6, 7], and some of the main differences are specified below: (1) structural conformation: more compact structure of nanometric order and different physicochemical properties; (2) high crystallinity: greater than 60%. Due to a more ordered chemical structure in which type I cellulose predominates (crystalline packaging, allomorph Iα predominates), while cellulose of plant origin is made up of cellulose type I (allomorph Iβ predominates); (3) high purity: contains no hemicellulose, lignin, or pectins, that is, no plant cell wall impurities; (4) high water retention capacity: it can absorb up to 200 times its weight, due to the 3D intertwined network formed by the nanometric size of its fibers; (5) biocompatibility: it is compatible with human tissue due to its high purity; (6) greater biodegradability; (7) high Young's modulus (>100 GPa) compared to all the biomaterials used; and (8) nanometric fibers and high transparency. As described before, Table 8.2 shows conditions and characteristics of cellulose obtained through fermentation.

8.6 BIOPROCESSES FOR OBTAINING MICROBIAL CELLULOSE

One of the most important challenges of biotechnology is the scaling of bioprocesses at an industrial level, as well as obtaining profitable returns. As mentioned above, there are many cellulose-producing microorganisms, but only some have been used in industrial processes, with bacteria

TABLE 8.2 Conditions and Characteristics of Cellulose Obtained Through Fermentation

Microorganism	Conditions of Fermentation	Characteristics of the Cellulose	Author
Three strains were evaluated: *Komagataeibacter* sp. FXV3, *Komagataeibacter* sp. NFXK3, and *Komagataeibacter intermedius* LMG 18909	Culture medium: Hestrin and Schramm medium (HS) (20 g/L glucose, 5 g/L yeast extract, 5 g/L peptone, 1.15 g/L citric acid, 2.7 g/L Na_2HPO_4, pH 6), incubated to 28°C, without agitation for seven days	• XRD showed the presence of type I cellulose, with up to 85% crystallinity • SEM, the surface morphology of cellulose changed depending on the strain and culture conditions (porosity, fiber size and surface area) • FTIR showed that the chemical composition of cellulose does not change, and there are no impurities such as hemicellulose and lignin • The permeability to N_2 and O_2 changed depending on the porosity of the cellulose membranes obtained	Fatima et al. [31]
Natural fermentation: Kombucha (symbiotic interaction of yeast, acetic and lactic acid bacteria) (SCOBY)	Culture medium: black tea, sucrose, hydrolyzed vinegar residues. Fermentation without agitation at 30°C for 15 days	• XRD demonstrated the presence of type I cellulose in all the cellulose obtained and allotrope Iα and showed a crystallinity index of up to 86.4% • TGA/DTGA showed high thermal stability since the cellulose degraded at 350°C • ^{13}C NMR demonstrated that the chemical composition of cellulose does not change, it also confirmed that only the allotrope Iα exists and did the presence of Iβ was not detected	Liu et al. [32]
Acetobacter xylinum	Static fermentation of coconut water at room temperature for seven days	• XRD showed bacterial cellulose of high crystallinity and purity • SEM microscopy showed high porosity in the prepared hydrogels, in addition, authors mention that the nanofibers are in the range of 40–80 nm	Phan et al. [33]
Gluconacetobacter hansenii, ATCC 23769	Fermentation in static conditions for 12 days at 30 °C, in the culture medium Luria Bertani Broth (LB)	• TGA/DTGA showed the degradation of cellulose at 321°C • SEM microscopy showed nanofibers from 52 to 98 nm. • High Young's modulus (64.6 MPa). • XRD showed that a semicrystalline structure is found the obtained cellulose	Ciftci [34]

largely standing out. The production of bacterial cellulose can be carried out in static culture or with shaking; in the first case, the film forms on the surface of the medium as a supernatant mainly due to the high oxygen pressure. Although fermentation technologies have advanced exponentially, static cultivation is still preferred for obtaining cellulose mainly due to performance and operating costs. In cultures with agitation, cellulose does not acquire the shape of a film but rather agglutinates into bodies a few millimeters in diameter whose appearance is not well defined and can vary between spheroid, cuboidal, or amorphous. It is worth mentioning that the choice of the type of culture can be influenced depending on the application of the product because the method can impact the physicochemical properties of the films [35, 36]. Evidence in the literature indicates that the *A. xylinum* species has been widely evaluated and has been classified as the cellulose producer par excellence because it synthesizes it in abundance. Recently, this species was reclassified taxonomically, first as *Gluconacetobacter xylinus* and finally named *K. xylinus* [37]. To implement a bacterial cellulose production process, it is necessary to consider several aspects, among which the following stand out: the origin of the biological agent, the type of bioreactor, nutrient sources, and automation [38]. Most commonly, the parameters evaluated to obtain the best performance are the carbon source, temperature, and pH. The culture media used to produce polysaccharides must be supplemented with a high proportion of carbon and other nutrients on a smaller scale. It is intended that most of the carbon source (>60%) can be metabolized to generate the polymer of interest; the most common, in order of use, are glucose, sucrose, fructose, maltose, xylose, starch, and glycerol. The type and initial concentration of the monosaccharide can impact cellulose production due to the appearance of metabolites that modify the conditions or act as inhibitors. Alternatively, nonconventional sources of carbon have been tested, including by-products of some industries, such as ethanol, citric acid, acetic acid, lactic acid, fructose, corn liquor, molasses, even derivatives of the confectionery industries, biodiesel, cognac, fruit juices, and wine, among others [39, 40]. Other growth medium components worth considering when producing bacterial cellulose are the nitrogen source, organic acids, and buffer salts. Many of the new processes are based on the medium originally developed by Scrhamm and Hestrin [35], which consists of a mixture of 2% glucose, 0.5% peptone, 0.5% yeast extract, citric acid, and Na_2HPO_4 to maintain a pH of 6. Temperature is another primary factor that can affect microbial growth, which is why it is not surprising that it must be controlled in detail during the fermentation processes. The great diversity of works on microbial cellulose production coincides with the first studies regarding this parameter and places the optimal temperature between 28°C and 30°C [41]. For its part, the best yields are achieved in a pH range between 4 and 6. Unlike the temperature, which is not affected mainly during the process, the pH can decrease due to the generation of metabolites, mainly of the acid type; for example, this factor must be controlled through automation of the bioreactor [42]. Once the microorganism produces cellulose, it must be recovered and purified since it contains materials from the culture medium, such as cells, components of the medium, and metabolites generated during fermentation [36]. Due to the acidic conditions of the bioprocess, a thermal–alkaline procedure with diluted alkali (100°C, 0.1 N NaOH/KOH) for 20 min is commonly used, followed by contact with organic acids (acetic acid) to level the pH and, finally, washed with water until obtaining a neutral pH [43]. The basic treatment dissolves the microorganisms and is the first attack on leveling the acidic pH of the culture medium, while the organic acids and water allow low molecular weight components to be carried away. Figure 8.1 schematizes the general process to produce bacterial cellulose in a medium without agitation.

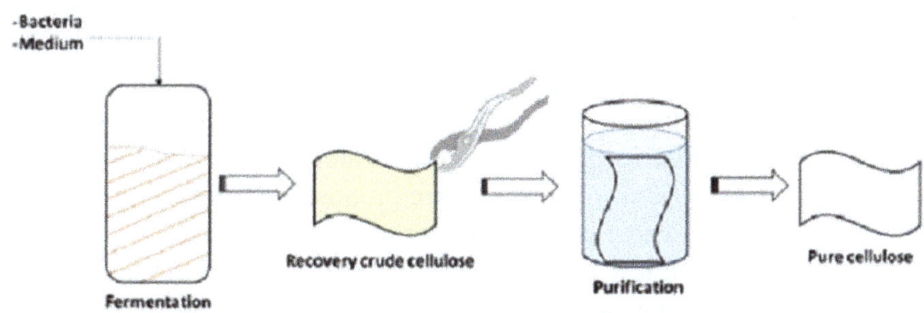

FIGURE 8.1 Process of bacterial cellulose production by static liquid culture.

It is considered that a bioprocess focused on bacterial cellulose can be scaled up to an industrial level when a performance comparable to the biosynthesis of plant cellulose is achieved, around 15 g/L in a minimum period of 50 h [44]. Bacterial cellulose has been produced industrially for more than three decades as a bioproduct to be used in multiple areas.

Jonas and Farah [45] mentioned that by the 1990s, the existence of three commercial brands was applied in the biomedical area.

8.7 POTENTIAL APPLICATION OF BACTERIAL CELLULOSE IN FOOD PACKAGING

The excellent functionality of bacterial cellulose, derived from its physicochemical properties, allows it to have enormous application potential. The most exploited have been from materials for biomedical use, food additives, ingredients in cosmetics, and even in the paint industry [38, 46]. In 1992, it was classified as generally recognized as safe by the FDA, the body responsible for the evaluation and approval of new substances with possible use in foods and medicines [36, 43]. As a food ingredient or additive, bacterial cellulose is used to maintain the viscosity of processed foods because it acts as a thickening and stabilizing agent. Furthermore, our body does not produce cellulase enzymes, so we cannot digest or absorb the biopolymer, that is, it acts as dietary fiber [36, 39]. As already explained, mainly using static culture, *G. xylinum* produces cellulose with better properties than that of plant origin, and the biopolymer is deposited in the form of a film on the surface of the culture. Regarding mechanical properties, we can highlight its thermal stability and barrier to water vapor, oxygen, and UV radiation; however, the films have little elasticity, which limits their use as food packaging [47]. To improve this condition, there are *in situ* (addition of supplementary materials during fermentation) and *ex situ* (direct treatments on the biopolymer) procedures to modify the structure and optimize its properties with the aim of a possible application in food packaging. *Ex situ* treatments are proposed as an excellent option to strengthen the scarce structural weaknesses of bacterial cellulose, mainly by combining it with plasticizing materials, which help mitigate its elastic weakness. Once modified, it can be used for food packaging; if the plasticizer is suitable, it even becomes an edible coating [48]. Likewise, cellulose films manage to impregnate themselves by capturing compounds in the empty spaces of their structure; these can be polymers or bioactive compounds, primarily antimicrobial agents have been tested [42]. In addition to requiring attention regarding the improvement of its mechanical properties, there are important challenges that still need to be overcome for the use of bacterial cellulose in food packaging. The main one is to achieve a production that allows better yields at a lower cost, current bioprocesses require a considerable economic investment, and the quantities of product obtained are destined for select markets, generally of high price and quality [8]. It is mainly for these reasons that we can currently only visualize it in the food packaging industry as an edible coating that increases the added value of the product. Under these conditions, it would not be profitable to think about the production of this biopolymer to be used as a temporary and waste reservoir such as chemically produced containers, bags, and packaging. Figure 8.2 shows the bioprocess of cellulose production and its application in cheese.

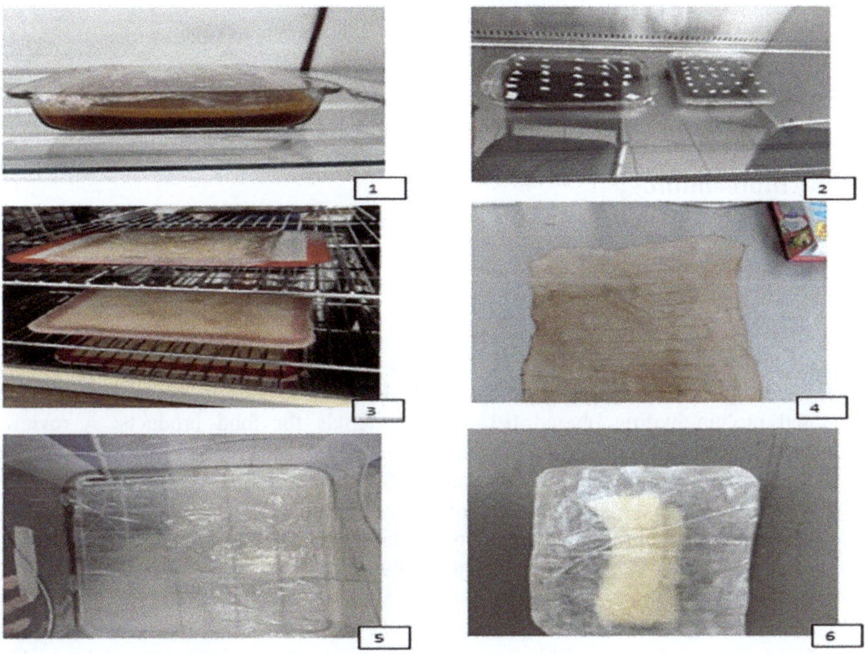

FIGURE 8.2 Bacterial cellulose production. Numbers 1 and 2 shows the process of fermentation, number 3 drying of cellulose, number 4 film manufacture, number 5 bleaching process, and number 6 application of the film in cheese.

8.8 CONCLUSION

Food packaging is an important source of environmental pollution; plastics are the main materials used for its production; due to how poorly biodegradable they are, these must be replaced by more environmentally friendly packaging. Cellulose can be used in the production of food packaging as it is the most abundant macromolecule on the planet, in addition to being able to be extracted from agroindustrial waste or produced with microorganisms. However, more research is required to ensure that sustainable packaging is affordable, more accessible, and with better functional characteristics than the current packaging.

KEYWORDS

- **microbial cellulose**
- **food packaging**
- **polymer**
- **biodegradable**
- **biotechnology**

REFERENCES

1. Sudheer, S., Bandyopadhyay, S., Bhat, R. (2023). Sustainable polysaccharide and protein hydrogel-based packaging materials for food products: A review. *International Journal of Biological Macromolecules*, 248, 125845.
2. Barone, A.S., Matheus, J.R.V., de Souza, T.S.P., Moreira, R.F.A., Fai, A.E.C. (2021). Green-based active packaging: Opportunities beyond COVID-19, food applications, and perspectives in circular economy—A brief review. *Comprehensive Reviews in Food Science and Food Safety*, 20, 4881–4905.
3. Amin, U., Khan, M.U., Majeed, Y., Rebezov, M., Khayrullin, M., Bobkova, E., Shariati, M.A., Chung, I.M., Thiruvengadam, M. (2021). Potentials of polysaccharides, lipids and proteins in biodegradable food packaging applications. *International Journal of Biological Macromolecules*, 183, 2184–2198.
4. Moghaddam, M.K., Gheshlagh, F.G., Moezzi, M. (2024). Extraction and characterization of cellulose microfibers from cornhusk for application as reinforcing agent in biocomposite. *International Journal of Biological Macromolecules*, 264, 130669.
5. Tuya, S.J., Gutiérrez, M.S., Merino, R.F. (2021). Ensayos preliminares para producción de celulosa por bacterias aisladas de caña de azúcar. *Revista de la Sociedad Química del Perú*, 87, 128–136.

6. Ramírez Tapias, Y.A., Di Monte, M.V., Peltzer, M.A., Salvay, A.G. (2022). Bacterial cellulose films production by Kombucha symbiotic community cultured on different herbal infusions. *Food Chemistry*, 372, 131346.
7. Agüero, A., Lascano D., Ivorra-Martinez J., Gómez-Caturla J., Arrieta P., Balart R. (2023). Use of bacterial cellulose obtained from kombucha fermentation in spent coffee grounds for active composites based on PLA and maleinized linseed oil. *Industrial Crops & Products*, 202, 116971.
8. Azeredo, H.M.C., Barud, H., Farinas, C.S., Vasconcellos, V.M., Claro, A.M. (2019). Bacterial cellulose as a raw material for food and food packaging applications. *Frontiers in Sustainable Food Systems*, 3, 7.
9. Al-Tayyar, N.A., Youssef, A.M., Al-Hindi, R. (2020). Antimicrobial food packaging based on sustainable bio-based materials for reducing foodborne Pathogens: A review. *Food Chemistry,* 310, 125915.
10. Porta, R., Sabbah, M., Di Pierro, P. (2020). Biopolymers as food packaging materials. *International Journal of Molecular Sciences*, 21, 4942.
11. Tan, C., Han, F., Zhang, S., Li, P., Shang, N. (2021). Novel bio-based materials and applications in antimicrobial food packaging: Recent advances and future trends. *International Journal of Molecular Sciences*, 22, 9663.
12. Mujtaba, M., Lipponen, J., Ojanen, M., Puttonen, S., Vaittinen, H. (2022). Trends and challenges in the development of bio-based barrier coating materials for paper/ cardboard food packaging; a review. *Science of the Total Environment*, 851, 158328.
13. Chen, S., Brahma, S., Mackay, J., Cao, C., Aliakbarian, B. (2020). The role of smart packaging system in food supply chain. *Journal of Food Science*, 85, 517–525.
14. Asgher, M., Qamar, S.A., Bilal, M., Iqbal, H.M. (2020). Bio-based active food packaging materials: Sustainable alternative to conventional petrochemical-based packaging materials. *Food Research International*, 137, 109625.
15. Muddasar, M., Beaucamp, A., Culebras, M., Collins, M.N. (2022). Cellulose: Characteristics and applications for rechargeable batteries. *International Journal of Biological Macromolecules*, 219, 788–803.
16. Perez, S., Samain, D. (2010). Structure and engineering of celluloses. *Advances in Carbohydrate Chemistry and Biochemistry*, 64, 25–116.
17. Lin, Z., Li, S., Huang, J. (2021). Natural cellulose substance based energy materials. *Chemistry–An Asian Journal*, 16, 378–396.
18. Ilangovan, M., Guna, V., Prajwal, B., Jiang, Q., Reddy, N. (2020). Extraction and characterization of natural cellulose fibers from *Kigelia africana*. *Carbohydrate Polymers*, 236, 115996.
19. Peter, Z. (2021). Order in cellulosics: Historical review of crystal structure research on cellulose. *Carbohydrate Polymers*, 254, 117417.
20. Mischnick, P., Momcilovic, D. (2010). Chemical structure analysis of starch and cellulose derivatives. *Advances in Carbohydrate Chemistry and Biochemistry*, 64, 117–210.
21. Kontturi, E., Tammelin, T., Österberg, M. (2006). Cellulose—model films and the fundamental approach. *Chemical Society Reviews*, 35, 1287–1304.
22. Miao, X., Lin, J., Tian, F., Li, X., Bian, F., Wang, J. (2016). Cellulose nanofibrils extracted from the byproduct of cotton plant. *Carbohydrate Polymers*, 136, 841–850.

23. Longaresi, R.H., de Menezes, A.J., Pereira-da-Silva, M.A., Barona, D., Mathias, S.L. (2019). The maize stem as a potential source of cellulose nanocrystal: Cellulose characterization from its phenological growth stage dependence. *Industrial Crops and Products*, 133, 232–240.
24. Rahmi., Patra, A., Lelifajri. (2024). Fabrication of coconut dregs residue derived nanocellulose film for food packaging. *South African Journal of Chemical Engineering*, 48, 71–79.
25. Randis, R., Darmadi, D.B., Gapsari, F., Sonief, A.A. (2024). Isolation and characterization of microcrystalline cellulose from oil palm fronds biomass using consecutive chemical treatments. *Case Studies in Chemical and Environmental Engineering*, 9, 100616.
26. Kumar, A., Dutta, D., Kalita, D., Majumdar, B., Saikia S.P., Banik, D. (2024). Extraction, physicochemical and structural characterisation of palm grass leaf fibres for sustainable and cleaner production of textile and allied cellulosic applications. *Journal of Cleaner Production*, 448, 141733.
27. Chamathka, W.W.D.R., Chai, T-T., Phuah T.T., Wong, J.X., Chen, S.N., Yassaralipour, A. (2024). Extraction and characterization of cellulose nanoparticles from palm kernel meal for potential application in active food packaging. *International Journal of Biological Macromolecules*, 260, 129637.
28. Brahma, R., Ray, S. (2024). Optimization of extraction conditions for cellulose from jackfruit peel using RSM, its characterization and comparative studies to commercial cellulose. *Measurement: Food*, 13, 100130.
29. Abzan, N., Abbasian, A., Jonoobi, M., Ghasemi, I. (2023). Cellulose microfiber extraction from leftover celery pulp: Chemomechanical treatments, structural, morphological, and thermal characterization. *International Journal of Biological Macromolecules*, 253, 126834.
30. Divakaran, D., Suyambulingam, I., Sanjay, M.R., Raghunathan, V., Ayyappan, V., Siengchin, S. (2023). Isolation and characterization of microcrystalline cellulose from an agro-waste tamarind (*Tamarindus indica*) seeds and its suitability investigation for biofilm formulation. *International Journal of Biological Macromolecules*, 254, 127687.
31. Fatima, A., Ortiz-Albo, P., Neves, L.A., Nascimento, F.X. (2023). Biosynthesis and characterization of bacterial cellulose membranes presenting relevant characteristics for air/gas filtration. *Journal of Membrane Science*, 674, 121509.
32. Liu, Z., Wang, Y., Guo, S., Liu, J., Zhu, P. (2024). Preparation and characterization of bacterial cellulose synthesized by kombucha from vinegar residue. *International Journal of Biological Macromolecules*, 258, 128939.
33. Phan, H.T., Nguyen, K.D., Nguyen, H.H.M., Dao, N.T., Le, P.T.K., Le, H.V. (2023). Nata de coco as an abundant bacterial cellulose resource to prepare aerogels for the removal of organic dyes in water. *Bioresource Technology Reports*, 24, 101603.
34. Ciftci, F. (2024). Bioadhesion, antimicrobial activity, and biocompatibility evaluation bacterial cellulose based silver nanoparticle bioactive composite films. *Process Biochemistry*, 137, 99–110.
35. Schramm, B.M., Hestrin, S. (1954). Factors affecting production of cellulose at the air/liquid interface of a culture of *Acetobacter xylinum*. *Journal of General Microbiolgy*, 11, 123–129.

36. Zhong, C. (2020). Industrial-scale production and applications of bacterial cellulose. *Frontiers in Bioengineering Biotechnology,* 8, 605374.
37. Gorgieva, S., Trček, J. (2019). Bacterial cellulose: Production, modification and perspectives in biomedical applications. *Nanomaterials*, 9, 1352.
38. Mona, S., Bajar, S., Deepak, B., Kiran, B., Kaushik, A. (2019). Microbial cellulose: production and application. In: Grumezescu, V., Mihai, G.A. (Eds.), *Materials for Biomedical Engineering: Absorbable Polymers* (pp. 316). Elsevier: Amsterdam.
39. Chawla, P.R., Bajaj, I.B., Survase, S.A., Singhal, R.S. (2009). Fermentative production of microbial cellulose. *Food Technology and Biotechnology*, 42, 107–124.
40. Tsouko, E., Kourmentza, C., Ladakis, D., Kopsahelis, N., Mandala, I., Papanikolaou, S., Paloukis, F., Alves, V., Koutinas, A. (2015). Bacterial cellulose production from industrial waste and by-product streams. *International Journal of Molecular Sciences*, 16, 14832–14849.
41. Gromet-Elhanan, Z., Hestrin, S. (1963). Synthesis of cellulose by *Acetobacter xylinum*. VI. Growth on citric acid-cycle intermediates. *Journal of Bacteriology*, 85, 284–292.
42. Masaoka, S., Ohe, T., Sakota, N. (1993). Production of cellulose from glucose by *Acetobacter xylinum*. *Journal of Fermentation and Bioengineering*, 75, 18–22.
43. Moniri, M., Boroumand Moghaddam, A., Azizi, S., Abdul Rahim, R., Bin Ariff, A., Zuhainis Saad, W., Navaderi, M., Mohamad, R. (2017). Production and status of bacterial cellulose in biomedical engineering. *Nanomaterials*, 7, 257–282.
44. Donini, I.A.N., de Salvi, D.T.B.D., Fukumoto, F.K., Lustri, W.R., Barud, H.S., Marchetto, R., Messaddeq, Y., Ribeiro, S.J.L. (2010). Biossíntese e recentes avanços na produção de celulose bacteriana. *Ecletica Quimica*, 35, 165–178.
45. Jonas, R., Farah, L.F. (1998). Production and application of microbial cellulose. *Polymer Degradation and Stability*, 59, 101–106.
46. Ha, J.H., Shehzad, O., Khan, Lee, S.Y., Park, J.W., Khan, T., Park, J.K. (2008). Production of bacterial cellulose by a static cultivation using the waste from beer culture broth. *Korean Journal of Chemical Engineering*, 25, 812–815.
47. Retegi, A., Gabilondo, N., Peña, C., Zuluaga, R., Castro, C., Gañan, P., de la Caba, K., Mondragon, I. (2010). Bacterial cellulose films with controlled microstructure–mechanical property relationships. *Cellulose*, 17, 661–669.
48. Cazón, P., Vázquez, M. (2021). Improving bacterial cellulose films by ex-situ and in-situ modifications: a review. *Food Hydrocolloids*, 113, 106514.

CHAPTER 9

Agave By-Products as a Source of Biodegradable Packaging Materials

ROGELIO SÁNCHEZ-VEGA,[1] MIROSLAVA CARRERA-ROMERO,[1] JUAN MANUEL TIRADO-GALLEGOS,[1] RUTH LECHUGA-VALLES,[1] and MARÍA JANETH RODRÍGUEZ-ROQUE[2*]

[1]Faculty of Zootechnics and Ecology, Autonomous University of Chihuahua, Chihuahua, Mexico

[2]Faculty of Agrotechnological Sciences, Autonomous University of Chihuahua, Chihuahua, Mexico

*Corresponding author

ABSTRACT

Agrifood waste can be used to produce biodegradable polymers to minimize losses and wastes along the agroindustrial chain. The use of renewable resources to obtain bio-based polymers generates a minimal impact on the environment compared to the use of synthetic plastics. Agricultural wastes such as pomace, skin, and peel of fruits, as well as husks and straw of cereals, and *Agave* can be used as carbon-rich precursors to produce bio-based polymers through physical, chemical, or biological methods. *Agave* waste, primarily generated during tequila, mezcal, or "pulque" production, can be upcycled to create environmentally safe, biodegradable, and compostable products for food packaging and other applications with notable properties such as tensile strength, water permeability, and ultraviolet (UV) absorbance. *Agave* waste can also be used as a source of prebiotic polymers, which have technological applications in food and can provide beneficial health effects. *Agave* products aim to minimize

Sustainable Food Packaging Materials and Technology. Juan Manuel Tirado-Gallegos, Juan Daniel Rivaldi Chávez, and José Juan Buenrostro-Figueroa (Eds)

DOI: 10.1201/9781779641434-9

traditional single-use plastics and support a low-carbon circular economy. Some companies have developed a line of products made from upcycled *Agave* materials, including biodegradable and durable eco-friendly plastic replacements for food service.

9.1 INTRODUCTION

Plastic pollution is widely distributed in farms, soils, deserts, oceans, and sediments. The presence of plastic waste in rivers, seas, and oceans is estimated to range between 9 and 23 million metric tons per year, without taking into account soil contamination, which reaches similar amounts (13 and 25 million metric tons yearly) [1, 2]. The rapid proliferation of plastic waste is due to a growth in production and the lack of measures to prevent pollution and efficient recycling, as well as the sharp increase in the consumption of single-use plastic, and a growing "throw-away" culture has increased the severity of the problem. In addition, the increment in plastic pollution is due to poor waste management capacity at a global level [3].

Due to the negative effect of the presence of plastics on the environment, there is a need to propose sustainable alternatives that contribute to the reduction of pollution and its effects on climate change. Plastic pollution requires several targeted strategies, such as reduction, reuse, and recycling besides novel polymer material design [2–4]. *Agave* by-products have emerged as a promising source of sustainable alternatives to conventional plastics, particularly in the production of biodegradable packaging materials. *Agave* is a plant native to Mexico and the southwestern United States, commonly used to produce tequila and other spirits [5]. The *Agave* industry generates significant amounts of by-products, including leaves, bagasse, and fibers, which can be revalorized for various applications, including packaging materials [5, 6]. *Agave* fibers are rich in cellulose, which can be transformed into cellulose nanoparticles, a key component in the production of biodegradable packaging materials. The transition of cellulose crystalline structure in biodegradable packaging materials is a promising approach for utilizing *Agave* by-products in sustainable packaging [7, 8]. Recent studies have focused on the potential application of *Agave*-based packaging materials with the potential to replace single-use plastics in various applications, such as flexible films, rigid materials, coatings, and adhesives [9–11].

The use of *Agave* by-products in the production of biodegradable packaging materials not only provides a sustainable alternative to conventional plastics but also helps address the environmental concerns associated with the disposal of agrowaste. By revalorizing agrowaste to generate bioplastics, the food packaging production chain can be made more sustainable, and, thus, the *Agave* industry can contribute to the reduction of plastic waste, the mitigation of greenhouse gas emissions, and the promotion of sustainable food packaging solutions [12].

The purpose of this document is to review the perspective of *Agave* by-products as a potential source of biodegradable packaging materials involving exploring the potential of utilizing *Agave* waste, a by-product of the spirit industry, in creating sustainable and eco-friendly packaging solutions.

9.2 AGAVE PLANTS AND THEIR BY-PRODUCTS

Agave plants are native to the arid regions of America, and they belong to the genus of monocotyledons of the order *Asparagales* within the *Agavaceae* family. The genus name *Agave* comes from the ancient Greek *agauos*, *agavos*, meaning "admirable, illustrious, noble" associated with spikes of large flowers distributed in many species [13], and it consists of around 200 species and 247 taxa, of which 150 species or 75% are found in Mexico. *Agave* plants are formed of large leaves with spines (used to extract sisal-type fibers) and "piña" (the raw material to produce tequila and mezcal, among other alcoholic beverages) [14–17].

The preparation of fermented beverages, as well as a source of foods, dietary supplements, antimicrobial agents, anti-inflammatory agents, biofuel, and fiber, are some examples of the uses and applications of *Agave*. *Agave* plants are also used for ornamental purposes in dry climates. *Agave* fibers are extracted for making rope, cloth, and netting, while thorns are used as sewing needles and for medicinal applications [17–20]. Pre-Hispanic cultures in Mexico used *Agave* derivatives for medicinal purposes, thanks to their therapeutic benefits [21].

Agave has also been used for creating handcrafted items, scaffolding, stools, and ladders and to reinforce cement in the construction industry, which influences its mechanical and durability properties [20].

Agave is a raw material for bioethanol production. Bioethanol can be produced from agroindustrial residues, which is carried out through the digestion of lignocelluloses to release sugars, a process that allows for the fermentation of cellulose and hemicellulose to fuels obtention [6, 22].

Another use of *Agave* plants includes the preparation of different Mexican dishes. Maguey leaves are toasted and used as material to wrap roast meat. The cuticle commonly called mixiote is carefully separated from the leaves, and it is used in various preparations based on different meats with sauce and steamed [6, 23]. The *Agave* plant produces edible flowers during its final season, and the stalks, leaves, and sap (aguamiel) are used as food sources [17]. In addition, *Agave americana* leaves powder was added to yogurt to improve its color, texture, and viscosity [24].

The sweet sap of the plant is fermented to get pulque, an alcoholic beverage used in celebrations. Mezcal, the first distilled spirit made from *Agave*, held a similar place in Mexican culture, figuring prominently in celebrations too [17]. The *Agave*-based industries, particularly tequila and mezcal, play a significant role in Mexico's economy and culture. These industries have experienced substantial growth in recent years, with tequila production increasing by 210% in the last decade and mezcal production reaching 8.1 million liters during 2021 [25]. However, the by-products generated during the production of *Agave* spirits, such as the solid and liquid residues, are highly polluting, resistant to decomposition, and potentially toxic to aquatic life if not properly managed. To address these environmental challenges, farmers and researchers are developing creative and sustainable solutions to reuse the waste generated during *Agave* spirit production. These solutions aim to protect the environment, create jobs along the value chain, and make mezcal production environmentally friendly [18, 26].

Agave has been used for food and fiber for thousands of years, and it was vital for ancient Mesoamerican people. *Agave* plants and their by-products have a diverse range of applications, including their use in packaging materials [14, 18, 19]. These by-products can be revalorized for their high-value components. For instance, cellulose from *Agave* is hydrolyzed to obtain simple sugars, which can then be used to create nanoparticles and bioplastics. The leaves and bagasse of *Agave* have been utilized to produce cellulose-based materials, with leaves showing higher crystallinity than other tissues [27]. Nanocellulose is obtained from sources that contain cellulose, such as *Agave*. The nanocellulose-based biocomposites are then incorporated into a biopolymer matrix to improve the overall properties of the composite.

9.3 AGAVE BY-PRODUCTS

The *Agave* transformation industry, especially the *Agave*-based alcohol industry, generates a great quantity of organic residues, such as leaves, bagasse, and fibers, which offer a wide array of uses beyond traditional applications (Figure 9.1). These by-products can be revalorized for their high-value components. For instance, cellulose from *Agave* can be hydrolyzed to obtain simple sugars, which can be used to create nanoparticles and bioplastics. The leaves and bagasse of *Agave* have been utilized to produce cellulose-based materials, with leaves showing higher crystallinity than other tissues. Additionally, *Agave* by-products have been explored for their nutraceutical value, showcasing antioxidant, anti-inflammatory, antibacterial, and antifungal properties [6, 27].

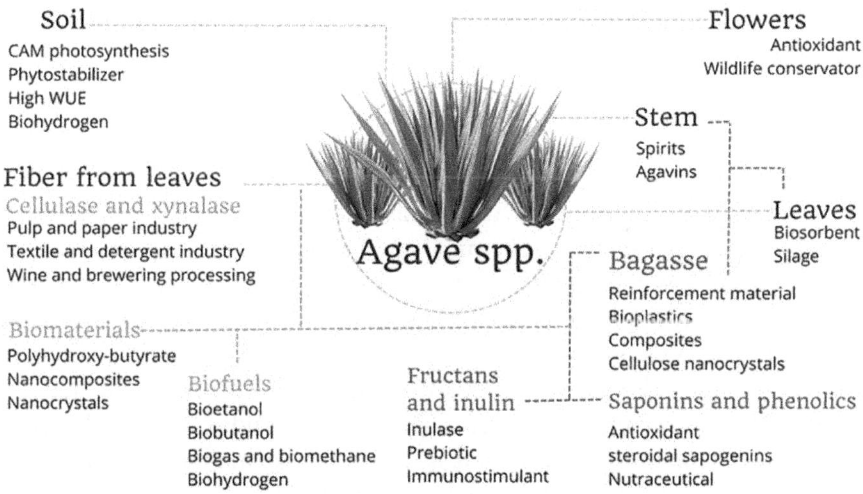

FIGURE 9.1 Common by-products obtained from *Agave* plant.

Agave fibers are composed of cellulose, hemicellulose, and lignin, among other polysaccharides. The chemical composition of natural fibers depends on various factors such as the type of soil used, weather conditions, and the age of the plants. Table 9.1 exhibits the chemical composition of some *Agave* plants. Bagasse and other *Agave* residues contain cellulose, hemicellulose, lignin, pectin, fructose, glucose, gums, sucrose, and saponins, with varying proportions depending on the *Agave* variety [6].

TABLE 9.1 Traditional Uses of *Agave* Species and Chemical Composition of Their Residues

Traditional Products	Agave Species	Type of Residue	Cellulose	Hemicellulose	Lignin	References
			Component (%)			
Tequila	*Agave tequilana*	Bagasse	42	22	18	[28]
		Leaves	38.9–54.5	15.2–19.7	9.8–16.3	[13, 29]
Mezcal	*Agave potatorum*	Bagasse	24 ± 1.2	15 ± 0.7	14 ± 0.9	[27]
	Agave karwinskii	Bagasse	30 ± 2	20 ± 1	14 ± 0.5	
	Agave americana var. *oaxacensis*	Bagasse	28 ± 1.5	16 ± 0.5	15 ± 1.1	
	Agave salmiana	Bagasse	35	4.6	13.0–19.1	[30]
		Leaves	20.7	3.7	9.5–26.1	[29, 31]
Pulque, fiber, saponins	*A. americana*	Leaves	65.2–68.5	5.6–18.4	2.7–9.1	[29, 32]
Fiber	*Agave lechuguilla*	Leaves	79.8	3–6	15.3	[33]
	Agave fourcroydes	Leaves	77.6	5–7	13.1	[34]
		Leaves	72	24	14	[35]

9.3.1 CELLULOSE

It is a polymer consisting of D-anhydro-glucopyranose units joined by β-1,4-glycosidic bonds. The chemical and physical properties of cellulose are determined by the molecular structure [36]. Cellulose is the principal component of *Agave* plants, which is mainly composed of glucans, second hemicellulose (xylans) and lignin. Cedeño-Cruz [37] reported a cellulose content of 43% in bagasse from *A. tequilana* (var. Weber, from the tequila industry). A lower cellulose content (between 20% and 30%) was observed in fibers of *A. salmiana*. Nonetheless, the cellulose augmented until 65% after heat pretreatment (121°C for 15–45 min) [38].

The length of the polymer chains varies according to the type of natural fiber, expressed as the degree of polymerization. In this sense, the type and geometric conditions of the cellulose affect the mechanical properties. Cellulose possesses various systems of hydrogen bonds, which could affect the low solubility in diverse solvents, the reactivity of the hydroxyl groups, and the crystallinity of cellulose samples. Moreover, cellulose also contains hydrophobic areas that have a certain influence on solubility [39].

Intermolecular hydrogen bonding is responsible for the strong interaction between cellulose chains. In this sense, it has been reported that there are three types of cellulose: I, II, and III. The difference between cellulose I and II is that in the latter, molecules are more densely packed and strongly inter-bonded, and, therefore, cellulose II is less reactive. Cellulose III is formed by treating cellulose I and II with liquid ammonia and certain amines [39].

Most authors determined higher characteristic values for type I than type II. Solid cellulose forms a microcrystalline structure with high-order regions (crystalline regions) and low-order regions (amorphous regions). Natural cellulose (cellulose I) crystallizes in monoclinic sphenodic structures [19, 36]. The crystallinity of cellulose in *Agave* leaves might be undesirable for some applications; however, through adequate pretreatment, nanocrystals with a high degree of crystallinity can be obtained. Cellulose can be hydrolyzed to obtain simple sugars that serve as building blocks for various applications, such as nanoparticles and bioplastics [6].

9.3.2 HEMICELLULOSE

Cellulose contains only 1,4-β-D-glucopyranose units, whereas hemicellulose contains several different sugar units; it exhibits a considerable degree of chain branching, and the degree of polymerization of native cellulose is 10–100 times higher than that of hemicellulose [36]. Hemicelluloses include xyloglucans, xylans, mannans, glucomannans, and β-(1→3, 1→4)-glucans [40]. The content of hemicellulose in *Agave* fibers has been reported between 10% and 13% [41].

Hemicellulose can be hydrolyzed to produce xylo-oligosaccharides (dietary fiber) that have prebiotic properties and improve the stability and texture of foods [42, 43]. Moreover, hemicellulose is used in the production of various materials (emulsifiers, films, and hydrogels) and some chemicals such as xylitol, ethanol, and furfural, among others for food applications [44]. Fructooligosaccharides (including oligosaccharides such as 1-ketose, nystose, and 1 F-β-fructofuranosyl nystose) are also present in *Agave* by-products, which are obtained from inulin hydrolysis [19, 43].

9.3.3 LIGNINS

They are complex hydrocarbon polymers with aliphatic and aromatic constituents [36]. Lignin has no regular repeating pattern, as in the case of cellulose. In lignin, the monolignol subunits are randomly connected by various linkages [45]. The content of lignins in *Agave* fibers ranged from 21% to 24% [37]. Spines are an important source of hemicellulose [35]. It has also been reported that lignins have antioxidant and prebiotic properties [19].

9.3.4 FUNCTIONAL PROPERTIES OF AGAVE BY-PRODUCTS

Scientific findings have demonstrated that *Agave* has beneficial effects in the treatment of diseases of bacterial etiology, gastrointestinal and wound infections, urologic disorders, dysentery, cancer, diabetes, anti-inflammatory, and hypertension, among others [46–51]. *Agave* and their subproducts are an important source of plant-derived compounds, defined as natural or synthetic compounds with biological activity in the living tissue producing beneficial effects [52].

As shown in Table 9.2, *Agave* by-products have been found to possess various functional properties due to their rich chemical composition. These by-products are composed of cellulose, hemicellulose, lignin, and fructans, as well as bioactive compounds, including phenolic compounds, carotenoids, and dietary fibers. In this respect, it has been demonstrated that bioactive compounds contained in *Agave* by-products exhibited antimicrobial, antioxidant, anticancer, and coadjuvant in mineral absorption, antifungal, cardioprotective, neuroprotective, and anti-inflammatory properties [46, 53]. The functional properties of *Agave* fibers depend on the particle size, extraction conditions, and structure of the plant polysaccharide [21].

Agave by-products have been also used in the production of biomethane, hydrogen, and methane through biological, enzymatic, chemical, and hydrothermal pretreatments [21].

9.4 TECHNOLOGIES USED FOR PROCESSING AND EXTRACTION OF AGAVE FIBERS

9.4.1 PRETREATMENT

Agave fibers can be extracted using various methods, including traditional and modern techniques. Pretreatments of *Agave* by-products are critical for cellulose depolymerization; it is due to pretreatment making cellulose more susceptible to be hydrolyzed [62].

TABLE 9.2 Major Applications in Biopolymers made of Diverse *Agave* By-Products

Agave By-Products	Materials	Application	References
Fibers (*A. tequilana* Weber var. Azul)	Composites	*Agave* fibers generate low-cost and compostable biocomposite, with lower disintegration rate and density useful for the fabrication of molds and three-dimensional (3D)-printing support materials	[54]
Fibers (*A. tequilana* Weber var. Azul)	Composites	*Agave* fiber addition to polyhydroxybutyrate (PHB) and poly(3-hydroxybutyrate-co-3-hydroxyvalerate) [P(HB-HV)] biopolymers produces composites with enhanced properties and potentially helps to cost reduction	[41]
Bagasse (*A. tequilana*)	Composites	Agave fiber/polylactic acid (PLA) composites produce consistent and repeatable tensile, flexural, impact, and water absorption properties	[55]
Fibers (*A. tequilana*)	Bioplastic	Hybridization reduced the materials cost and gave them better mechanical properties	[56]
Fibers (*A. tequilana*)	Automotive composites	The results from mechanical and thermal characterization suggested that *Agave* fiber fillers can be used to enhance desired properties in biocomposites	[57]
Fructan (*A. tequilana* Weber var. Azul)	Probiotic	Inulin-type fructan from chicory and mixed-type fructan from *Agave* exerted a significant growth enhancement	[58]
Dried boles (*Agave sisalana*)	Probiotic	Extract of boles from *A. sisalana* exerts a prebiotic effect linked to its significant sugar content; therefore, it could be an economically viable material for both the pharmaceutical and food industries	[59]
Fructan (*A. tequilana* Weber var. Azul)	Bioplastic	*Agave* fibers withstand high-stress levels being suitable for packaging heavy or fragile items and avoiding damage during transportation and storage	[60]
Fibers (*A. tequilana*)	Bioplastic	*Agave* fibers in conjunction with other botanical sources have been of interest for the development of biodegradable packaging with hydrophobic and resistant properties	[61]

The traditional process involves drying the leaves, boiling them (or water retting them), and then removing the nonfibrous materials by squashing leaves or mechanical scraping [63, 64]. A more recent method consists of immersing the blades of the *Agave* leaves in water for 10–13 days in a container [65]. The fibers can also be obtained without using water by burying the leaves in the ground for three months, resulting in better mechanical properties. Some authors have used a pectinase solution or a hydrolysis process for fiber extraction [63].

It has been observed that the chemical composition and mechanical properties of the fibers (fiber length and diameter) are significantly influenced by the extraction process [64].

9.4.2 TECHNOLOGIES APPLIED IN THE EXTRACTION OF AGAVE FIBERS

9.4.2.1 MECHANICAL EXTRACTION

Mechanical extraction consists of crushing and beating the leaves, so that only fibers remain. People in rural areas performed decortication by hand, crushing the leaves, and scraping the pulp with a knife, which is time-consuming and needs manpower. Mechanical extraction produces long and coarse fibers suitable for ethnic clothing and strings, and it does not require chemicals or water, making it environmentally friendly [66, 67].

When the decortication and washing are finished, the fibers are subjected to a drying process (in mechanical dryers or the sun). Mechanical extraction has the disadvantage of being inefficient in removing cementing compounds between the fibers, such as waxes, hemicellulose, and lignin, leading to lower yield [66, 67].

9.4.2.2 SOLVENT CASTING

Solvent casting is a simple technique based on dissolving the polymer (using water as solvent) matrix at low temperature. To produce homogeneous compounds, nanocellulose is added with mechanical stirring, combined with ultrasound to reduce bubble formation. Subsequently, by evaporation of the solvent, the film is obtained [68].

9.4.2.3 CHEMICAL EXTRACTION

Fiber extraction by chemical methods involves the use of acids, alkalis, and enzymes [66]. Chemical extraction can extract fibers efficiently and cost-effectively and improve fiber properties by removing cementing materials. Fiber extraction with alkaline solutions favors the dissolution of lignins present between the fibers. In this way, the contact surface and the degree of polymerization increase, and the breaking resistance of the fibers is reduced [66]. In this sense, Barcelos et al. [69] evaluated the effect of delignification pretreatment with dilute NaOH upon glucose yield. The authors reported that increasing concentrations of NaOH showed the positive effect of this pretreatment on enhancing cellulose digestibility.

A greater degree of crystallinity of the fibers is obtained when acid solutions are used, which hydrolyze lignin and hemicellulose into shorter-chain pentose molecules [66].

9.4.2.4 ULTRASOUND

Ultrasonication (US) technology is based on the cavitation effect caused by ultrasonic energy, an acoustic wave that oscillates at frequencies above 16 kHz. In US processing, two ranges of ultrasound are distinguished: (1) high frequency (2–10 MHz) and (2) low frequency (20 to around 100 kHz) [70, 71]. Most US pretreatment takes place at frequencies below 50 kHz, leading to delignification and extraction of high-added-value chemicals [72].

Solís-Rosales et al. [73] evaluated the influence of US processing on the thermal, morphological, and structural properties of *A. tequilana* fibers. A reduction of 35% in lignin content was registered as a consequence of hydrogen bonds breaking in the structural components of the fiber. In addition, after US treatment, a morphological change (high roughness and porosity) in fiber was observed. Sonication for 60 min and immersion time of 12 h caused the most effective method to enhance the fiber properties.

An ultrasound between 30 and 60 min at 42 kHz and 132 W was applied to *Agave durangensis* leaves. The time has a significant influence on the lignin and holocellulose content. This was justified by the fact that the US improved the fructans and phenolic compounds [33]. In this context, US pretreatment induced lignin elimination when US pretreatment (amplitude 60%, 30 min) was applied. In addition, lignin removal was significantly improved to 47% in comparison to that obtained without US pretreatment (30.92%) [74]. In general, US pretreatment requires lower process time and temperature, and higher delignification occurs [72].

Ríos-González et al. [75] evaluated the influence of pretreatments with alkaline solutions and ultrasound on the elimination of hemicellulose and *Agave* digestibility. The authors reported that 28% of hemicellulose was removed when 4.1% NaOH, 27 min, 60°C, and 50% sonication amplitude were applied.

9.4.2.5 *MICROWAVES*

Microwave is a form of electromagnetic wave energy, ranging from 300 MHz to 300 GHz in the electromagnetic wave spectrum. Microwave can be classified as an electrothermal process for pasteurization of food. Penetration in food is in function on the wavelength, depth of material, composition, geometry, and structure of food [70, 76].

Rosas-Hernández et al. [77] observed a cellulose content of 42.76% in bagasse of *Agave angustifolia* haw when organosolv fractionation assisted (40% ethanol for 1.5 h) with microwaves was applied.

It was reported that microwave-assisted pretreatment removed a higher content of *Agave* hemicellulose (77.5%) than that performed with US (28.2%) [75]. Furthermore, microwave pretreatment was more efficient than the US in reducing the leaves' recalcitrance and improved saccharification [6].

9.5 APPLICATIONS OF AGAVE-BASED PACKAGING

As was previously mentioned, *Agave* plants and their by-products have a diverse range of applications, including their use in packaging materials. The incorporation of compounds extracted from *Agave* led to the development of novel materials with structural potential. This approach reduces production costs through the revalorization of agroindustrial by-products for the design and reformulation of food packaging.

Agave fiber serves as a renewable resource that can be used to create bioplastic materials, offering a sustainable alternative to traditional plastics and reducing CO_2 emissions. Overall, utilizing *Agave* for food packaging supports environmental conservation and promotes a more sustainable approach to packaging solutions [6, 78].

The properties of *Agave* films are significantly affected by processing techniques. Singha and Rana [79] used lignocellulosic fibers extracted from leaves of wild *Agave* plant (*A. americana*) as a reinforcement in the preparation of polystyrene matrix-based biocomposites using compression molding technique. They reported that composites reinforced with short *Agave* fibers show better tensile strength, better compressive strength, and a similar trend in flexural properties when compared to neat polystyrene resin. In this line, another study analyzed the effect of *Agave* fibers on the thermal properties of high-density polyethylene (HDPE), linear low-density polyethylene (LLDPE), and polypropylene (PP) matrices. The results showed that the addition of *Agave* fibers did not significantly affect the overall thermal properties of the composite materials, but the elastic modulus and yielding stress of the composite materials were generally proportional to the fiber content [57].

Muñiz-Márquez et al. [80] found that thermal processing can affect the sugar composition of Mexican *Agave* syrup. Specifically, they found that the thermal effect can lead to the formation of new sugars in the syrup, which can affect its physicochemical properties. In addition, it was reported that *Agave* and its derivatives, such as *Agave* syrup, were reported to be a source of nutritional compounds, including polyphenols and antioxidants. However, the extraction and processing methods used can affect the concentration and composition of these bioactive compounds in the final product [81].

9.5.1 BENEFITS OF USING AGAVE AS A FOOD PACKAGING MATERIAL

Cellulose fibers can absorb water due to the network of fibers linked by intermolecular and intramolecular bonds, formed by the hydroxyl groups of cellulose [82].

The advantages of using *Agave*-based packaging materials are numerous and contribute to generating sustainable and ecological packaging.

- *Reduced carbon footprint*: By incorporating *Agave* into packaging materials, companies can reduce waste and lower their carbon footprint. *Agave*-based packaging helps in minimizing greenhouse gas emissions, contributing to a healthier environment and offering a reduced carbon footprint compared to traditional packaging materials [6, 4].
- *Biodegradability*: *Agave* fibers used in packaging materials are biodegradable, unlike synthetic fibers, which help reduce waste

and do not contribute to microplastic pollution. This biodegradability supports a circular economy, promotes waste reduction, and offers superior strength compared to traditional packaging materials [82].

- *Strength and durability*: *Agave*-based food packaging products, such as bagasse trays and containers, offer durability and resistance to moisture and grease, making them suitable for a variety of food products [68].
- *Cost-effectiveness*: *Agave*-based packaging solutions can be cost-effective due to their lightweight nature, which reduces shipping costs and can lead to immediate savings for businesses [68].
- *Versatility*: *Agave* fiber, used in packaging materials, offers versatility in applications, making it a flexible and practical choice for various food packaging needs [83].
- *Compostability*: *Agave*-based food packaging products are compostable within a short period, contributing to waste reduction and supporting a circular economy model [84].

9.5.2 PROPERTIES OF AGAVE FILMS

The mechanical properties of *Agave* films include the elastic modulus, yielding stress, and mechanical strength. Studies have shown that the addition of *Agave* fibers to thermoplastic composite films can enhance the mechanical properties of the films. Specifically, the elastic modulus and yielding stress of the composite films are generally proportional to the fiber content, indicating that the mechanical resistance of the films increases with higher fiber content [57, 85]. *Agave* fibers have been found to increase the mechanical strength and elasticity of biocomposite materials, contributing to improved mechanical properties of the films. Additionally, the morphology of the cross-section of *Agave* fiber-reinforced films can impact their mechanical properties, with factors such as fiber length and orientation influencing the overall mechanical strength of the films [57].

9.5.3 PHYSICAL CHARACTERISTICS OF AGAVE FIBERS

Generally, natural fibers are hydrophilic, such as observed in fibers of *A. americana*. Natural cellulosic fibers have characteristic physicochemical properties, such as high moisture absorbency (up to 9%), low density (1.36

g/cm^3), and high tenacity (from 16 to 41 cN/tex). Cellulosic fibers are flexible, smooth, and shiny [66, 67].

Some fiber characteristics were described by Hulle et al. [66] given as follows:

- *Fiber morphology*: The average diameter of fibers is around 3.1 μm, and the length of a spiral side is about 10.1 μm. The thickness is in the range of 100–150 μm. The average diameter of *Agave* fibers is smaller than those of flax, sisal, and alfa. This structure is used to explain the mechanical properties of fibers [66].
- *Tensile properties*: The tenacity of *A. americana* fiber ranges between 16 and 41 cN/tex. The elongation of the fiber when subjected to a force is in the range of 2%–4%. *A. americana* fiber is rigid and has low elongation [66].

9.5.4 THERMAL AND TENSILE PROPERTIES OF AGAVE FILMS

Agave films, being biopolymers, exhibit specific thermal characteristics that impact their applications. The inclusion of *Agave* gel in starch-based films can alter their thermal resistance, which is essential for various applications such as medication administration, tissue engineering, biomedicine, and food packaging [86]. Biopolymers such as cellulose, starch, chitosan, and PLA have been extensively studied for their thermal properties, including thermal stability, degradation steps, glass transition temperature, and crystallinity, which are key factors in determining their performance in different environments [86].

Kestur et al. [12] evaluated the thermal stability of blue *Agave* bagasse fiber, which took place between 100 and 220°C. The authors associated this high thermal stability with the high value of crystallinity (70%) and high lignin content of *Agave* fiber.

The thermal properties of *Agave* fiber-reinforced thermoplastic composite films have been investigated, revealing that the addition of *Agave* fibers generally does not significantly affect the thermal properties of the composite materials. However, the elastic modulus and yielding stress of these composite films are typically proportional to the fiber content, indicating a relationship between fiber content and mechanical properties rather than thermal properties [57].

The specific thermal properties can vary depending on the type of biopolymer used and the processing conditions. Annandarajah et al. [57] investigated the effect of *Agave* fibers on the thermal properties of HDPE, LLDPE, and PP matrices. The authors reported that the addition of *Agave* fibers did not significantly affect the overall thermal properties of the composite materials, but the elastic modulus and yielding stress of the composite materials were generally proportional to the fiber content. Therefore, the specific thermal properties can vary depending on the type of biopolymer used and the processing conditions.

In this respect, Syafri et al. [60] studied the influence of ultrafine grinding and ultrasonic treatment on the mechanical and thermal properties of polyvinyl alcohol (PVA) films with *Agave gigantea* biocomposite. The findings demonstrated that the ultrafine grinding and ultrasonic treatment improved the mechanical and thermal properties of the PVA films.

The thermal properties of *Agave* films can be affected by factors such as processing conditions (ultrafine grinding and ultrasonic treatment can alter the mechanical and thermal properties of the films), the type of biopolymer used, the presence of *Agave* fibers, and the composition of the composite materials. Understanding these thermal properties is essential for optimizing the performance and applications of *Agave*-based films in various industries [57].

The potential drawbacks of using *Agave* films for thermal applications include their limited thermal stability and potential degradation at high temperatures. *Agave* fibers have been shown to increase the mechanical strength and elasticity of biocomposite materials, but the addition of natural fillers such as *Agave* fibers does not affect its thermal properties. Therefore, *Agave* films may not be suitable for applications that require high thermal stability or resistance to heat [57].

9.6 CHALLENGES AND FUTURE TRENDS

The current challenges in the development and commercialization of *Agave*-based packaging materials include the need to ensure that these materials meet the quality standards required for food packaging (durability, barrier properties, and food safety, among others). Despite theoretical advances in the use of *Agave* derivatives, there are important challenges regarding how these materials can be used effectively in commercial products. Additionally, of the global production of plastics, the generation of biodegradable packaging barely represents 1%, which indicates the need for more studies to help improve and optimize the use of materials of biological origin for the manufacture of packaging.

The challenges in scaling up *Agave*-based packaging production include the requirement for significant investment in infrastructure and technology to enhance production capacity and the necessity to address supply chain complexities related to sourcing *Agave* feedstocks and ensuring consistent quality. In addition, regulatory compliance, market acceptance, and the need for further research and development to optimize the performance and properties of *Agave*-based packaging materials for widespread commercial viability are required. The most promising applications of *Agave*-based packaging materials in the future include the development of biodegradable and edible packaging materials for food and beverage products, which can help reduce waste and improve sustainability.

9.7 CONCLUSION

Agave by-products, such as *Agave* leaves, bagasse, and fibers, are often considered waste and discarded, causing serious environmental pollution problems. These by-products can be revalued for various applications, especially its use for packaging design. Studies were cited that reported the use of *Agave* fibers to reinforce and produce polymer-based packaging. An example of this was a bioplastic made from tequila waste, specifically *Agave* fibers, which demonstrates the versatility and potential of *Agave* by-products as a sustainable source of bioactive compounds for various industrial applications. These functional properties make *Agave* a valuable resource for developing innovative packaging solutions. Another potential application of *Agave* by-products in packaging is the production of nanoparticles. The abundance of cellulose in *Agave* makes it a suitable source for developing nanoparticles that can be integrated into packaging materials.

Overall, *Agave* by-products offer a sustainable and versatile source of bioactive compounds and functional properties that can be harnessed for the development of innovative and eco-friendly packaging materials with additional health benefits and improved sustainability profiles. The use, application, and development of *Agave* by-products in packaging materials must also consider regulatory compliance, including standards on the effect that these packagings have when coming into contact with food.

KEYWORDS

- *Agavaceae*
- **biodegradable packaging**
- **bioplastics**

REFERENCES

1. MacLeod, M., Arp, H.P.H., Tekman, M.B., Jahnke, A. (2021). The global threat from plastic pollution. *Science*, 373, 61–65. doi: 10.1126/science.abg5433.
2. Thushari, G.G.N., Senevirathna, J.D.M. (2020). Plastic pollution in the marine environment. *Heliyon*, 6, e04709. doi: 10.1016/j.heliyon.2020.e04709.
3. Lau, W.W.Y., Shiran, Y., Bailey, R.M., Cook, E., Stuchtey, M.R., Koskella, J., Velis, C.A., Godfrey, L., Boucher, J., Murphy, M.B., Thompson, R.C., Jankowska, E., Castillo, A.C., Pilditch, T.D., Dixon, B., Koerselman, L., Kosior, E., Favoino, E., Gutberlet, J., Baulch, S., Atreya, M.E., Fischer, D., He, K.K., Petit, M.M., Sumaila, U.R., Neil, E., Bernhofen, M.V., Lawrence, K., Palardy, J.E. (2020). Evaluating scenarios toward zero plastic pollution. *Science*, 369, 1455–1461. https://www.science.org.
4. Singh, N., Ogunseitan, O.A., Wong, M.H., Tang, Y. (2022). Sustainable materials alternative to petrochemical plastics pollution: A review analysis. *Sustainable Horizons*, 2, 100016. doi: 10.1016/j.horiz.2022.100016.
5. Robles, E., Fernández-Rodríguez, J., Barbosa, A.M., Gordobil, O., Carreño, N.L.V., Labidi, J. (2018). Production of cellulose nanoparticles from blue agave waste treated with environmentally friendly processes. *Carbohydrate Polymers*, 183, 294–302. doi: 10.1016/j.carbpol.2018.01.015.
6. Álvarez-Chávez, J., Villamiel, M., Santos-Zea, L., Ramírez-Jiménez, A.K. (2021). Agave by-products: An overview of their nutraceutical value, current applications, and processing methods. *Polysaccharides*, 2, 720–743. doi: 10.3390/polysaccharides2030044.
7. Espino, E., Cakir, M., Domenek, S., Román-Gutiérrez, A.D., Belgacem, N., Bras, J. (2014). Isolation and characterization of cellulose nanocrystals from industrial by-products of *Agave tequilana* and barley. *Industrial Crops and Products*, 62, 552–559. doi: 10.1016/j.indcrop.2014.09.017.
8. Robles, E., Salaberria, A.M., Herrera, R., Fernandes, S.C.M., Labidi, J. (2016). Self-bonded composite films based on cellulose nanofibers and chitin nanocrystals as antifungal materials. *Carbohydrate Polymers*, 144, 41–49. doi: 10.1016/j.carbpol.2016.02.024.
9. Chatterjee, B., Mazumder, D. (2024). A critical review of the advances in valorizing agro-industrial wastes through mixed culture fermentation. *Journal of Environmental Chemical Engineering*, 12, 111838.

10. Hernández, J., Romero, V., Escalante, A., Toríz, G., Rojas, O.J., Sulbarán, B. (2018). *Agave tequilana* bagasse as source of cellulose nanocrystals via organosolv treatment. *BioResources*, 13, 3603–3614. doi: 10.15376/biores.13.2.3603-3614.
11. Schutz, G.F., de Ávila Gonçalves, S., Alves, R.M.V., Vieira, R.P. (2024). A review of starch-based biocomposites reinforced with plant fibers. *International Journal of Biological Macromolecules*, 261, 129916. doi: 10.1016/j.ijbiomac.2024.129916.
12. Kestur G., S., Flores-Sahagun, T.H.S., Dos Santos, L.P., Dos Santos, J., Mazzaro, I., Mikowski, A. (2013). Characterization of blue agave bagasse fibers of Mexico. *Composites Part A: Applied Science and Manufacturing*, 45, 153–161. doi: 10.1016/j.compositesa.2012.09.001.
13. Rijal, D., Vancov, T., McIntosh, S., Ashwath, N., Stanley, G.A. (2016). Process options for conversion of *Agave tequilana* leaves into bioethanol. *Industrial Crops and Products*, 84, 263–272. doi: 10.1016/j.indcrop.2016.02.011.
14. López-Romero, J.C., Ayala-Zavala, J.F., González-Aguilar, G.A., Peña-Ramos, E.A., González-Ríos, H. (2018). Biological activities of Agave by-products and their possible applications in food and pharmaceuticals. *Journal of the Science of Food and Agriculture,* 98, 2461–2474. doi: 10.1002/jsfa.8738.
15. Martínez-Herrera, R.E., Rutiaga-Quiñones, O.M., Alemán-Huerta, M.E. (2021). Integration of Agave plants into the polyhydroxybutyrate (PHB) production: A gift of the ancient Aztecs to the current bioworld. *Industrial Crops and Products*, 174, 114188. doi: 10.1016/j.indcrop.2021.114188.
16. García-Mendoza, A. (2002). *Distribution of Agave (Agavaceae) in Mexico. Cactus and Succulent Journal*, 74, 177–187.
17. Nava-Cruz, N.Y., Medina-Morales, M.A., Martínez, J.L., Rodríguez, R., Águilar, C.N. (2015). Agave biotechnology: An overview. *Critical Reviews in Biotechnology,* 35, 546–559. doi: 10.3109/07388551.2014.923813.
18. Davis, S.C., Ortiz-Cano, H.G. (2023). Lessons from the history of Agave: Ecological and cultural context for valuation of CAM. *Annals of Botany*, 132, 819–833. doi: 10.1093/aob/mcad072.
19. Márquez-Rangel, I., Cruz, M., Ruíz, H.A., Rodríguez-Jasso, R.M., Loredo, A., Belmares, R. (2023). Agave waste as a source of prebiotic polymers: Technological applications in food and their beneficial health effect. *Food Bioscience*, 56, 103102.
20. Onuaguluchi, O., Banthia, N. (2016). Plant-based natural fibre reinforced cement composites: A review. *Cement and Concrete Composites*, 68, 96–108. doi: 10.1016/j.cemconcomp.2016.02.014.
21. Bermúdez-Bazán, M., Castillo-Herrera, G.A., Urias-Silvas, J.E., Escobedo-Reyes, A., Estarrón-Espinosa, M. (2021). Hunting bioactive molecules from the agave genus: An update on extraction and biological potential. *Molecules*, 26, 6789. doi: 10.3390/molecules26226789.
22. Davis, S.C., Kuzmick, E.R., Niechayev, N., Hunsaker, D.J. (2016). Productivity and water use efficiency of *Agave americana* in the first field trial as bioenergy feedstock on arid lands. *GCB Bioenergy*, 9, 314–325. doi: 10.1111/gcbb.12324.
23. Pérez-España, V.H., Cuervo-Parra, J.A., Paz-Camacho, C., Morales-Ovando, M.A., Gomez-Aldapa, C.A., Rodriguez-Jimenes, G.C., Robles-Olvera, V.J., Lopez-Pérez, P.A., Romero-Cortés, T. (2019). General characterizaton of cuticular membrane

isolated from Agave salmiana. *International Journal of Bio-Resource and Stress Management*, 10, 46–52.
24. Bouaziz, M.A., Bchir, B., Chalbi, H., Sebii, H., Karra, S., Smaoui, S., Attia, H., Besbes, S. (2021). Techno-functional characterization and biological potential of *Agave americana* leaves: Impact on yoghurt qualities. *Journal of Food Measurement and Characterization*, 15, 309–326. doi: 10.1007/s11694-020-00632-9.
25. Ruiz-Camou, C., Núñez, J., Musule, R. (2023). Evaluating the environmental performance of mezcal production in Michoacán, México: A life cycle assessment approach. *International Journal of Life Cycle Assessment*, 28, 1658–1671. doi: 10.1007/s11367-023-02237-y.
26. Campilho, R.D.S.G. (2016). *Natural Fiber Composites* (1st edition). CRC Press.
27. Stewart, J.R. (2015). Agave as a model CAM crop system for a warming and drying world. *Frontiers in Plant Science*, 6, 684. doi: 10.3389/fpls.2015.00684.
28. Corona-González, R.I., Varela-Almanza, K.M., Arriola-Guevara, E., Martínez-Gómez, Á.de J., Pelayo-Ortiz, C., Toriz, G. (2016). Bagasse hydrolyzates from *Agave tequilana* as substrates for succinic acid production by *Agave tequilana* in batch and repeated batch reactor. *Bioresource Technology*, 205, 15–23. doi: 10.1016/j.biortech.2015.12.081.
29. Flores-Gómez, C.A., Escamilla Silva, E.M., Zhong, C., Dale, B.E., Da Costa Sousa, L., Balan, V. (2018). Conversion of lignocellulosic agave residues into liquid biofuels using an AFEXTM-based biorefinery. *Biotechnology for Biofuels*, 11, 7. doi: 10.1186/s13068-017-0995-6.
30. Delfín-Ruíz, M.E., Calderón-Santoyo, M., Ragazzo-Sánchez, J.A., Gómez-Rodríguez, J., López-Zamora, L., Aguilar-Uscanga, M.G. (2020). Acid pretreatment optimization for xylose production from *Agave tequilana* Weber var. azul, *Agave americana* var. oaxacensis, *Agave karwinskii*, and *Agave potatorum* bagasses using a Box-Behnken design. *Biomass Conversion and Biorefinery*, 10, 949–958. doi: 10.1007/s13399-019-00497-z.
31. Jones, A.M., Zhou, Y., Held, M.A., Davis, S.C. (2020). Tissue composition of *Agave americana* L. yields greater carbohydrates from enzymatic hydrolysis than advanced bioenergy crops. *Frontiers in Plant Science*, 11, 654. doi: 10.3389/fpls.2020.00654.
32. Perez-Pimienta, J.A., Flores-Gómez, C.A., Ruiz, H.A., Sathitsuksanoh, N., Balan, V., da Costa Sousa, L., Dale, B.E., Singh, S., Simmons, B.A. (2016). Evaluation of agave bagasse recalcitrance using AFEXTM, autohydrolysis, and ionic liquid pretreatments. *Bioresource Technology*, 211, 216–223. doi: 10.1016/j.biortech.2016.03.103.
33. Contreras-Hernández, M.G., Ochoa-Martínez, L.A., Rutiaga-Quiñones, J.G., Rocha-Guzmán, N.E., Lara-Ceniceros, T.E., Contreras-Esquivel, J.C., Prado Barragán, L.A., Rutiaga-Quiñones, O.M. (2018). Effect of ultrasound pre-treatment on the physicochemical composition of *Agave durangensis* leaves and potential enzyme production. *Bioresource Technology*, 249, 439–446. doi: 10.1016/j.biortech.2017.10.009.
34. Vieira, M.C., Heinze, T., Antonio-Cruz, R., Mendoza-Martinez, A.M. (2002). Cellulose derivatives from cellulosic material isolated from *Agave lechuguilla* and *fourcroydes*. *Cellulose*, 9, 203–212. doi: 10.1023/A:1020158128506.
35. Morán-Velázquez, D.C., Monribot-Villanueva, J.L., Bourdon, M., Tang, J.Z., López-Rosas, I., Maceda-López, L.F., Villalpando-Aguilar, J.L., Rodríguez-López,

L., Gauthier, A., Trejo, L., Azadi, P., Vilaplana, F., Guerrero-Analco, J.A., Alatorre-Cobos, F. (2020). Unravelling chemical composition of agave spines: News from agave fourcroydes Lem. *Plants*, 9, 1–15. doi: 10.3390/plants9121642.

36. Bledzki, A.K., Gassan, J. (1999). Composites reinforced with cellulose based fibres. *Progress in Polymer Science*, 24, 221–274.
37. Cedeño-Cruz, M. (2003). Tequila production from agave: historical influences and contemporary processes. In: KA. Jacques, TP. Lyons, DR. Kelsall. (Eds.), *The Alcohol Textbook* (pp. 223–245). Nottingham University Press.
38. Medina-Morales, M.A., Contreras-Esquivel, J.C., De la Garza-Toledo, H., Rodriguez, R., Aguilar, C.N. (2011). Enzymatic bioconversion of agave leaves fiberhydrolysis using plackett-burman design. *American Journal of Agricultural and Biological Science*, 6, 480–485. doi: 10.3844/ajabssp.2011.480.485.
39. Heinze, T. (2016). Cellulose: Structure and properties. In: Rojas, O. J. (Eds.), *Cellulose Chemistry and Properties: Fibers, Nanocelluloses and Advanced Materials* (pp. 1–52). Springer.
40. Gao, Y., Guo, M., Wang, D., Zhao, D., Wang, M. (2023). Advances in extraction, purification, structural characteristics and biological activities of hemicelluloses: A review. *International Journal of Biological Macromolecules*, 225, 467–483. doi: 10.1016/j.ijbiomac.2022.11.099.
41. Torres-Tello, E.V., Robledo-Ortíz, J.R., González-García, Y., Pérez-Fonseca, A.A., Jasso-Gastinel, C.F., Mendizábal, E. (2017). Effect of agave fiber content in the thermal and mechanical properties of green composites based on polyhydroxybutyrate or poly(hydroxybutyrate-co-hydroxyvalerate). *Industrial Crops and Products*, 99, 117–125. doi: 10.1016/j.indcrop.2017.01.035
42. Akpinar, O., Erdogan, K., Bostanci, S. (2009). Production of xylooligosaccharides by controlled acid hydrolysis of lignocellulosic materials. *Carbohydrate Research*, 344, 660–666. doi: 10.1016/j.carres.2009.01.015.
43. Martins, M., Ávila, P.F., Paim de Andrade, C.C., Goldbeck, R. (2020). Synergic recombinant enzyme association to optimize xylo-oligosaccharides production from agricultural waste. *Biocatalysis and Agricultural Biotechnology*, 28, 101747. doi: 10.1016/j.bcab.2020.101747.
44. Rao, J., Lv, Z., Chen, G., Peng, F. (2023). Hemicellulose: Structure, chemical modification, and application. *Progress in Polymer Science*, 140, 101675. doi: 10.1016/j.progpolymsci.2023.101675.
45. Luo, J., Liu, T.L. (2023). Electrochemical valorization of lignin: Status, challenges, and prospects. *Journal of Bioresources and Bioproducts*, 8, 1–14. doi: 10.1016/j.jobab.2022.11.003.
46. Ahumada-Santos, Y.P., Montes-Avila, J., Uribe-Beltrán, Mde J., Díaz-Camacho, S.P., López-Angulo, G., Vega-Aviña, R., López-Valenzuela, J.Á., Heredia, J.B., Delgado-Vargas, F. (2013). Chemical characterization, antioxidant and antibacterial activities of six Agave species from Sinaloa, Mexico. *Industrial Crops and Products*, 49, 143–149. doi: 10.1016/j.indcrop.2013.04.050.
47. Chen, P.Y., Kuo, Y.C., Chen, C.H., Kuo, Y.H., Lee, C.K. (2009). Isolation and immunomodulatory effect of homoisoflavones and flavones from *Agave sisalana* Perrine Ex Engelm. *Molecules*, 14, 1789–1795. doi: 10.3390/molecules14051789.

48. Cornara, L., La Rocca, A., Marsili, S., Mariotti, M.G. (2009). Traditional uses of plants in the Eastern Riviera (Liguria, Italy). *Journal of Ethnopharmacology*, 125, 16–30. doi: 10.1016/j.jep.2009.06.021.
49. Da Silva, B.P., De Sousa, A.C., Silva, G.M., Mendes, T.P., Parente, J.P. (2002). A New Bioactive Steroidal Saponin from Agave attenuata. *Zeitschrift für Naturforschung C. Journal of Biosciences*, 57, 423–428.
50. Semenya, S., Potgieter, M., Tshisikhawe, M., Shava, S., Maroyi, A. (2012). Medicinal utilization of exotic plants by Bapedi traditional healers to treat human ailments in Limpopo province, South Africa. *Journal of Ethnopharmacology*, 144, 646–655. doi: 10.1016/j.jep.2012.10.005.
51. Verástegui, Á., Verde, J., García, S., Heredia, N., Oranday, A., Rivas, C. (2008). Species of Agave with antimicrobial activity against selected pathogenic bacteria and fungi. *World Journal of Microbiology and Biotechnology*, 24, 1249–1252. doi: 10.1007/s11274-007-9563-8.
52. Messinese, E., Pitirollo, O., Grimaldi, M., Milanese, D., Sciancalepore, C., Cavazza, A. (2024). By-products as sustainable source of bioactive compounds for potential application in the field of food and new materials for packaging development. *Food and Bioprocess Technology*, 17, 606–627. doi: 10.1007/s11947-023-03158-2.
53. Ribeiro, B.D., Alviano, D.S., Barreto, D.W., Coelho, M.A.Z. (2013). Functional properties of saponins from sisal (*Agave sisalana*) and juá (*Ziziphus joazeiro*): Critical micellar concentration, antioxidant and antimicrobial activities. *Colloids and Surfaces A: Physicochemical and Engineering Aspects*, 436, 736–743. doi: 10.1016/j.colsurfa.2013.08.007.
54. Figueroa-Velarde, V., Diaz-Vidal, T., Cisneros-López, E.O., Robledo-Ortiz, J.R., López-Naranjo, E.J., Ortega-Gudiño, P., Rosales-Rivera, L.C. (2021). Mechanical and physicochemical properties of 3d-printed agave fibers/poly(Lactic) acid biocomposites. *Materials (Basel)*, 14, 3111. doi: 10.3390/ma14113111.
55. Huerta-Cardoso, O., Durazo-Cardenas, I., Longhurst, P., Simms, N.J., Encinas-Oropesa, A. (2020). Fabrication of *Agave tequilana* bagasse/PLA composite and preliminary mechanical properties assessment. *Industrial Crops and Products*, 152, 112523. doi: 10.1016/j.indcrop.2020.112523.
56. Pérez-Fonseca, A.A., Robledo-Ortíz, J.R., Ramirez-Arreola, D.E., Ortega-Gudiño, P., Rodrigue, D., González-Núñez, R. (2014). Effect of hybridization on the physical and mechanical properties of high density polyethylene-(pine/agave) composites. *Materials and Design*, 64, 35–43. doi: 10.1016/j.matdes.2014.07.025.
57. Annandarajah, C., Li, P., Michel, M., Chen, Y., Jamshidi, R., Kiziltas, A., Hoch, R., Grewell, D., Montazami, R. (2018). Study of agave fiber-reinforced biocomposite films. *Materials (Basel)*, 12, 99. doi: 10.3390/ma12010099.
58. Mueller, M., Reiner, J., Fleischhacker, L., Viernstein, H., Loeppert, R., Praznik, W. (2016). Growth of selected probiotic strains with fructans from different sources relating to degree of polymerization and structure. *Journal of Functional Foods*, 24, 264–275. doi: 10.1016/j.jff.2016.04.010.
59. Conceição Apolinário, A., Silva Vieira, A.D., Marta Isay Saad, S., Converti, A., Pessoa, A., da Silva, J.A. (2020). Aqueous extracts of *Agave sisalana* boles have prebiotic potential. *Natural Product Research*, 34, 2367–2371. doi: 10.1080/14786419.2018.1536129.

60. Syafri, E., Jamaluddin, J., Sari, N.H., Mahardika, M., Suryanegara, L., Sinaga, R., Yudhanto, F., Zainul, R., Nugroho, A., Khan, A., Wazzan, M.A. (2023). Effect of ultrafine grinding and ultrasonication duration on the performance of polyvinyl alcohol (PVA) *Agave gigantea* cellulose micro fiber (CMF) bio-composite film. *Journal of Natural Fibers*, 20, 2192545. doi: 10.1080/15440478.2023.2192545.
61. Sifuentes-Nieves, I., Yáñez Macías, R., Neira Velázquez, G., Velázquez, G., Garcia Hernandez, Z., Gonzalez Morones, P., Hernández-Hernández, E. (2021). Biobased sustainable materials made from starch and plasma/ultrasound modified Agave fibers: Structural and water barrier performance. *International Journal of Biological Macromolecules*, 193, 2374–2381. doi: 10.1016/j.ijbiomac.2021.11.071
62. Gupta, R., Khasa, Y.P., Kuhad, R.C. (2011). Evaluation of pretreatment methods in improving the enzymatic saccharification of cellulosic materials. *Carbohydrate Polymers*, 84, 1103–1109. doi: 10.1016/j.carbpol.2010.12.074.
63. Ortega, Z., Castellano, J., Suárez, L., Paz, R., Díaz, N., Benítez, A.N., Marrero, M.D. (2019). Characterization of *Agave americana* L. plant as potential source of fibres for composites obtaining. *SN Applied Sciences*, 1, 987. doi: 10.1007/s42452-019-1022-2.
64. Quattrocchi, U. (2017). CRC World Dictionary of Plant Names. In: (Eds.), *Common Names, Scientific Names, Eponyms, Synonyms, and Etymology*. Routledge.64. Suvitha, L. (2021). *Fabrication and Evaluation of Agave americana Leaf Fiber*. Proceedings of the First International Conference on Combinatorial and Optimization, ICCAP 2021, December 7-8 2021, Chennai, India. doi: 10.4108/eai.7-12-2021.2314639.
65. Bezazi, A., Belaadi, A., Bourchak, M., Scarpa, F., Boba, K. (2014). Novel extraction techniques, chemical and mechanical characterisation of *Agave americana* L. Natural fibres. *Composites Part B: Engineering*, 66, 194–203. doi: 10.1016/j.compositesb.2014.05.014.
66. Hulle, A., Kadole, P., Katkar, P. (2015). *Agave americana* leaf fibers. *Fibers*, 3, 64–75. doi: 10.3390/fib3010064.
67. Kolte, P.P., Daberao, A.M. (2012). *Agave americana*: The natural leaf fiber. *Textile Review*, 4, 23.
68. Aigaje, E., Riofrio, A., Baykara, H. (2023). Processing, properties, modifications, and environmental impact of nanocellulose/biopolymer composites: A review. *Polymers (Basel)*, 15, 1219. doi: 10.3390/polym15051219.
69. Barcelos, C.A., Maeda, R.N., Betancur, G.J.V., Pereira, N. (2013). The essentialness of delignification on enzymatic hydrolysis of sugar cane bagasse cellulignin for second generation ethanol production. *Waste and Biomass Valorization*, 4, 341–346. doi: 10.1007/s12649-012-9137-3.
70. Barbosa-Cánovas, G., Bermúdez-Aguirre, D. (2010). Other novel milk preservation technologies: Ultrasound, irradiation, microwave, radio frequency, ohmic heating, ultraviolet light and bacteriocins. In: Griffiths, M.W. (Eds.), *Improving the Safety and Quality of Milk* (pp. 420–450). Elsevier: Cambridge.
71. Rastogi, N.K. (2011). Opportunities and challenges in application of ultrasound in food processing. *Critical Reviews in Food Science and Nutrition*, 51, 705–722. doi: 10.1080/10408391003770583.
72. Gallego-García, M., Moreno, A.D., Manzanares, P., Negro, M.J., Duque, A. (2023). Recent advances on physical technologies for the pretreatment of food waste and

lignocellulosic residues. *Bioresource Technology*, 369, 128397. doi: 10.1016/j.biortech.2022.128397.
73. Solís-Rosales, S.G., Naranjo-Naranjo, L., Fonseca-Florido, H.A., González-Morones, P., Hernández, Z.G., Macías, R.Y., Romero, G.Y., Gallardo-Vega, C.A., Sifuentes-Nieves, I., Hernández-Hernández, E. (2022). Alkali/Ultrasound treatment as alternative to modify structural and thermal properties of *Agave tequilana* fibers. *Journal of Natural Fibers*, 19, 9309–9322. doi: 10.1080/15440478.2021.1982822.
74. Ong, V.Z., Wu, T.Y., Chu, K.K.L., Sun, W.Y., Shak, K.P.Y. (2021). A combined pretreatment with ultrasound-assisted alkaline solution and aqueous deep eutectic solvent for enhancing delignification and enzymatic hydrolysis from oil palm fronds. *Industrial Crops and Products*, 160, 112974. doi: 10.1016/j.indcrop.2020.112974.
75. Ríos-González, L.J., Medina-Morales, M.A., Rodríguez-De la Garza, J.A., Romero-Galarza, A., Medina, D.D., Morales-Martínez, T.K. (2021). Comparison of dilute acid pretreatment of agave assisted by microwave versus ultrasound to enhance enzymatic hydrolysis. *Bioresource Technology*, 319. doi: 10.1016/j.biortech.2020.124099.
76. Tsubaki, S., Azuma, J.-I., Yoshimura, T., Maitani, M.M., Suzuki, E., Fujii, S., Wada, Y. (2016). Microwave-Induced Biomass Fractionation. *Biomass Fractionation Technologies for a Lignocellulosic Feedstock Based Biorefinery*, 103–126. doi: 10.1016/B978-0-12-802323-5.00005-0.
77. Rosas-Hernández, Y., García-Serrano, L.A., Tapia-Maruri, D., Jiménez-Aparicio, A.R., Camacho-Díaz, B.H., Arenas-Ocampo, M.L. (2018). Optimization of the microwave-assisted ethanosolv extraction of lignocellulosic compounds from the bagasse of Agave angustifolia haw using the response methodology. *Journal of Agricultural and Food Chemistry*, 66, 3533–3540. doi: 10.1021/acs.jafc.7b04627.
78. Pérez-Zavala, M.de L., Hernández-Arzaba, J.C., Bideshi, D.K., Barboza-Corona, J.E. (2020). Agave: A natural renewable resource with multiple applications. *Journal of the Science of Food and Agriculture*, 100, 5324–5333. doi: 10.1002/jsfa.10586.
79. Singha, A.S., Rana, R.K. (2013). Preparation and properties of agave fiber-reinforced polystyrene composites. *Journal of Thermoplastic Composite Materials*, 26, 513–526. doi: 10.1177/0892705711425848.
80. Muñiz-Márquez, D.B., Contreras, J.C., Rodríguez, R., Mussatto, S.I., Wong-Paz, J.E., Teixeira, J.A., Aguilar, C.N. (2015). Influence of thermal effect on sugars composition of Mexican Agave syrup. *CYTA: Journal of Food*, 13, 607–612. doi: 10.1080/19476337.2015.1028452.
81. Santos-Zea, L., María Leal-Díaz, A., Cortés-Ceballos, E., Alejandra Gutiérrez-Uribe, J. (2012). Agave (*Agave* spp.) and its traditional products as a source of bioactive compounds. *Current Bioactive Compounds*, 8, 218–231.
82. Chang, S., Weng, Z., Zhang, C., Jiang, S., Duan, G. (2023). Cellulose-based intelligent responsive materials: A review. *Polymers (Basel)*, 15, 3905. doi: 10.3390/polym15193905.
83. Saraiva, A., Carrascosa, C., Ramos, F., Raheem, D., Raposo, A. (2022). Agave syrup: Chemical analysis and nutritional profile, applications in the food industry and health impacts. *International Journal of Environmental Research and Public Health*, 19, 7022. doi: 10.3390/ijerph19127022.
84. Crespo-González, M.R., González-Eguiarte, D.R., Rodríguez-Macías, R., Rendón-Salcido, L.A., Del Real-Laborde, J.I., Torres-Morán, J.P. (2013). Evaluation of

agave bagasse compost as a component of substrates to produce seedling blue agave. *Revista Mexicana de Ciencias Agrícolas*, 4, 1161–1173.
85. Chowdhury, I.H., Abdelwahab, M.A., Misra, M., Mohanty, A.K. (2021). Sustainable biocomposites from recycled bale wrap plastic and agave fiber: Processing and property evaluation. *ACS Omega*, 6, 2856–2864. doi: 10.1021/acsomega.0c05186.
86. Amrutha, S.R., Rejimon, P.K., Suja, N.R., Alson, M. (2023). Thermal properties of biopolymers. In: Rehm, B.H.A. (Eds.), *Handbook of Biopolymers* (pp. 269–296). Springer Nature: Singapore.

CHAPTER 10

Biodegradable Active Films with Antioxidants and Antimicrobials Extracted from Residual Biomass

AILEN M. POZA,[1,2] NATALIA GONZÁLEZ,[3,4] DAIANA V. TRAPÉ,[5,6] OLIVIA V. LÓPEZ,[4,5*] and MARCELO A. VILLAR[5,6]

[1]*Instituto de Investigaciones Bioquímicas de Bahía Blanca (INIBIBB) (UNS-CONICET), Bahía Blanca, Argentina*

[2]*Centro i~mar, Universidad de Los Lagos, Puerto Montt, Los Lagos, Chile*

[3]*Instituto de Química del Sur INQUISUR (UNS-CONICET), Bahía Blanca, Argentina*

[4]*Departamento de Química, Universidad Nacional del Sur (UNS), Bahía Blanca, Argentina*

[5]*Planta Piloto de Ingeniería Química, PLAPIQUI (UNS-CONICET), Bahía Blanca, Argentina*

[6]*Departamento de Ingeniería Química, Universidad Nacional del Sur (UNS), Bahía Blanca, Argentina*

*Corresponding author

ABSTRACT

This chapter is focused on the application of natural additives such as antioxidants and antimicrobials obtained from residual biomass. Waste and by-products derived from algae processing, wine and olive oil industries, and food processing, among others, are considered. Active bio-based

Sustainable Food Packaging Materials and Technology. Juan Manuel Tirado-Gallegos, Juan Daniel Rivaldi Chávez, and José Juan Buenrostro-Figueroa (Eds)

DOI: 10.1201/9781779641434-10

and biodegradable films with natural compounds for food packaging, highlighting their advantages with respect to synthetic ones, are presented. Effectiveness of packaging with bioactive components to protect different food products and extend their shelf life is described. The presence of antioxidants in the packaging, particularly natural extracts rich in phenolic compounds, such as tocopherols, flavonoids, and phlorotannins, delay oxidative processes and reduce the food deterioration induced by ultraviolet (UV) radiation. On the other hand, antimicrobial food packaging controls the growth of microorganisms potentially harmful. Oxidative and decomposition processes compromise the food sensory quality and contribute to the loss of nutrients and the formation of toxic compounds. Migration kinetics of antioxidant and antimicrobial agents from the packaging to the food product, which are crucial to ensure the safety and effectiveness of the packaging to preserve food quality while minimizing potential health risks, are also included.

10.1 FOOD PACKAGING

10.1.1 DEFINITION AND FUNCTIONS OF FOOD PACKAGING

Food packaging dates back to the 18th century [1], and its functions have been adapted to the specific characteristics of the foods (nutritional, organoleptic, and hygienic) that humans have been incorporating into their diet. Nowadays, packaging fulfills different conventional and innovative functionalities in order to minimize the food alteration caused by physical factors (light, temperature, humidity, etc.), chemical compounds (oxygen, ethylene, water vapor, etc.), and biological agents (bacteria, yeast, molds, etc.). The functions that food packaging must comply are physical and barrier protection, containment, information transmission, marketing, security, convenience, and portion control [2]. Table 10.1 lists the main food packaging functions, explaining the purposes that they should fulfill.

According to Han et al. [3], the key function of food packaging aims to extend the shelf life of the packed product, reducing food losses. Worldwide, 30% of food is lost throughout the entire production, distribution, and consumption chain; therefore, the packaging of food products surely helps to minimize food waste [4]. The fact that food products are protected by packages reduce undesirable changes in their organoleptic (color, flavor, and texture), hygienic (microorganism growth), and nutritional

TABLE 10.1 Food Packaging Functions and Purposes

Functions	Purposes
Physical protection	To protect from shock, vibration, compression, temperature, and microorganisms
Barrier protection	To act as a barrier to gasses, water vapor, and dust. To keep the packed food clean, fresh, and safe. Some packages also maintain modified or controlled atmospheres and others contain desiccants, oxygen absorbers, or ethylene absorbers to extend food shelf life
Containment	To provide containment to food products as liquids, powders, and granular to allow their efficient handling
Information transmission	To communicate through labels how to use, transport, recycle, or dispose of the package or product. Some types of information are required by governments
Marketing	To encourage consumers to purchase the product through aesthetically pleasing and eye-appealing food presentations
Security	To reduce the security risks of shipment. To indicate that the package and contents are not counterfeit through authentication seals. To avoid pilferage by including antitheft devices, such as dye packs, RFID tags, or electronic article surveillance tags, that can be activated or detected by devices at exit points
Convenience	To facilitate distribution, handling, stacking, display, sale, opening, reclosing, usc, and rcusc
Portion control	To allow control of the portion size of food single-serving packaging

(nutrient degradation) characteristics [5–7]. Besides, food packaging must meet the demands of today's consumers such as saving time in food preparation, easy to open, easy to reclose, and suitability for microwave heating, among others [8].

10.1.2 MATERIALS FOR FOOD PACKAGING

Historically, materials that have been used to manufacture food packaging are glass, metal, paper, and polymers. Packaging usually contains more than one type of these materials to have a product that meets the functional and aesthetic requirements demanded by consumers and market [9].

Among the mentioned materials employed for food packaging, polymers are the most used due to their low cost, easy processing, and versatility [10]. Polymers began to be manufactured industrially in the 1930s and, since then, have been essential in the food packaging industry.

Nowadays, Market Plastic Reports 2023 estimated that the global plastic market will reach USD 721.14 billon by 2027 with a compound annual growth rate (CAGR) of 4.5% during the period 2023–2027 [11]. According to the cost/benefit low ratio, packages based on polymers are able to transport, storage, and conservate many foods because of an equilibrate balance between their mechanical and barrier properties, cost, and lightweight [12]. Thereby, synthetic petroleum-based polymers such as polyethylene (PE), polypropylene (PP), polyethylene terephthalate (PET), and polystyrene (PS) have been widely employed as materials for food packaging due to their adequate properties [10, 13–15]. Despite the advantages of petroleum-based polymers, their use compromises the environment mainly due to its nonrenewable origin and its nonbiodegradable character remaining in the nature for more than 450 years [16, 17]. In this sense, they are responsible of pollution of soil and waterways and some health problems in humans. According to this, Leslie et al. [18] warned about the presence of microplastics in human blood that could affect immune regulation or predispose to immunological diseases. For these reasons, the massive use of petroleum-based polymers as food packaging materials leads to serious environmental problems and health risks, being single-use packaging the main responsible for these adverse effects [19]. Although there are alternatives that could minimize the consequences of the irresponsible use and final disposal of packaging from nonrenewable origin and nonbiodegradable nature, according to the reduce, reuse, and recycle (3R) concept for efficient waste management, they are not sufficient to solve this problem. Within this context, bio-based and biodegradable polymers appear as promising alternative materials to face these challenges associated with environmental preservation and health care. As it was stressed by Bilal and Iqbal [20], this proposal implies the search for renewable sources of environmentally friendly biopolymers as alternative materials to conventional nonbiodegradable fossil-based polymers. The renewable origin guarantees the preservation of the planet's nonrenewable resources, so that they are available for future generations, and the biodegradable nature avoids environmental pollution and possible health risks for human beings. In accordance with Ruggero et al. [21], bio-based polymers limit the extraction and use of fossil resources by using (partly) biomass; meanwhile, biodegradability is an add-on property, which depends not only of the polymer chemical structure but also of the environmental conditions under which the degradation occurs.

10.1.3 BIODEGRADABLE BIO-BASED POLYMERS FOR FOOD PACKAGING

Bio-based polymers are extracted from renewable sources and they are usually biodegradables. They have a low impact in carbon dioxide emissions and they are associated with the environmental sustainability. Biodegradability is the capacity for biological degradation of organic materials by enzymatic action of microorganisms down to water, carbon dioxide, methane, and biomass. The International Union of Pure and Applied Chemistry (IUPAC) defines the term biodegradation as the "breakdown of a substance catalyzed by enzymes *in vitro* or *in vivo*" [22]. Nevertheless, the time required for the material biodegradation is critical, and this process can take days or thousands of years [23]. According to Brdlík et al. [24], the ideal biodegradation time of a biodegradable material should be 1–2 months under composting conditions. Biodegradable polymers can be obtained from biological sources (bio-based polymers) or synthetized from petroleum-based resources with chemical structures that allows biodegradation [21].

Following these concepts, bio-based polymers possess enormous advantages such as sustainability and eco-friendliness, attractive to replace the conventional materials employed for food packaging. These polymers can be obtained from different natural sources such as polysaccharides, proteins, lipids, and microorganisms. There are several biopolymers that have been proposed as alternative materials for food packaging. The most studied bio-based polymers for the development of food packaging will be described below.

Polysaccharides are the most macromolecules produced by nature and they are easily extracted from diverse plants, fungi, and algae [25]. Among polysaccharides, the most employed for food packaging are starch, cellulose, chitosan, and alginate. They possess interesting properties for food packaging such as biocompatibility, biodegradability, and nontoxicity. They are usually used to develop multifunctional films and edible coatings, among other applications. Besides, by incorporating antimicrobial or antioxidant biocomponents, it is possible to obtain active packaging or develop smart materials with the ability to act as sensors when faced with an external stimulus [26]. However, despite the aforementioned advantages, polysaccharide-based materials present, in general, some limitations associated with the polymer's hydrophilic character. According to this, films and coatings usually present high water-vapor permeability and poor mechanical properties, restricting the polysaccharides applications as food packaging materials [27]. In order to get over polysaccharides-based materials limitations, these biopolymers can be modified by different chemical and/or physical methods. Janik et al. [28] published an interesting review reporting the possibilities of modification of polysaccharides such as starch, chitosan, sodium alginate, and cellulose to improve their hydrophilic, barrier, and mechanical properties. The introduction of new functional groups by esterification, etherification, or complexation modification to the base structure of polysaccharides gives different functionalities such as hydrophobicity, antimicrobial, or antioxidant capacity, depending on the characteristics of the introduced groups [29]. Besides, they can be combined with other biopolymers to obtain improved materials taking advantage of the individual properties of each material [30]. Another alternative to enhance polysaccharide-based materials properties is to develop bionanocomposites by adding to the biopolymer different organic (cellulosic whiskers, lignocellulosic, chitin whisker, starch nanocrystal, etc.) or inorganic nanofillers (metallic nanoparticles, metal oxides, carbon nanotubes, or layered silicates) [31].

Proteins are considered promising sustainable alternatives to traditional polymers for food packaging and can be divided into plant proteins (gluten, soy, and zein) and animal proteins (casein, whey, and gelatin). Besides, proteins can be obtained from agroindustrial residues or by-products giving added value to these no exploiting materials, generating economics returns and environmental benefits [32, 33]. Proteins are good candidates to be used as food packaging due to their relative abundance, good film-formation ability, biodegradability, and nutritive value. The ability to form films is highly dependent on their molecular characteristics such as molar mass, conformation, charge, flexibility, and thermal stability. Films based on proteins have barrier capacity to different gasses and aromas and present good mechanical properties. An attractive characteristic of proteins for food packaging application is their amphiphilic nature that allows proteins to be carriers for active compounds [34].

Nowadays, the use of lipids in food packaging is being taken into account because they provide proteins- and polysaccharides-based films and coatings several benefits, including gloss, hydrophobicity, water vapor barrier, mechanical, antimicrobial, and antioxidant properties [35]. Yousuf et al. [36] stressed that vegetable oils, edible waxes, and fatty acids, including glycerides, lecithin, and tocopherols, have been used to develop food films and coatings. These authors also highlight that lipids are fundamental constituents since the mechanical and barrier properties of films and coatings mainly depend on the polarity of the components. The nonpolar nature of lipids generally gives food packaging hydrophobicity enhanced barrier properties against water vapor. An application of this type of compounds as food packaging that dates back a long time is the application of waxes to fruits and vegetables to prolong their postharvest shelf life [37]. However, the enormous potential of lipids as food packaging materials is still under study, and there are many aspects to investigate, such as different types of lipids and the properties and functionality that lipids give to films and coatings.

Another group of biodegradable bio-based polymers employed to develop food packaging are those obtained from microorganisms. Among them, polyhydroxyalkanoates (PHAs) are considered as an interesting alternative to replace synthetic nonbiodegradable polymers since they are hydrophobic with gas barrier properties similar to low-density polyethylene (LDPE) [19]. These biopolymers are linear polyesters of hydroxyalkanoic acids produced by bacterial fermentation under unbalanced growth conditions. PHAs have been used to develop packaging for different foods such as oils, cheese, and meat [38]. The main disadvantage is the relatively high cost in comparison with petroleum-based materials. However, there are many alternatives to reduce the production cost of PHAs, such as agroindustrial by-products as carbon source, among others [39–41].

The use of biodegradable bio-based polymers to (partially) replaces nowadays conventional materials employed for food packaging has benefits and limitations (Figure 10.1). Then, it is important to deepen the study of biopolymers and improvement strategies for the preparation of new materials that can be truly considered more environmentally friendly alternatives to materials based on synthetic polymers.

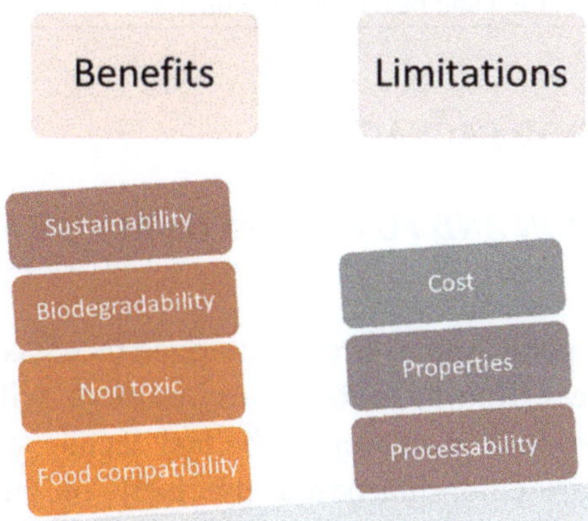

FIGURE 10.1 Benefits and limitations of biodegradable bio-based food packaging.

10.1.4 TYPES OF BIODEGRADABLE BIO-BASED FOOD PACKAGING

Packaging for foodstuffs comes in many different forms, based on technical requirements throughout the supply chain, as well as marketing needs (brand identity or consumer information) and other criteria [42]. Nowadays, it can be found different types of biodegradable bio-based food packaging that are worldwide commercial. Some of them are shown in Table 10.2, indicating the company responsible for the development and marketing.

TABLE 10.2 Types of Biodegradable Bio-Based Food Packaging Worldwide Commercial

Biodegradable Bio-Based Polymer	Packaging Type	Company
Corn starch	Trays, lids, bags, plates, bowls, and containers	PakFactory (https://pakfactory.com/blog/category/knowledge-base/)
Corn starch	Meat trays, containers, deli containers, lids, and plates	Servous (https://atyourservous.com/benefits-of-cornstarch-packaging/)
Polyhydroxyalkanoates	Food containers, tableware, cups and straws, juice bottles, and bags	Good Start Packaging (https://www.goodstartpackaging.com/guide-to-pha-bioplastic-polyhydroxyalkanoates/)
Polylactic acid	Salad bowl, deli containers, and cups	Bio Futura (https://www.biofutura.com/en/materials/pla)
Cellulose	Coffee and tea packaging, snacks and dried food packaging, and films for fresh food and bakery, among others	NatureFlex (https://www.natureflex.com/packaging-solutions/)

10.2 ACTIVE FOOD PACKAGING

10.2.1 DEFINITION AND FUNCTIONS OF ACTIVE FOOD PACKAGING

Traditionally, food packaging was characterized to act as simple containers and isolating barriers from the outside environment, with minimal impact on the packaged product. However, in the last decade, new food preservation technologies are emerging based, precisely, on enhancing or taking advantage of the possible interactions of the packaging with the product and/or the environment that surrounds the food [43, 44]. Thus, active packaging can be defined as a food/package/environment system that acts in a coordinated way to improve food safety, extending its shelf life and preserving its quality. Accordingly, packaging goes from being a mere container to playing an active role in maintaining or even improving the quality of the packaged food products. Considering the redefinition of the concept of active packaging, research and development must be done on designing packaging and packaging technologies tailored to the needs of different products and the consumers, enabling new forms of food preservation and marketing. These packaging systems can incorporate components that absorb chemicals that are part of the food or release substances into the environment surrounding the food [45, 46]. Main active functions of food packaging are shown in Figure 10.2. They can absorb different compounds such as oxygen, carbon dioxide, ethylene, water, flavors, contaminants, and ultraviolet (UV) light. Regarding the release functions, active packaging can contain antimicrobial compounds, antioxidants, inorganic preservatives, aromas, and so on [47].

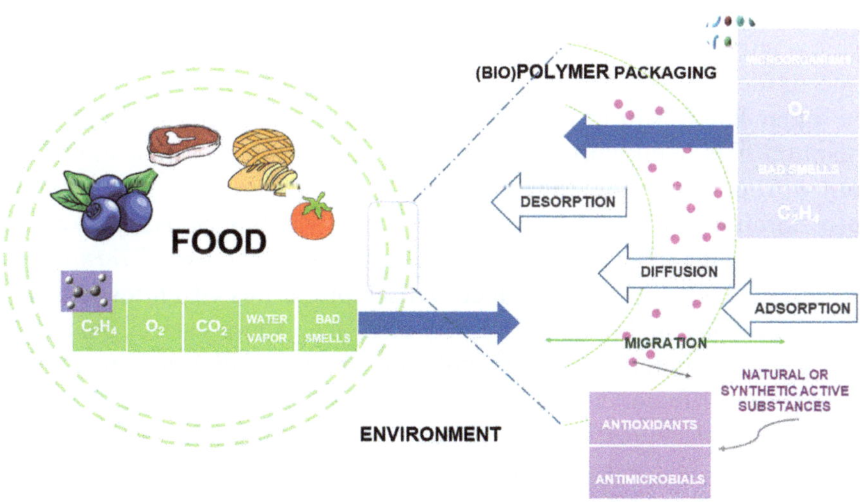

FIGURE 10.2 Food/packaging/environment interactions.

Regarding absorption function of active food packaging, several recent researches can be found. For example, strawberries suffer deterioration due to the release of ethylene; Fan et al. [48] proposed the use of depolymerized zein films (D-zein) that contain amino, carboxyl, and thiol groups that can react with ethylene, resulting in a good adsorbent for this compound. Similarly, Nian et al. [49] incorporated a metal–organic compound to a chitosan quaternary ammonium salt–gelatin matrix to develop active food packaging with ethylene adsorption capacity and improved water stability. These authors demonstrated the effectiveness of this active packaging on the preservation of a climacteric fruit (banana). Kumar et al. [50] proposed the use of sillimanite and bentonite clay as an ethylene scavenging material to preserve fruits and vegetables. On the other hand, Zúniga et al. [51] developed an active food packaging based on composites from avocado seed starch modified with tert-butyl acetoacetate and incorporated in a polylactic acid matrix that have light and oxygen blocking characteristics. Besides, numerous articles reporting packaging whose main active function is to absorb UV light using inorganic compounds (zinc oxide, titanium oxide, ferric oxide), carbonous materials (carbon dot, graphene), or natural compounds (lignin, tannic acid, polyphenols, flavonoids, essential oils) can be found in the literature [52]. Hence, Gasti et al. [53] developed active packaging with UV-blocking capacity for meat products, which were composed by biopolymers (chitosan and pullulan) and chitosan–zinc oxide hybrid nanoparticles loaded with clove essential oil. On the other hand, Sun et al. [54] proposed the preparation of graphene quantum dots using corn starch as green precursor and their modification with polyethyleneimine. These nanofillers were then incorporated into starch/polyvinyl alcohol (PVA) films to obtain active packaging with UV light barrier capacity. It was also studied the use of *Terminalia catappa* leaves waste to enhance barrier properties of chitosan-based films achieving materials with low wettability. These active films were employed to maintain the organoleptic characteristics and to extend shrimp shelf life under refrigeration [55].

As it was aforementioned, active packaging with the function of release of compounds inside the packages, in order to maintain an adequate storage microenvironment in terms of food preservation, is another type of studied packaging. Currently, most of these released compounds are bioactive compounds with different functions such as antibacterial, antifungal, and antioxidants, among others.

However, the efficiency of those packages mainly depends on the release of bioactive compounds from both sides of the films. According to this, Feng et al. [56] obtained an active packaging with unidirectional release based on glutenin/tamarind gum loaded with a binary microemulsion of melatonin/grapefruit essential oil. This active packaging presented an inner layer with bacteriostatic and antioxidant activities and it was employed for the packaging of white mushrooms. Chandrasekar et al. [57] proposed the use of films of two biothermoplastic layers based on starch for meat packaging. The material employed had an external layer rich in FeO nanoparticles with high oxygen scavenging capacity and an internal layer rich in ZnO nanoparticles with antimicrobial activity. Regarding multilayer materials, it can be mentioned the three-layer film proposed by Wang et al. [58], obtained by the layer-by-layer casting method. This system was composed by a loading layer (polyquaternium-10 and carboxymethyl chitosan) containing thymol and carvacrol nanoemulsions, an adhesive layer (polyquaternium-10 and zein), and a barrier layer (zein solution). This material, proposed for the preservation of pork meat, showed good interlaminar bonding strength, suitable water retention, satisfactory biological safety, as well as great antioxidant and antibacterial activities. He et al. [59] developed interesting bilayer films constituted by an external hydrophobic layer based on zein and an internal hydrophilic layer composed by K-carrageenan, gelatin, and curcumin as matrix, containing nisin as antibiotic. These active films were proposed as packaging for fish and fulfilled the function of releasing nisin; besides, they can also be considered smart packaging since the color of curcumin changes due to changes in the surrounding pH. Previous examples of active packaging refer to materials that can release active compounds to preserve perishable meat products. In the case of active packaging used for fruits and vegetables, in general, they are focused on minimizing the changes induced by products processing (peeling, cutting, slicing) that generates tissue damage, loss of texture and color, development of browning and unpleasant odors, and allows growth of microorganisms [60, 61]. Then, this chapter is focused on bio-based food packaging with active compounds, principally antioxidants and antimicrobials. This approach is based on the observed reports of the high economic losses generated by spoiled food waste mainly caused by microbial deterioration and oxidation phenomena. Therefore, it is relevant to develop new active packaging systems that avoid these losses and achieve adequate food preservation with a lower environmental impact [46].

10.2.2 ACTIVE COMPOUNDS FOR FOOD PACKAGING

In recent decades, the use of nonbiodegradable synthetic polymers for food packaging has shown a notable increase, generating a large amount of waste, with only a small part being recycled [14]. Considering this and taken into account the concern about extending the food shelf life to minimize food losses, the use of biodegradable bio-based active packaging with synthetic or natural active agents (antimicrobials, antioxidants, colorants, deodorizers, acidulants or acidity correctors, etc.) became a possible alternative [62, 63]. Active compounds help, in general, to extend food shelf life while preserving their organoleptic characteristics such as flavor, color, and texture that can be modified by the oxidation of lipids and proteins or by the contamination with different microorganisms. In this way, Mansour et al. [64] proposed an alginate-based packaging with lauric arginate as an active compound to avoid organoleptic modifications in chicken meat such as discoloration, meat oxidation, fat rancidity, and high microbial load.

10.2.2.1 SYNTHETIC ACTIVE COMPOUNDS

Active compounds used in food packaging can be natural or synthetic. Among synthetic antioxidants, usually used to prevent lipid deterioration, are butylated hydroxyanisole (BHA) and butylated hydroxytoluene (BHT). Likewise, as synthetic antimicrobials, it can be cited sulfur compounds such as sulfites, widely used to inhibit bacterial growth in wines, brines, dried fruits, and sorbic acid and sorbates that inhibit the growth of molds and yeasts. Some of these synthetic compounds fulfill both antioxidant and antimicrobial functions, as in the case of nitrites used in meats and sulfites employed in wines. Yahaya et al. [65] incorporated BHA as a synthetic antioxidant in a biopolymeric matrix based on semirefined carrageenan with glycerol as plasticizer and cellulose nanofiber as reinforcement. Incorporation of BHA within the packaging materials avoids possible health problems that this compound could cause if it is used directly in food. Wang et al. [66] employed propyl gallate as a synthetic antioxidant in multilayer packaging and demonstrated a slowly release of this compound from this material. Shi et al. [67] developed a packaging for Chinese giant salamander fillets using polylactic acid electrospun fibers containing octyl gallate encapsulated in β-cyclodextrin as bactericidal agent and demonstrated that this active packaging extended the food shelf life 15 days at 4°C. On the other hand, several authors have proposed the use of sorbic acid or sorbate and its salts as active antimicrobial substances to inhibit the growth of molds, yeasts, and fungi. Fasihnia et al. [68] designed a material for food packaging using polypropylene as polymer matrix and sorbic acid as antimicrobial. As it was stressed by Oliviero et al. [69], sorbate is generally used as an active agent for fresh foods. For example, Wangprasertkul et al. [70] developed films of polybutylene adipate-*co*-terephthalate (PBAT) and thermoplastic starch with potassium sorbate for fresh noodles. Likewise, Viscusi et al. [71] obtained a material to pack bread to be stored at room temperature, which was composed by polypropylene with a double hydroxide coating based on Zn/Al as a nanocarrier of the antimicrobial sorbate.

Other synthetic compounds used as antimicrobials are sulfite compounds; they allow releasing sulfur dioxide to inhibit nonenzymatic and enzymatic browning and to prevent the growth of microorganisms. San and Harnkarnsujarit [72] used sulfite compound in the production of antimicrobial and antibrowing packaging based on thermoplastic cassava starch and PBAT, while the incorporation of this synthetic compound directly into the food could cause health problems for consumers. Similarly, it has been observed that propyl gallate, used to prevent acidification in foods, could have negative consequences on male reproduction, thus being considered a toxic compound since it prevents the normal growth of testicular cells [73]. Despite sulfite compounds, usually used in fruits, vegetables, and seafood preservation, are considered "generally recognized as safe" (GRAS) by the Food and Drug Administration, it has been shown that they can cause respiratory disorders, allergies, and fatal anaphylactic shock in people who suffer asthma [74]. Sulfites in wines and juices fulfill various functions and, therefore, in this type of beverages, their control is essential to avoid undesirable organoleptic properties and adverse effects, for example, in asthmatic people [75]. It also had been reported that synthetic antioxidants BHA and BHT could be considered possible carcinogens for humans and also produce certain types of allergies, affecting the consumer's health [76,77].

Another issue that has to be taken into account about synthetic active compounds is related to the environmental pollution associated with the synthesis methods employed [78]. In this sense, some companies that produce these compounds proposed a reduction in the synthesis steps, along with a decreasing in both the use of solvents and the generation of chemical waste. Bolan et al. [79] stressed that environmental impact caused by synthetic nanoparticles used as a preservative generates ecotoxicity since these compounds reach the soil and water courses through discarded food products that contain them and due to the excretion generated by the intake of these nanocomposites. Besides, synthetic antioxidants such as BHA and BHT have been reported as hazardous substances for marine animals [80]. Li et al. [81] proposed an interesting study about microplastics in water, highlighting the importance of the compounds added to them, such as synthetic antioxidants, and the potential dangers that can cause in ecosystems. Authors indicated that physiological activity of organisms that inhabit different water bodies can be affected by the release of compounds that make up microplastics. Release of compounds noncovalently linked to microplastics could occur, thus producing leaching from the microplastic surfaces, generating adverse effects in freshwater bodies. Figure 10.3 shows the main synthetic active substances used in food packaging, indicating their negative impact on human health and the environment.

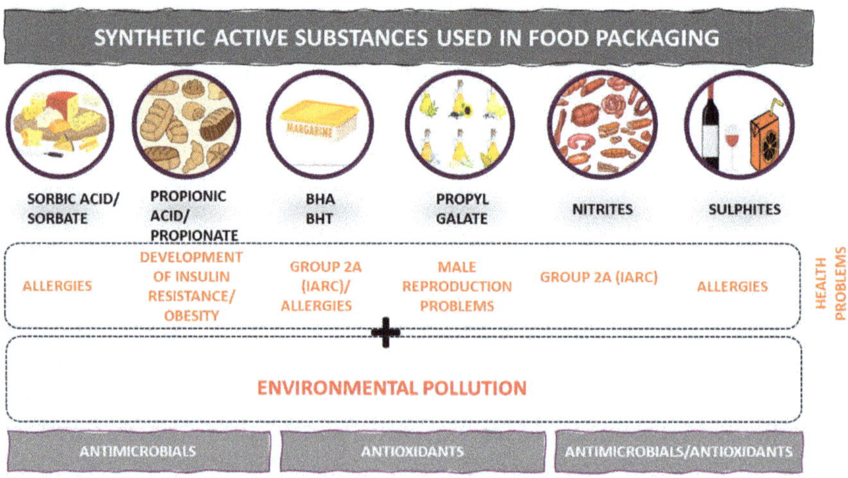

FIGURE 10.3 Synthetic active substances used in food packaging.

Taking into account both the consumers health and environmental problems caused by the synthesis and use of synthetic compounds added to food packaging, alternatives are currently being sought to replace them.

In accordance with this, nowadays the use of natural active substances in food packaging is proposed.

10.2.2.2 NATURAL ACTIVE COMPOUNDS

Currently, it is a growing trend toward developing active packaging with natural instead of synthetic substances. Natural active compounds are generally classified according to their origin: plants, animals, bacteria, algae, and fungi.

Active compounds derived from microorganisms are called biotics and are widely known for their prophylaxis in chronic diseases. Gervasoni et al. [82] proposed the use of nanocellulose combined with biotics to development active packaging with potential application in the food industry. Likewise, some bacteria synthesize proteins and peptides known as bacteriocins that inhibit Gram-positive bacteria growth and can be used as antimicrobial substances in food packaging instead of synthetic compounds [83]. Furthermore, Hong et al. [84] synthetized bacteria (*Pseudomonas aeruginosa*)-derived carbon dots doped with sulfur and they were then incorporated to films based on PVA. These materials were employed to package pork meat and apple slices, demonstrating their antimicrobial and antioxidant capacities.

Regarding active substances of animal origin, hydroalcoholic extracts of bee pollen and honey were incorporated to K-carrageenan-based packaging. Honey has antimicrobial activity (against *P. aeruginosa*, *Escherichia coli*, *Salmonella typhi*, *Staphylococcus aureus*, and *Vibrio cholerae*, among others), and extracts derived from pollen are considered potential antioxidants since they contain a high phenolic content [85]. On the other hand, Ananthi et al. [86] obtained carbon dots derived from cow milk doped with silver as an antibacterial agent. These systems were incorporated to alginate to obtain packaging for tomato slices. Antimicrobial enzymes such as lysozyme, lactoperoxidase, chitinase, and glucose oxidase can be obtained from different sources of animal origin. For example, lysozyme, which can be extracted from eggs, inhibits Gram-positive growth hydrolyzing the peptidoglycan of the bacteria cell wall. According to this, lysozyme was used to obtain coatings on the surface of packaging polymers with good barrier properties and their antimicrobial efficiency was tested with pasteurized milk [87]. Besides, Zhang and Rhim [88] proposed the use of natural enzymes such as lysozyme and lactoperoxidase, obtained from milk colostrum, for food packaging.

Among active compounds derived from fungi, it can be mentioned ergosterol and vitamin D2 obtained from mushroom waste as antimicrobial and antioxidant for food packaging [89]. Besides, Gomma et al. [90] developed edible antioxidant films based on alginates and fucoidans extracted from seaweeds and chitosans extracted from filamentous fungus.

Regarding to active compounds derived from plants, currently there are many researches that can be found in the literature. For example, antioxidant extracts obtained from green tea leaves and chañar fruits were used to improve ductility of pectin films, demonstrating that these materials extended shelf life of fatty foods [91]. Other interesting research reported the use of hydroalcoholic extract obtained from a green algae as active compound of chitosan edible coating to extend shelf life of green chilli [92]. It is important to highlight that many active compounds possess antimicrobial and antioxidant properties. In this way, it can be cited the work of Severo et al. [93] about the use of cranberry extracts as active compounds of chitosan films. These extracts have antioxidant and antimicrobial activities against *E. coli* ATCC 25922 and *S. aureus* ATCC 25923. On the other hand, it had also been used oregano essential oil to load multilayer films based on sodium alginate, K-carrageenan, and ZnO nanoparticles [94]. Priya et al. [95] proposed the use of *Plectranthus amboinicus* essential oil for functionalized chitosan films and demonstrated that obtained active materials presented antimicrobial capacity against food pathogens, enhancing their mechanical properties. In order to give antimicrobial and antioxidant properties to gum-guar films, it was incorporated ethanolic extract of algae *Dunaliella tertiolecta*, and the presence of this active compounds also enhanced surface hydrophobicity and mechanical and barrier properties [96]. Abdo et al. [97] obtained a chloroform extract of microalgae *Scenedesmus obliquus* and incorporated this active compound to a polyurethane matrix. Final materials presented antimicrobial activity toward Gram-positive and Gram-negative bacteria as well as to fungi.

10.3 NATURAL ANTIMICROBIAL AND ANTIOXIDANTS OBTAINED FROM WASTES AND BY-PRODUCTS

As it was aforementioned, the use of synthetic polymers in food packaging materials is a growing concern due to its serious environmental issues.

As an alternative, biodegradable films and coatings are gaining attention as eco-friendly substitutes for synthetic packaging. Moreover, many studies also reported the efficacy of active films and coatings in preserving the quality and extending the shelf life of various food products, including vegetables, fruits, meats, and dairy [98]. Therefore, there is an intensive growth in the development of biodegradable bio-based materials with bioactive compounds, which can be obtained from low-cost materials such as agrofood by-products [99]. Waste and by-products derived from the food, agricultural, forestry, fishing, and aquaculture industries, all of which generate substantial quantities, could be used to obtain bioactive compounds [100]. Obtaining added value products using agrofood by-products is an interesting attraction from the economic and environmental points of view [101], and there are diverse biomass reservoirs extensively explored for this purpose [102]. Nowadays, waste produced by the food industry is destined for various activities, such as composting, use as agricultural fertilizers, anaerobic digestion, bioenergy generation, cogeneration, incineration, disposal in sewers, landfills, or even disposed of in the sea. These practices have economic consequences since the material is underexploited or undervalued and in addition the excess of waste generates serious environmental damage, since it contributes to greenhouse gas emissions and can cause contamination of groundwater due to its dumping in landfills [103]. Food waste is considered a valuable raw material for so-called biorefineries due to its high organic content and the presence of valuable compounds such as oils, sugars, proteins, and phenolic compounds. Biorefineries are industries capable of taking advantage of all available biomass fractions (including industrial waste, organic fraction of municipal solid waste, lignocellulose, and algae biomass) and transforming them into a wide range of renewable bioproducts such as biofuels, bioplastics, fertilizers, nutraceuticals, and/or phytochemicals [103]. In this scenario, utilizing food waste as a primary resource for biorefineries will aid in decreasing the amount of waste disposed of, presenting significant potential and opportunities for enhancing both the economic and environmental aspects of the food industry. Additionally, this approach will support the establishment of a "zero waste" bioeconomy, where industrial symbiosis and repurposing waste from one process as feedstock for another play a pivotal role.

In this way, the idea of using waste raw materials to obtain bioactive compounds and then formulate bioactive packaging is an innovative and promising strategy in the field of sustainable and functional packaging. Table 10.3 shows natural compounds from plants, animals, and microorganisms used as antioxidant or antimicrobial compounds in polymers. In particular, by-products derived from algae processing, wine production, and olive oil extraction receive significant attention due to their widespread availability and low cost; consequently, their use for the formulation of bio-based and biodegradable active films is discussed below.

TABLE 10.3 Natural Compounds from Plants, Animals, and Microorganisms Used as Antioxidant or Antimicrobial Compounds in Polymers

Antioxidant/Antimicrobial Agent	Polymeric Matrix	References
Petit Verdot Red Grape Pomace Extract	Jute fibers	[104]
Oregano essential oil, rosemary extract, and green tea extract	Poly(3-hydroxybutyrate-co-3-hydroxyvalerate) (PHBV) fibers	[105]
Murta fruit extract	Biocomposite films based on methyl cellulose (MC)	[106]
Carvacrol and pomegranate peel extract	Chitosan-based active films	[107]
Corn cobs, almond shells, Eucalyptus wood, and grape pomace	Sodium–alginate matrix to formulate edible films	[108]
Phellodendron extract	Film with soybean protein isolate (SPI)	[109]
Mangosteen rind powder	Chitosan film	[110]
Ethanolic propolis extract	Cassava starch films	[111]
Natural extract of Ocimum tenuiflorum	Chitosan (CS)- and PVA-based films	[112]
Extracts of sage and rosemary	Chitosan films	[113]

10.3.1 OLIVE OIL INDUSTRY: WASTE AND BIOMASS VALORIZATION FOR FOOD PACKAGING

During olive oil production, solid and liquid wastes are produced, being olive pomace and olive mill wastewater (OMW) the main ones [114]. The olive oil processing schemes with the main streams are shown in Figure 10.4. The composition of waste fluctuates depending on various factors such as climatic conditions, olive provenance, cultivation practices, and extraction techniques [115]. It is marked by a notable presence of lipids, organic, and phenolic compounds, which have a negative impact on the environment. As a result, waste results in a significant environmental problem in the primary olive oil manufacturing process. According to the International Olive Council, global olive oil production for the 2022/2023 campaign reached 2.6 million tons [116]. This production is predominantly concentrated in the Mediterranean region, including Spain, Portugal, Italy, Greece, Turkey, Morocco, and Tunisia, which collectively contribute to over 90% of the world's olive oil output. Then, the discharge of olive mill wastewater presents a significant ecological concern, particularly in the Mediterranean area. Annually are generated approximately 2.8 million tons of olive pomace, while global olive mill wastewater production ranges from 7 to over 30 million cubic meters per year [117].

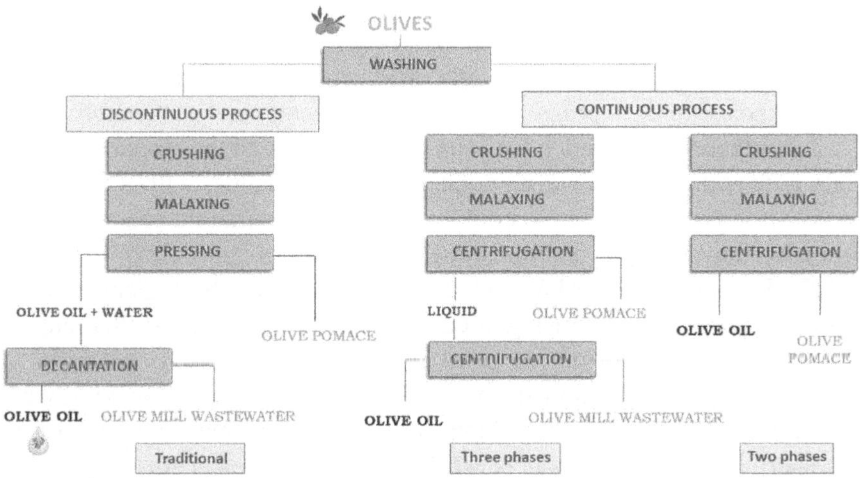

FIGURE 10.4 Processes of olive oil production.

Olive pomace is a semisolid by-product composed mainly of olive skins, pulp residues, and pit fragments [118]. Pomace is considered a pollutant due to its high acidity (pH = 5.5) and elevated polyphenol content [119]. Olive pomace contains approximately 98% of olive fruit phenolic compounds and encompasses a rich array of constituents, notably polysaccharides, proteins, fatty acids, pigments, and polyphenols. Polyphenols include simple phenols such as tyrosol (2-(4-hydroxyphenyl) ethyl alcohol) and hydroxytyrosol (HT) (2-(3,4-dihydroxyphenyl) ethyl alcohol); cinnamic acid derivatives such as caffeic acid and verbascoside;

flavonoids such as apigenin, luteolin, and rutin (quercetin-3-rutinoside); and secoiridoids exemplified by oleuropein, oleuropein aglycone, and decarboxymethyl oleuropein aglycone isomers [120].

Additionally, a significant volume of wastewater, known as OMW, is generated during the extraction process constituted by sugars, polyalcohols, lipids, pectins, and polyphenols being HT the principal phenolic compound, complemented by significant levels of tyrosol, oleuropein, caffeic acid, vanillic acid, gallic acid, luteolin, and verbascoside, among other noteworthy constituents [121]. Due to the substantial volumes of wastewater released into the natural environment, characterized by high chemical oxygen demand (COD) and polyphenol content, OMW has been regarded as a hazardous waste with a negative impact on the environment and an economic load on the olive oil industry for a long period. However, in the last decade, this view has changed, and OMW is now being recognized as a potential low-cost starting material rich in bioactive compounds. As environmental awareness grows, the olive oil industry is actively pursuing innovative methods to utilize these wastes efficiently and sustainably [118].

A fundamental approach to optimizing the value of these residual materials is to extract valuable phytochemical compounds, with a particular focus on polyphenols. These compounds offer remarkable properties that are advantageous for pharmaceuticals, cosmetics, and food industries [122]. Transforming OMW and olive pomace into biologically active materials presents a promising alternative for valorizing these challenging waste streams.

As it was aforementioned, the food industry is actively seeking ways to incorporate natural antioxidants into packaging materials to prolong the shelf life of food products while preserving their sensory qualities. To this end, the development of active packaging materials using olive oil by-products has gained attention. For instance, Akretche et al. [123] conducted the valorization of OMW to produce bio-based polymer films with antioxidant properties through environmentally friendly processes. Their approach involved directly bonding phenolic compounds into pectin films using OMW. These efforts aimed to create packaging materials that could help prevent oxidation and preserve products in various food systems. Moreover, Nunes et al. [124] studied the potential application of lignocellulosic fibers derived from olive pomace, on the development of biocomposites for food packaging with promissory results. However, further studies are necessary to achieve the expected total efficacy on food packaging functionalities (mainly, food safety and maintenance of organoleptic properties) and at the same time be sustainable. Studies have also investigated the integration of olive leaf extract into biodegradable films to give antioxidant properties. However, weak interactions between the polymer matrix and antioxidants resulted in their migration toward food products, limiting the suitability of such materials to items with shorter shelf life. Additionally, Luzi et al. [125] explored the use of hydroxytyrosol- and oleuropein-enriched extracts from olive oil by-products as active antioxidants in PVA-based films. Their research aimed to enhance the antioxidant capabilities of PVA films, thereby improving their oxidative stability for a wide range of applications. Moreover, Apicella et al. [126] developed multilayer active films containing extracts from OMW polyphenols, demonstrating that the addition of OMW extract did not compromise the oxygen and water vapor barrier properties of the films. The OMW extract exhibited strong compatibility with the film matrix, particularly when combined with 95% ethanol, and effective release of the antioxidant compounds. On the other hand, Galanakis et al. [127] investigated the potential use of phenols extracted from OMW as a UV barrier in cosmetics, aiming to enhance the UV protection properties.

Antimicrobial properties of extracts derived from OMW against various bacterial strains, including both Gram-positive (*Streptococcus pyogenes* and *S. aureus*) and Gram-negative (*E. coli* and *K. pneumonia*) bacteria were investigated by Tafesh et al. [128]. Their findings suggested that OMW extracts, particularly when combined with natural phenols, could effectively inhibit the growth of pathogenic bacteria. Moreover, additional research has indicated that OMW may be a valuable source of antifungal agents, offering a potential natural alternative to synthetic fungicides, as demonstrated by Yangui et al. [129].

10.3.2 WINE INDUSTRY: WASTE AND BIOMASS VALORIZATION FOR FOOD PACKAGING

This industry has considerable socioeconomic importance worldwide, with global production reaching 260 million hL in 2022 [130].

However, this industry is one of those that contributes the most to the accumulation of agricultural waste [131]. Approximately 750 L of wine are obtained from 1000 kg of grapes processed, along with 1650 L of wastewater and around 200 kg of solid waste [132]. Winery wastewater is characterized by having an acidic, phytotoxic pH, and a high organic load. These residues contain a high load of carbohydrates, lipids, proteins, and phenolic compounds [133]. Without adequate treatment, through aerobic processes (aeration), the discharge of wastewater can have a very negative impact on the environment since eutrophic environments can cause the death of aquatic organisms and the generation of bad odors as a consequence of a reduction of oxygen levels and high COD values [132]. Residual water generally contains two fractions: (1) the *wine lees* that are obtained after the clarification process of wine making and contains the solid sediments that settle at the bottom of the containers after fermentation and maturation of the wine and (2) the *vinasse*, which is the wastewater generated in the distillery. Both phases can be used separately, generating added value to the production process. The *wine lees* can be recovered from the bottom of the container after fermentation, as well as collected in the form of residue after filtration or centrifugation. This fraction represents 5% v/v of the total wine production [134]. This residue is considered a microbial biomass and contains a fraction of phenolic compounds, inorganic matter, organic acids, anthocyanins, and microorganisms [135]. On the other hand, *vinasse* contains part of the seeds, skin, and dead yeast. Some products such as tartaric acid can be produced from lees and vinasse, which have great economic importance [134]. Athanasiou et al. [136] performed an enzymatic treatment of wine lees with high anthocyanin and flavonoid content to produce polymeric compounds with improved antioxidant properties. These synthesized polymeric products proved to be effective as additives in food packaging films, reinforcing their initial antioxidant activity. In the same way, the use of vinasse extract as a component for the production of PVA/gelatin films was also investigated [137]. The incorporation of these extracts gave the films antioxidant activity, so they have great potential to form an active and degradable packaging material that contains powerful antioxidants. On the other hand, the presence of tartaric acid in these wastes is useful as a pH adjuster and as an effective antimicrobial agent by creating an environment that is too acidic for most microorganisms to grow. This makes it a potential candidate for use in active packaging, which is designed to improve food safety and quality and extend shelf life [138].

The main solid waste generated during the wine making process is known as grape pomace, which represents between 10% and 30% w/w of the initial raw material [139–141]. This residue contains the skins (47%), the stems (28%), and the seeds (25%) [140]. It is well known that grape pomace contains a variety of phenolics, pigments, and antioxidants [139, 142, 143, 144]. Among the phenolic compounds can be found, phenolic acids such as gallic acid, monomeric flavan-3-ols such as catechin, epicatechin, gallocatechin, epigallocatechin, and epicatechin 3-*O*-gallate, as well as dimers and trimers of procyanidin and highly polymerized compounds [145]. The extract exhibits significantly higher antioxidant potential than vitamins E and C [145]. However, the rich content of polyphenols, especially proanthocyanidins, which contribute to the red color and astringent taste, could affect the color and sensory attributes of the products when used at higher concentrations [146].

Several studies used different solid fractions from the wine making process to formulate bioactive packaging. For example, Cejudo-Bastante et al. [104] discussed the use of natural jute fibers impregnated with a Petit Verdot red grape pomace extract as a new active material for food packaging. The impregnation conditions were optimized by varying the pressure, the percentage of modifier, and the pomace dry mass. As a result, a material with antioxidant and antibacterial capacities against *E. coli*, *S. aureus,* and *P. aeruginosa* was obtained. Saurabh et al. [147] developed guar gum-based active packaging films using grape pomace extract. The resulting active films had adequate mechanical and water vapor barrier properties. They further demonstrated significant antimicrobial activity against several food pathogens, including *E. coli*, *S. aureus*, *Bacillus cereus*, and *Salmonella typhimurium*, and films were stable to radiation up to doses of 5 kGy. Another study combines the mechanical and antimicrobial properties of films based on chitosan or starch by incorporating cellulose nanocrystals and grape pomace extract, rich in phenolic compounds and antioxidants [148]. The results showed that the incorporation of cellulose nanocrystals increased the tensile and Young's modulus of the films, while the addition of grape pomace extract had an opposite effect on the mechanical properties by decreasing the tensile strength but increasing the percentage of elongation at break. Nanocomposite films exhibited significant antimicrobial activity against several foodborne pathogens, including *E. coli*, *S. aureus*, and *Listeria monocytogenes*. Also, films with high concentrations of cellulose nanocrystals significantly reduced water vapor permeability but increased the release of the phenolic compound.

Among the solid residues, grape stems are rich in dietary fibers (up to 77% of the dry weight), which are mainly composed of lignin and structural carbohydrates (cellulose and hemicelluloses) [149]. Engel et al. [150] developed biodegradable foams using starch, grape stems rich in lignocellulosic fibers, and glycerol as a plasticizer. The addition of grape stem waste improved the tensile strength and elasticity of the foam by promoting interfacial interaction between the matrix and the fibers, resulting in foams with favorable mechanical properties, low density, and higher resistance to cracking.

10.3.3 HYDROCOLLOID INDUSTRY: WASTE AND ALGAE BIOMASS VALORIZATION FOR FOOD PACKAGING

The main source of algae exploitation to date derives from the hydrocolloid industry, such as alginate, carrageenan, and agar, which processes more than 100,000 t of harvested or cultivated algae annually around the world [151]. Unlike terrestrial plant biomass, residual biomass derived from the production of biopolymers from algae is highly underexploited and its potential is largely unknown. However, this residual biomass constitutes a vast renewable source of different derived products with economic importance for the food industry, biomedical ingredients, cosmetics, specialized chemicals, and materials in future biorefineries. Currently, although many policy efforts have been made to promote circular bioeconomy practices and new forms of sustainable exploitation of the oceans, little has changed in the industrial sector [152–154].

10.3.3.1 ALGINATE INDUSTRY

Alginate is a linear heteropolysaccharide composed mainly of β-D-mannuronic acid (M) and α-L-guluronic acid (G), and it is the main component of the cell wall of many brown algae, representing 17–45% of their dry weight [155]. The main properties of alginates, including their gel-forming, emulsifying, and film-forming abilities, have made them widely used ingredients in various fields such as pharmaceutical, cosmetic, textile, and food industries [90]. Particularly, industrial alginate production only uses 13–55% of the total dry mass of the algae, while the rest is discarded or drained as residual effluent, which is an environmental and economic disadvantage [156]. The extraction process at an industrial level involves the collection, transportation, and storage of algae, with the eventual addition of formaldehyde to avoid chemical or enzymatic reactions that could reduce the quality of the alginate. It also involves acidic pretreatment at room temperature to remove pigments, other carbohydrates, and low molecular weight compounds. Then, alkaline extraction is performed, followed by solid/liquid separation, drying, and grinding, to produce high-quality alginate [157, 158]. However, the surplus or residual biomass could be used to obtain other high-value products such as phlorotannins, fucoidans, and cellulose, among others [156]. During the alginate extraction process, the algal biomass is pretreated to be clarified, removing pigments and phlorotannins. Phlorotannins are a type of tannins found specifically in brown algae. These compounds are formed by the oligomeric dehydrogenative coupling of phloroglucinol (1,3,5-trihydroxybenzene) [159] and possess multiple resonance stabilization sites and a variety of condensed units, giving them great potential. Furthermore, phlorotannins also show promising antibacterial and antifungal properties [160–162]. The biological activity of these compounds is largely influenced by the extraction method, which determines the chemical structure of extracted phenolic compounds [163]. These properties give them great economic importance for several industries [164, 165].

Recently, there has been increasing interest in the incorporation of natural phlorotannins into polymeric packaging materials with the aim of improving a variety of properties, including antimicrobial, antioxidant, and physicochemical properties of the resulting films [166]. These compounds have been the subject of numerous studies that highlight them as natural active agents for the development of active packaging films. For example, Yu et al. [167] studied the immobilization of phlorotannins on nanochitin to improve its stability and effectiveness as a biopreservative. Here, the resulting nanochitin–phlorotannin complex was applied to chilled sea bass fillets to test its ability to inhibit bacterial growth and maintain quality. The study found that the nanochitin–phlorotannin complex was effective in reducing bacterial growth and improving the overall quality of sea bass filets during refrigerated storage. Another study analyzed the structural and physical properties, antioxidant and antimicrobial activity of konjac glucomannan (KGM)/cellulose nanocrystal bionanocomposite films incorporated with phlorotannin from *Sargassum* [165]. The incorporation of phlorotannin from *Sargassum* significantly improved the antioxidant and antimicrobial activities of the KGM/cellulose nanocrystal films.

Also, it was tested the use of carboxymethyl chitosan (CMQ), a derivative of chitosan that is biocompatible, nontoxic, and biodegradable, together with gliadin/phlorotannin nanoparticles in order to obtain new composite active packaging films [164]. The findings suggest that CMQ-based films have great potential for use in food packaging due to their enhanced functional properties and ability to inhibit microbial growth, extending the shelf life of food products and reducing food waste. The incorporation of phlorotannins provides the ability to impart antimicrobial properties to packaging films, which helps to prolong the shelf life of packaged products by inhibiting the growth of unwanted microorganisms. Its antioxidant activity is also beneficial, as it helps protect foods from oxidation and spoilage. However, the use of phlorotannins derived from waste from the alginate industry to produce active packaging has not currently been tested [90, 168, 169]. On the other hand, fucoidans, highly valuable water-soluble sulfated heteropolysaccharides, are another by-product derived from algae, which are usually discarded during the alginate extraction process. During the acid treatment of the algal biomass, fucoidans are retained in the supernatant. Its structure is based on L-fucose and sulfate residues with α-1,3 and/or α-1,4 bonds, along with variable amounts of other sugars such as D-galactose, D-mannose, D-xylose and, at times, D-glucuronic acid [170]. This polysaccharide has been widely studied for its antioxidant, anticoagulant, antithrombotic, immunoregulatory, antiviral, and anti-inflammatory properties. But unlike alginate, fucoidan cannot form gels or films on its own. However, this polysaccharide can be mixed with other polymers, such as alginate to form bio-based films. For example, Gomaa et al. [90] designed films based on alginate–fucoidan with antioxidant properties, evaluating and modeling the kinetic absorption and release of moisture and polyphenols. In this case, they formulate the films with raw alginate obtained from the brown macroalgae *Sargassum*, which maintains the remnants of fucoidans and polyphenols as it is not purified. This study could be a clear example of how waste products from the alginate industry can have added value to develop edible packaging with beneficial properties for food preservation.

On the other hand, the waste generated during the extraction of alginate contains the insoluble cellulose fraction, which has great commercial potential for the production of cellulose nanofibrils. Furthermore, the acid and alkali treatment of the extraction of alginate and fucoidans is a pretreatment that favors the extraction of the cellulose fraction [171].

Nanofibrils have received significant attention in recent years due to their potential biomedical applications, including controlled drug delivery, utilization as a formulation excipient, wound dressing materials in wound healing, transdermal drug delivery, and skin tissue engineering [172]. Moreover, there is growing interest in leveraging nanofibrils to enhance the properties of packaging materials, such as biodegradability, mechanical strength, transparency, and oxygen barrier properties. Nanocellulose, in particular, stands out as a highly promising material for such applications. Notably, it is renewable, biocompatible, and biodegradable while also exhibiting exceptional mechanical properties. These attributes render nanocellulose an appealing choice for sustainable and environmentally friendly packaging solutions. Cebrián-Lloret et al. [173] used the residual biomass of alginate for the development of cellulose-based packaging films. Then, the use of cellulose appears as a material with high potential for the production of biodegradable and sustainable packaging. The alkaline treatment applied to the residual biomass produced films with an optimal balance between functional properties of cellulose, with economic and environmental efficiency. In particular, crystalline cellulose content in the *Saccharina* residue led to stiffer and less permeable films, suggesting that algal residues may be a valuable source for food packaging applications [174].

10.3.3.2 CARRAGEENAN AND AGAR INDUSTRY

Regarding the extraction of biopolymers, other important industries correspond to those that are mainly dedicated to the production of agar and carrageenan. Biofilms of these polysaccharides usually have good mechanical resistance, moderate gas barrier properties and, most importantly, are edible and easily degradable in most cases.

Agar, formed by alternating β-D-galactose and α-L-galactose, can represent up to 57% of dry weight of some algae and it has application in the food and pharmaceutical industries, while agar oligosaccharides present interesting therapeutic properties [174, 175]. Carrageenan, which is a linear polysaccharide formed by a sequence of α-D-galactopyranose and β-D-galactopyranose, represents up to 75% of the dry weight of algae and it is used for its thickening and stabilizing properties and for its biological activity [174].

The yield of agar varies between 27% and 33%, depending on the harvest time and the species. Furthermore, during the manufacturing process, between 15% and 40% of the initial dry biomass is generated as solid waste [174, 176]. The residual algal biomass obtained after agar extraction is frequently used as fertilizer [177], plant defense stimulator [178], or is discarded as waste [179]. However, this residual biomass has the potential to be transformed into packaging materials due to its composition rich in polysaccharides derived from agar remains and other biocompatible substances with antioxidant properties. Agar residues in biomass have gelling and stabilizing properties, which could improve the characteristics of packaging materials, such as their mechanical resistance and their ability to retain moisture [176, 180].

Some of the antioxidant substances in red algae are a consequence of the stressful conditions to which these organisms are exposed. Red algae have evolved sophisticated defense mechanisms to protect themselves from UV radiation, which can be detrimental to their health and survival. Some of the photoprotective molecules against UV radiation are pigmented compounds (i.e., chlorophyll *a*, phycobiliproteins, and carotenoids) and amino acids similar to mycosporine [181, 182]. In a recent research, the extraction of mycosporine-like amino acids (MAAs), such as shinorine, P334, palythine, asterin, and palithinol, from the residual biomass of the agar industry biorefinery, has been a topic of interest [180]. These compounds absorb UV radiation and have been in the spotlight in recent years for their photoprotective role. MAAs disperse harmful UV radiation by converting it into thermal energy and dissipating it into the environment inside the cell [183]. MAAs also have strong antioxidant capacity by quenching reactive oxygen species [184]. The use of red algae mycosporine in packaging could provide an additional layer of protection against degradation of packaging materials due to exposure to sunlight. Incorporating these compounds into wrapping films or packaging coatings could help extend the shelf life of light-sensitive products, such as food and cosmetics, among others, by reducing the impact of UV radiation [180]. Gan et al. [185] prepared anti-UV films based on fish gelatin and sodium alginate containing MAAs. They study the influence of different concentrations of MAAs (0–5%) on the physical, antioxidant, antibacterial, and anti-UV properties of composite films. Results indicate that the content of MAAs affected the mechanical properties of the films but did not influence their water content, swelling, or water vapor permeability.

Films with MAAs showed antioxidant and antibacterial activities and protection against UV rays, suggesting their potential for active and antioxidant packaging applications.

Furthermore, recently, red algae waste derived from the agar extraction industry has been used to produce PHAs through a fermentation process through hydrolysates mediated by a species of halophilic bacteria, *Halomonas boliviensis* [180]. These wastes, rich in cellulose and with traces of agar, were used as sugar raw materials for the production of these biodegradable polymers.

Carrageenan is the seaweed hydrocolloid with the highest sales volume, more than double that of alginate and triple that of agar [174]. The yield of carrageenan can represent 32% of the dry weight of the algae, but the amount of residual fractions generated represents 30% of the initial material, and the increasing production of carrageenan makes it necessary to valorize the residual fractions [186–188].

Reddy et al. [189] proposed a biorefinery process for the extraction of carrageenan using *Kappaphycus*. Authors propose starting with the aqueous extraction of fresh algae to extract pigments, followed by precipitation with ammonium sulfate. The extracted algal matter is subjected to lipid extraction using a solvent mixture of chloroform and methanol. If *Kappaphycus* is processed, the extracted solid is treated with a KOH solution to dissolve it and precipitate the carrageenan, while cellulose can be obtained from the residual solid left after recovery of the carrageenan. The remaining solid residues following carrageenan extraction typically consist predominantly of cellulose, along with minor quantities of polymers derived from mannose and xylose, as well as residual galactose and anhydrogalactose, which vary depending on the effectiveness of the carrageenan extraction procedure [190]. Sedayu et al. [191] explores the use of nanocellulose fibrils (NCFs) to reinforce semirefined carrageenan (SRC) and refined carrageenan (RC) films for food packaging. The study evaluates the impact of varying levels of NCF reinforcement on the water sensitivity, physicomechanical, and thermal properties of SRC and RC films. Results indicate that NCF reinforcement enhances the overall properties of both SRC and RC films, particularly their water and moisture barrier. However, the improvement in properties is less pronounced in SRC films compared to RC films. The enhancement is attributed to the stronger polymer structure and better adhesion between NCF and the carrageenan matrix.

10.4 SCIENTIFIC INTEREST IN NATURAL ANTIOXIDANTS AND ANTIMICROBIALS FOR FOOD PACKAGING

The data analyzed in this chapter were mainly obtained from the Scopus database, Research Rabbit, and Google Patents, using [TITLE-ABS-KEY (food packaging + antioxidant packaging + antimicrobial packaging + bio-based packaging)]. Figure 10.5 shows the evolution of the number of documents related to the production of food packaging that incorporates specific characteristics such as antioxidants, antimicrobial properties, and bio-based materials over the last 10 years, reaching a total of 80 manuscripts. This evolution can be fitted to an exponential function with a coefficient of 0.56 per year, higher than that of the general number of publications. Therefore, there is a growing interest in this topic, due to the emerging and current need to manufacture food packaging from sustainable and biodegradable materials. The transfer of this knowledge to the food industry is also undoubtedly relevant, taking into account the number of patents granted for the production of food packaging and biofilms with antimicrobial and antioxidant properties that can be reinforced with biodegradable products such as nanocellulose, lignin, and chitosan, among others, which exceed the number of manuscripts (Figure 10.5A). Of a total of 152 patents related to this field, China and Japan account for the majority, followed by the United States (Figure 10.5B).

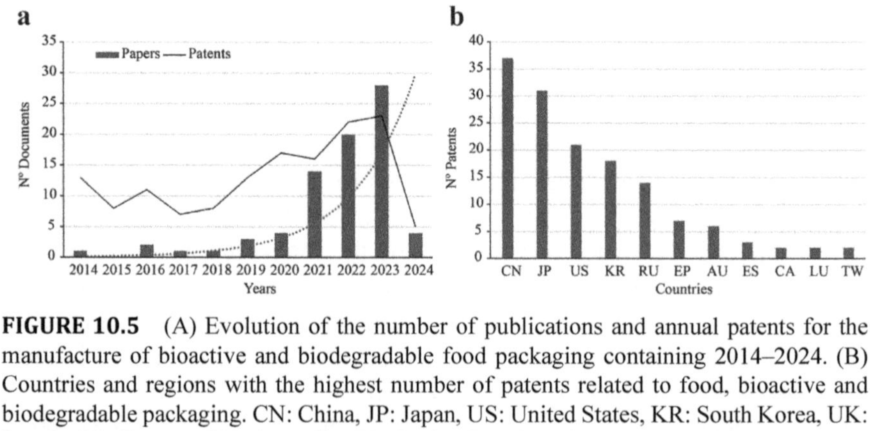

FIGURE 10.5 (A) Evolution of the number of publications and annual patents for the manufacture of bioactive and biodegradable food packaging containing 2014–2024. (B) Countries and regions with the highest number of patents related to food, bioactive and biodegradable packaging. CN: China, JP: Japan, US: United States, KR: South Korea, UK: Russia, EP: Europe, AU: Australia, ES: Spain, CA: Canada, LU: Luxembourg, and TW: Taiwan.

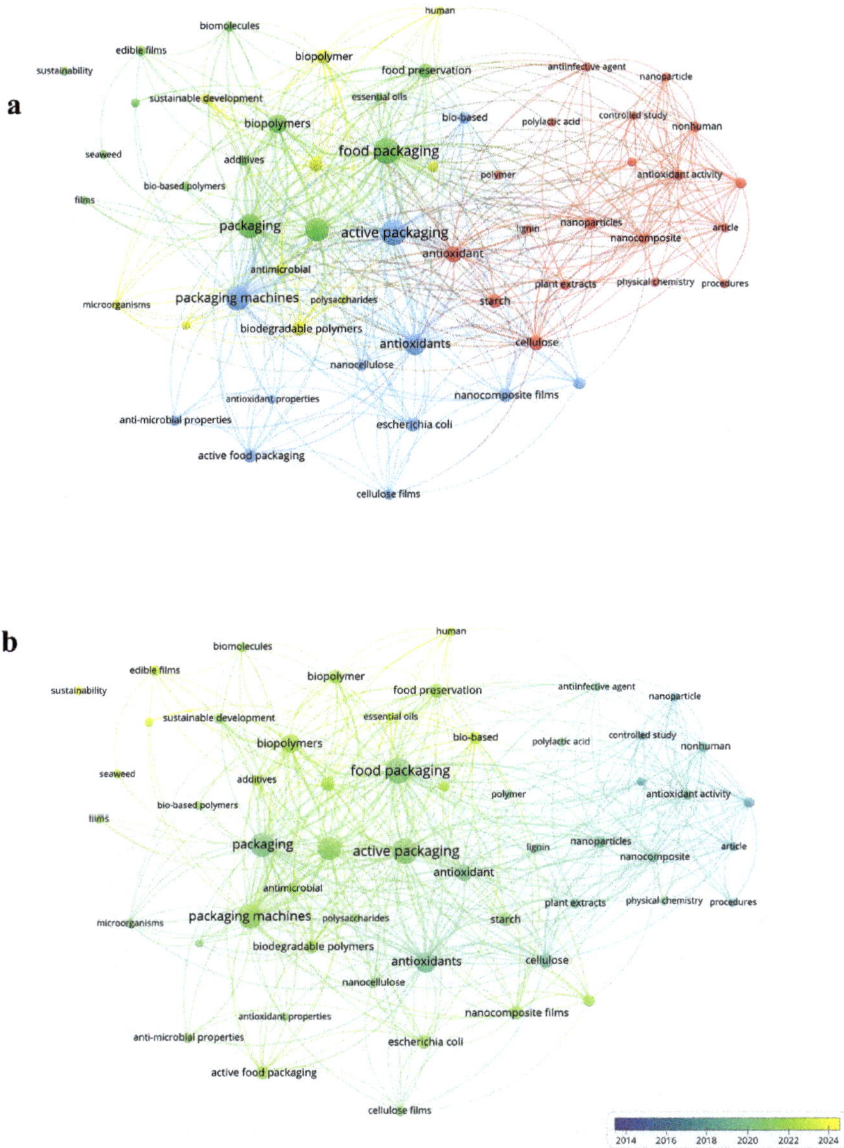

FIGURE 10.6 (A) Keyword groups in publications on the manufacturing of food, bioactive, and biodegradable packaging. (B) Evolution of these keywords throughout the period 2014–2024.

Figure 10.6A shows the distribution of keywords according to their co-occurrence. They were identified by themes or communities of keywords with different colors, establishing the size of the label according to its abundance in the literature using the VOSviewer software. The groups of keywords in publications on the manufacture of food packaging with bioactive properties comprise four fundamental aspects: (1) as a central axis, they start from the idea of active packaging from which the terms that relate to the materials for its production arise, mainly biopolymers; (2) the sustainability and biodegradability of the materials; (3) the bioactive properties of packaging such as antioxidants and antimicrobials; and (4) finally the reinforcing materials such as cellulose and lignin. It is important to highlight that the most recent research is related to the development of sustainable and bio-based materials, through the use of renewable and biodegradable materials in the manufacture of packaging, instead of traditional materials derived from petroleum (Figure 10.6B). These materials appear to come primarily from biopolymers derived from algae or chitosan. The oldest publications refer to the active properties of packaging using mainly plant extracts. In this way, the use of biodegradable and bioactive materials is a hot topic today (Figure 10.6B). This aspect is also reflected in the fact that the number of patents produced related to the manufacture of bioactive food packaging exceeds the number of publications in scientific journals.

10.5 CONCLUSION

Olive oil production process generates substantial solid and liquid waste, notably olive pomace and OMW, presenting significant environmental challenges due to their high lipid, organic, and phenolic compounds content. Efforts to utilize olive oil by-products in active packaging materials have shown promise, with studies exploring their incorporation into bio-based polymer films and biocomposites for food packaging. Additionally, research indicates the antimicrobial properties of extracts derived from OMW against various bacterial and fungal strains, suggesting their potential as natural antimicrobials. However, further studies are necessary to optimize the efficacy and sustainability of these materials in food packaging applications, particularly in terms of food safety and preservation of organoleptic properties.

On the other hand, the development of packaging using waste from the wine industry, supplemented with cellulose or other additives, also presents promising prospects for the future. These initiatives could significantly contribute to reducing waste and promoting sustainability in the food and packaging industry. By incorporating waste from the wine industry, such as grape pomace or grape skins, into packaging manufacturing, a resource that would otherwise be wasted is used, thereby reducing the environmental footprint of the wine industry. Additionally, the addition of cellulose or other complementary materials can improve the mechanical and functional properties of packaging, such as strength, flexibility, and moisture and gas barrier. This allows the creation of more robust and versatile packaging, suitable for a wide range of applications in the food and packaging industry.

Finally, the use of biopolymers such as alginate, agar, and carrageenan, together with by-products derived from algae, such as polyphenols and pigments with antioxidant properties and cellulose waste, in the development of active packaging, represents an innovative strategy within the framework of a circular economy. Algae biopolymers are biodegradable and compostable, making them more environmentally friendly and reducing the accumulation of plastic waste. Furthermore, biopolymers in combination with by-products derived from algae, such as molecules with antioxidant properties and cellulose, can add functional value to active packaging.

By adopting a circular economy approach, the life cycle of these materials can be closed, promoting the collection, recycling, and reuse of active packaging made from algae biopolymers. This involves designing waste management systems that allow the recovery of materials for subsequent use, thus creating a continuous and sustainable cycle of production and consumption.

KEYWORDS

- **antioxidants**
- **antimicrobials**
- **food packaging**
- **polyethylene**
- **polypropylene**
- **polyethylene terephthalate**

REFERENCES

1. Gupta, R.K., Dudeja, P. (2017). Food packaging. In: *Food safety in the 21st century* (pp. 547-553). Academic Press. doi: 10.1016/B978-0-12-801773-9.00046-7.
2. Biji, K.B., Ravishankar, C.N., Mohan, C.O., Srinivasa Gopal, T.K. (2015). Smart packaging systems for food applications: A review. *Journal of Food Science and Technology*, 52, 6125–6135. doi: 10.1007/s13197-015-1766-7.
3. Han, J.W., Ruiz-Garcia, L., Qian, J.P., Yang, X.T. (2018). Food packaging: A comprehensive review and future trends. *Comprehensive Reviews in Food Science and Food Safety*, 17, 860–877. doi: 10.1111/1541-4337.12343.
4. Petkoska, A.T., Daniloski, D., D'Cunha, N.M., Naumovski, N., Broach, A.T. (2021). Edible packaging: Sustainable solutions and novel trends in food packaging. *Food Research International,* 140, 109981. doi: 10.1016/j.foodres.2020.109981.
5. Daniloski, D., Petkoska, A.T., Ščetar, M., Kurek, M., Vaskoska, R., Kalevska, T., Nedelkoska, D.N. (2019). The effect of barrier properties of polymeric films on the shelf life of vacuum packaged fresh pork meat. *Meat Science*, 158, 107880. doi: 10.1016/j.meatsci.2019.107880.
6. Jeevahan, J., Anderson, A., Sriram, V., Durairaj, R.B., Britto Joseph, G., Mageshwaran, G. (2021). Waste into energy conversion technologies and conversion of food wastes into the potential products: A review. *International Journal of Ambient Energy*, 42, 1083–1101. doi: 10.1080/01430750.2018.1537939.
7. Umaraw, P., Munekata, P.E., Verma, A.K., Barba, F.J., Singh, V.P., Kumar, P., Lorenzo, J.M. (2020). Edible films/coating with tailored properties for active packaging of meat, fish and derived products. *Trends in Food Science Technology*, 98, 10–24. doi: 10.1016/j.tifs.2020.01.032.
8. Salgado, P.R., Di Giorgio, L., Musso, Y.S., Mauri, A.N. (2021). Recent developments in smart food packaging focused on biobased and biodegradable polymers. *Frontiers in Sustainable Food Systems*, 5, 630393. doi: 10.3389/fsufs.2021.630393.
9. Marsh, K., Bugusu, B. (2007). Food packaging—roles, materials, and environmental issues. *Journal of Food Science*, 72, R39–R55. doi: 10.1111/j.1750-3841.2007.00301.x.
10. Asgher, M., Qamar, S.A., Bilal, M., Iqbal, H.M. (2020). Bio-based active food packaging materials: Sustainable alternative to conventional petrochemical-based packaging materials. *Food Research International*, 137, 109625. doi: 10.1016/j.foodres.2020.109625.
11. https://www.reportlinker.com (accessed April 01, 2024).
12. Vallejos, S., Trigo-López, M., Arnaiz, A., Miguel, Á., Muñoz, A., Mendía, A., García, J.M. (2022). From classical to advanced use of polymers in food and beverage applications. *Polymers,* 14, 4954. doi: 10.3390/polym14224954.
13. Fahmy, H.M., Eldin, R.E.S., Serea, E.S.A., Gomaa, N.M., AboElmagd, G.M., Salem, S.A., Elsayedf, Z.A., Edreesa, A., Shams-Eldin, E., Shalan, A.E. (2020). Advances in nanotechnology and antibacterial properties of biodegradable food packaging materials. *RSC Advances*, 10, 20467-20484. doi: 10.1039/d0ra02922j.
14. Ncube, L.K., Ude, A.U., Ogunmuyiwa, E.N., Zulkifli, R., Beas, I.N. (2020). Environmental impact of food packaging materials: A review of contemporary

development from conventional plastics to polylactic acid based materials. *Materials,* 13, 4994. doi: 10.3390/ma13214994.

15. Deshwal, G.K., Panjagari, N.R., Alam, T. (2019). An overview of paper and paper based food packaging materials: Health safety and environmental concerns. *Journal of Food Science and Technology*, 56, 4391–4403. doi: 10.1007/s13197-019-03950-z.
16. Cazón, P., Vázquez, M. (2021). Bacterial cellulose as a biodegradable food packaging material: A review. *Food Hydrocolloids,* 113, 106530. doi: 10.1016/j.foodhyd.2020.106530.
17. Hahladakis, J.N. (2020). Delineating the global plastic marine litter challenge: Clarifying the misconceptions. *Environmental Monitoring and Assessment*, 192, 1–11. doi: 10.1007/s10661-020-8202-9.
18. Leslie, H.A., Van Velzen, M.J., Brandsma, S.H., Vethaak, A.D., Garcia-Vallejo, J.J., Lamoree, M.H. (2022). Discovery and quantification of plastic particle pollution in human blood. *Environment International*, 163, 107199. doi: 10.1016/j.envint.2022.107199.
19. Cheng, J., Gao, R., Zhu, Y., Lin, Q. (2024). Applications of biodegradable materials in food packaging: A review. *Alexandria Engineering Journal*, 91, 70–83. doi: 10.1016/j.aej.2024.01.080.
20. Bilal, M., Iqbal, H.M. (2018). Bio-based biopolymers and their potential applications for bio-and non-bio sectors. In: (Eds.), *Handbook of biopolymers: Advances and multifaceted applications.* CRC Press.
21. Ruggero, F., Carretti, E., Gori, R., Lotti, T., Lubello, C. (2020). Monitoring of degradation of starch-based biopolymer film under different composting conditions, using TGA, FTIR and SEM analysis. *Chemosphere*, 246, 125770. doi: 10.1016/j.chemosphere.2019.125770.
22. https://goldbook.iupac.org/terms/view/B00656 (accessed April 01, 2024).
23. Silva, R.R.A., Marques, C.S., Arruda, T.R., Teixeira, S.C., de Oliveira, T.V. (2023). Biodegradation of polymers: Stages, measurement, standards and prospects. *Macromol*, 3, 371–399. doi: 10.3390/macromol3020023.
24. Brdlík, P., Borůvka, M., Běhálek, L., Lenfeld, P. (2021). Biodegradation of poly (lactic acid) biocomposites under controlled composting conditions and freshwater biotope. *Polymers*, 13, 594. doi: 10.3390/polym13040594.
25. Ramesh, H.P., Tharanathan, R.N. (2003). Carbohydrates—The renewable raw materials of high biotechnological value. *Critical Reviews in Biotechnology*, 23, 149–173. doi: 10.1080/713609312.
26. Nešić, A., Cabrera-Barjas, G., Dimitrijević-Branković, S., Davidović, S., Radovanović, N., Delattre, C. (2019). Prospect of polysaccharide-based materials as advanced food packaging. *Molecules*, 25, 135. doi: doi:10.3390/molecules25010135.
27. Youssef, A.M., El-Sayed, S.M. (2018). Bionanocomposites materials for food packaging applications: Concepts and future outlook. *Carbohydrate Polymers*, 193, 19–27. doi: 10.1016/j.carbpol.2018.03.088.
28. Janik, W., Jakubski, Ł., Kudła, S., Dudek, G. (2023). Modified polysaccharides for food packaging applications: A review. *International Journal of Biological Macromolecules*, 128916. doi: 10.1016/j.ijbiomac.2023.128916.
29. Zhang, B., Lan, W., Xie, J. (2022). Chemical modifications in the structure of marine polysaccharide as serviceable food processing and preservation assistant: A review.

International Journal of Biological Macromolecules, 223, 1539–1555. doi: 10.1016/j.ijbiomac.2022.11.034.
30. Hiremani, V.D., Gasti, T., Masti, S.P., Malabadi, R.B., Chougale, R.B. (2022). Polysaccharide-based blend films as a promising material for food packaging applications: Physicochemical properties. *Iranian Polymer Journal*, 31, 503–518. doi: 10.1007/s13726-021-01014-8.
31. Olayil, R., Prabu, V.A., DayaPrasad, S., Naresh, K., Sreekanth, P.R. (2022). A review on the application of bio-nanocomposites for food packaging. *Materials Today: Proceedings*, 56, 1302–1306. doi: 10.1016/j.matpr.2021.11.315.
32. Ramos, M., Valdés, A., Beltrán, A., Garrigós, M.C. (2016). Gelatin-based films and coatings for food packaging applications. *Coatings,* 6, 41. doi: 10.3390/coatings6040041.
33. Oymaci, P., Altinkaya, S.A. (2016). Improvement of barrier and mechanical properties of whey protein isolate based food packaging films by incorporation of zein nanoparticles as a novel bionanocomposite. *Food Hydrocolloids*, 54, 1–9. doi: 10.1016/j.foodhyd.2015.08.030.
34. Calva-Estrada, S.J., Jiménez-Fernández, M., Lugo-Cervantes, E. (2019). Protein-based films: Advances in the development of biomaterials applicable to food packaging. *Food Engineering Reviews,* 11, 78–92. doi: 10.1007/s12393-019-09189-w.
35. Mohamed, S.A., El-Sakhawy, M., El-Sakhawy, M.A.M. (2020). Polysaccharides, protein and lipid-based natural edible films in food packaging: A review. *Carbohydrate Polymers,* 238, 116178. doi: 10.1016/j.carbpol.2020.116178.
36. Yousuf, B., Sun, Y., Wu, S. (2022). Lipid and lipid-containing composite edible coatings and films. *Food Reviews International*, 38, 574–597. doi: 10.1080/87559129.2021.1876084.
37. Devi, L.S., Kalita, S., Mukherjee, A., Kumar, S. (2022). Carnauba wax-based composite films and coatings: Recent advancement in prolonging postharvest shelf life of fruits and vegetables. *Trends in Food Science Technology*, 129, 296–305. doi: 10.1016/j.tifs.2022.09.019.
38. Kumari, S.V.G., Pakshirajan, K., Pugazhenthi, G. (2022). Recent advances and future prospects of cellulose, starch, chitosan, polylactic acid and polyhydroxyalkanoates for sustainable food packaging applications. *International Journal of Biological Macromolecules*, 221, 163–182. doi: 10.1016/j.ijbiomac.2022.08.203.
39. Trapé, D.V., López, O.V. Villar, M.A. (2021). Vinasse: From a residue to a high added value biopolymer. *Bioresources and Bioprocessing*, 8, 130. doi: 10.1186/s40643-021-00476-1.
40. Trapé, D.V., López, O.V., Villar, M.A. (2024). Exploitation of Biomass to the Integrated Production of Bioethanol and Poly (hydroxyalkanoate)s. *BioEnergy Research,* 17, 27–41. doi: 10.1007/s12155-023-10661-z.
41. Trapé, D.V., Ramos, F.D., López, O.V., Diaz, M.S., Villar, M.A. (2023). Sustainable biopolymers production using a bioethanol residue as raw material. *Journal of Polymers and the Environment*, 31, 804–811. doi: 10.1007/s10924-022-02639-3.
42. https://www.foodpackagingforum.org/food-packaging-health/food-packaging-materials (accessed April 01, 2024).
43. Golmakani, M. T., Hajjari, M.M., Kiani, F., Sharif, N., Hosseini, S.M.H. (2024). Application of electrospinning to fabricate phycocyanin-and Spirulina extract-loaded

gliadin fibers for active food packaging. *Food Chemistry: X*, 22, 101275. doi: 10.1016/j.fochx.2024.101275.

44. Zhang, Y., Min, T., Zhao, Y., Cheng, C., Yin, H., Yue, J. (2024). The developments and trends of electrospinning active food packaging: A review and bibliometrics analysis. *Food Control,* 110291. doi: 10.1016/j.foodcont.2024.110291.
45. Jagoda, S.U.M., Gamage, J.R., Karunathilake, H.P. (2023). Environmentally sustainable plastic food packaging. *Journal of Cleaner Production*, 400, 136680. doi: 10.1016/j.jclepro.2023.136680.
46. Du, H., Sun, X., Chong, X., Yang, M., Zhu, Z., Wen, Y. (2023). A review on smart active packaging systems for food preservation: Applications and future trends. *Trends in Food Science Technology*, 104200. doi: 10.1016/j.tifs.2023.104200.
47. Vilela, C., Kurek, M., Hayouka, Z., Röcker, B., Yildirim, S., Antunes, M.D.C., Freire, C.S. (2018). A concise guide to active agents for active food packaging. *Trends in Food Science Technology*, 80, 212–222. doi: 10.1016/j.tifs.2018.08.006.
48. Fan, X., Yang, T., Li, Y., Zhao, J., Rong, L., Feng, L., Huang, J. (2023). Fabrication of depolymerized zein films via high-intensity ultrasound for ethylene adsorption packaging. *Food Packaging and Shelf life,* 39, 101159. doi: 10.1016/j.fpsl.2023.101159.
49. Nian, L., Wang, M., Zeng, Y., Jiang, J., Cheng, S., Cao, C. (2023). Modified HKUST-1-based packaging with ethylene adsorption property for food preservation. *Food Hydrocolloids*, 135, 108204. doi: 10.1016/j.foodhyd.2022.108204.
50. Kumar, P., Tripathi, S., Ramakanth, D., Gaikwad, K.K. (2024). Novel ethylene scavenger based on sillimanite and bentonite clay for packaging applications: A sustainable alternative for preservation of fresh produce. *Sustainable Chemistry and Pharmacy*, 38, 101516. doi: 10.1016/j.scp.2024.101516.
51. Zúniga, M.M.G., Ding, R., Oh, E., Nguyen, T.B., Trung, T.T., Nam, J.D., Suhr, J. (2024). Avocado seed starch utilized in eco-friendly, UV-blocking, and high-barrier polylactic acid (PLA) biocomposites for active food packaging applications. *International Journal of Biological Macromolecules*, 265, 130837. doi: 10.1016/j.ijbiomac.2024.130837.
52. Roy, S., Ramakrishnan, R., Goksen, G., Singh, S., Łopusiewicz, Ł. (2023). Recent progress on UV-light barrier food packaging films–a systematic review. *Innovative Food Science Emerging Technologies*, 103550. doi: 10.1016/j.ifset.2023.103550.
53. Gasti, T., Dixit, S., Hiremani, V.D., Chougale, R.B., Masti, S.P., Vootla, S.K., Mudigoudra, B.S. (2022). Chitosan/pullulan based films incorporated with clove essential oil loaded chitosan-ZnO hybrid nanoparticles for active food packaging. *Carbohydrate Polymers*, 277, 118866. doi: 10.1016/j.carbpol.2021.118866.
54. Sun, Q., Zhang, L., Huang, M., Ma, M., Zeng, J., Le, T. (2024). Modification of starch-derived graphene quantum dots as multifunctional nanofillers to produce polymer starch/polyvinyl alcohol composite films for active packaging. *LWT,* 198, 115953. doi: 10.1016 /j.lwt.2024.115953.
55. Haridevamuthu, B., Raj, D., Chandran, A., Murugan, R., Seetharaman, S., Dhanaraj, M., Almutairi, B.O., Arokiyaraj, S., Arockiaraj, J. (2024). Sustainable food packaging: Harnessing biowaste of *Terminalia catappa* L. for chitosan-based biodegradable active films for shrimp storage. *Carbohydrate Polymers*, 329, 121798. doi: 10.1016/j.carbpol.2024.121798.

56. Feng, L., Jiang, X., Han, J., Li, L., Kitazawa, H., Wang, X., Liu, H. (2023). Properties of an active film based on glutenin/tamarind gum and loaded with binary microemulsion of melatonin/pummelo essential oil and its preservation for Agaricus bisporus. *Food Chemistry*, 429, 136901. doi: 10.1016/j.foodchem.2023.136901.
57. Chandrasekar, C.M., Nespoli, L., Bellesia, T., Ghaani, M., Farris, S., Romano, D. (2024). Fabrication of double layer nanoparticle infused starch-based thermoplastic food packaging system for meat preservation. *International Journal of Biological Macromolecules*, 254, 127689. doi: 10.1016/j.ijbiomac.2023.127689.
58. Wang, L., Yuan, M., Sun, E., Wu, J., Lv, A., Zhang, X., Wang, K. (2023). Antibacterial food packaging capable of sustained and unidirectional release carvacrol/thymol nanoemulsions for pork preservation. *Food Hydrocolloids*, 145, doi: 10.1016/109169. 10.1016/j.foodhyd.2023.109169.
59. He, F., Kong, Q., Jin, Z., Mou, H. (2020). Developing a unidirectionally permeable edible film based on κ-carrageenan and gelatin for visually detecting the freshness of grass carp fillets. *Carbohydrate Polymers*, 241, 116336. doi: 10.1016/j.carbpol.2020.116336.
60. Torales, A.C., Gutiérrez, D.R., Rodríguez, S.D.C. (2020). Influence of passive and active modified atmosphere packaging on yellowing and chlorophyll degrading enzymes activity in fresh-cut rocket leaves. *Food Packaging and Shelf life*, 26, 100569. doi: 10.1016/j.fpsl.2020.100569.
61. Liu, K., Dong, H., Peng, J., Liao, W., Yang, X., He, Q. (2024). Design of equilibrium modified atmosphere packaging for postharvest cabbages preservation based on introducing available active sites into film materials as gas transport channels. *Food Research International*, 177, 113900. doi: 10.1016/j.foodres.2023.113900.
62. Abedi-Firoozjah, R., Parandi, E., Heydari, M., Kolahdouz-Nasiri, A., Bahraminejad, M., Mohammadi, R., Garavand, F. (2023). Betalains as promising natural colorants in smart/active food packaging. *Food Chemistry*, 424, 136408. doi: 10.1016/j.foodchem.2023.136408.
63. Manzoor, A., Yousuf, B., Pandith, J.A., Ahmad, S. (2023). Plant-derived active substances incorporated as antioxidant, antibacterial or antifungal components in coatings/films for food packaging applications. *Food Bioscience*, 53, 102717. doi: 10.1016/j.fbio.2023.102717.
64. Mansour, H.A., Abdelrahman, H.A., Zayed, N.E., Abdel-Naeem, H.H. (2023). The effects of novel alginate-lauric arginate coatings with temperature on bacterial quality, oxidative stability, and organoleptic characteristics of frozen stored chicken drumsticks. *International Journal of Biological Macromolecules*, 239, 124242. doi: 10.1016/j.ijbiomac.2023.124242.
65. Yahaya, W.A.W., Chik, S.M.S.T., Azman, N.A.M., Nor, A.M., Hamid, K.H.A., Ajit, A. (2023). Mechanical properties and antioxidant activity of carrageenan-cellulose nanofiber incorporated butylated hydroxyanisole as active food packaging. *Materials Today: Proceedings*, 107, 128–135. doi: 10.1016/j.matpr.2023.08.180.
66. Wang, P., Li, Y., Zhang, C., Que, F., Weiss, J., Zhang, H. (2020). Characterization and antioxidant activity of trilayer gelatin/dextran-propyl gallate/gelatin films: Electrospinning versus solvent casting. *LWT*, 128, 109536. doi: 10.1016/j.lwt.2020.109536.

67. Shi, Y.G., Li, D.H., Kong, Y.M., Zhang, R.R., Gu, Q., Hu, M.X., Jin, W.G. (2022). Enhanced antibacterial efficacy and mechanism of octyl gallate/beta-cyclodextrins against *Pseudomonas fluorescens* and *Vibrio parahaemolyticus* and incorporated electrospun nanofibers for Chinese giant salamander fillets preservation. *International Journal of Food Microbiology*, 361, 109460. doi: 10.1016/j.ijfoodmicro.2021.109460.
68. Fasihnia, S.H., Peighambardoust, S.H., Peighambardoust, S.J., Oromiehie, A. (2018). Development of novel active polypropylene based packaging films containing different concentrations of sorbic acid. *Food Packaging and Shelf life*, 18, 87–94. doi: 10.1016/j.fpsl.2018.10.001.
69. Oliviero, M., Lamberti, E., Cafiero, L., Pace, B., Cefola, M., Gorrasi, G., Sorrentino, A. (2023). Biodegradable cellulose acetate/layered double-hydroxide composite film for active packaging of fresh food. *Materials Chemistry and Physics*, 310, 128469. doi: 10.1016/j.matchemphys.2023.128469.
70. Wangprasertkul, J., Siriwattanapong, R., Harnkarnsujarit, N. (2021). Antifungal packaging of sorbate and benzoate incorporated biodegradable films for fresh noodles. *Food Control*, 123, 107763. doi: 10.1016/j.foodcont.2020.107763.
71. Viscusi, G., Bugatti, V., Vittoria, V., Gorrasi, G. (2021). Antimicrobial sorbate anchored to layered double hydroxide (LDH) nano-carrier employed as active coating on Polypropylene (PP) packaging: Application to bread stored at ambient temperature. *Future Foods*, 4, 100063. doi: 10.1016/j.fufo.2021.100063.
72. San, H., Harnkarnsujarit, N. (2023). Sulfite incorporated thermoplastic cassava starch blended PBAT blown films as antimicrobial and antibrowning packaging. *Industrial Crops and Products*, 206, 117610. doi: 10.1016/ j.indcrop.2023.117610.
73. Ham, J., Lim, W., Park, S., Bae, H., You, S., Song, G. (2019). Synthetic phenolic antioxidant propyl gallate induces male infertility through disruption of calcium homeostasis and mitochondrial function. *Environmental Pollution*, 248, 845–856. doi: 10.1016/j.envpol.2019.02.087.
74. Devaramani, S., Kumar, K.S., Suma, B.P., Pandurangappa, M. (2022). Rhodamine B phenylhydrazide as a new chemosensor for sulfite quantification: Application to food samples. *Materials Today: Proceedings*, 49, 748–755. doi: 10.1016/j.matpr.2021.05.227.
75. Villar, A., Merino, S., Boto, R.A., Aizpurua, J., Garcia-Etxarri, A., Azkune, M., Zubia, J. (2023). Sulfites detection by surface-enhanced Raman spectroscopy: A feasibility study. *Spectrochimica Acta Part A: Molecular and Biomolecular Spectroscopy*, 300, 122899. doi: 10.1016/j.saa.2023.122899.
76. Manoranjitham, J.J., Narayanan, S.S. (2021). Electrochemical sensor for determination of butylated hydroxyanisole (BHA) in food products using poly O-cresolphthalein complexone coated multiwalled carbon nanotubes electrode. *Food Chemistry*, 342, 128246. doi: 10.1016/j.foodchem.2020.128246.
77. Felter, S.P., Zhang, X., Thompson, C. (2021). Butylated hydroxyanisole: Carcinogenic food additive to be avoided or harmless antioxidant important to protect food supply? *Regulatory Toxicology and Pharmacology*, 121, 104887. doi: 10.1016/j.yrtph.2021.104887.
78. Kumar, V., Bansal, V., Madhavan, A., Kumar, M., Sindhu, R., Awasthi, M.K., Saran, S. (2022). Active pharmaceutical ingredient (API) chemicals: A critical

review of current biotechnological approaches. *Bioengineered*, 13, 4309–4327. doi: 10.1080/21655979.2022.2031412.

79. Bolan, S., Sharma, S., Mukherjee, S., Zhou, P., Mandal, J., Srivastava, P., Bolan, N. (2024). The distribution, fate, and environmental impacts of food additive nanomaterials in soil and aquatic ecosystems. *Science of the Total Environment*, 916, 170013. doi: 10.1016/j.scitotenv.2024.170013.
80. Ácsová, A., Hojerová, J., Hergesell, K., Hideg, É., Csepregi, K., Bauerová, K., Martiniaková, S. (2022). Antioxidant and Anti-Pollution Effect of Naturally Occurring Carotenoids Astaxanthin and Crocin for Human Skin Protection. *Chemistry Select*, 7, e202201595. doi: 10.1002/slct.202201595.
81. Li, X., Pu, Q., Xu, Y., Yang, H., Wu, Y., Wang, W., Li, Y. (2024). The masking phenomenon of microplastics additives on oxidative stress responses in freshwater food chains. *Science of The Total Environment*, 927, 172156. doi: 10.1016/j.scitotenv.2024.172156.
82. Gervasoni, L.F., Gervasoni, K., de Oliveira Silva, K., Mendes, M.E.F., Maddela, N.R., Prasad, R., Winkelstroter, L.K. (2023). Postbiotics in active food packaging: The contribution of cellulose nanocomposites. *Sustainable Chemistry and Pharmacy*, 36, 101280. doi: 10.1016/j.scp.2023.101280.
83. Correia, L.F., da Silva Pinho, G., da Cruz Neves, T.J., de Oliveira Vieira, K.C., Maddela, N.R., Prasad, R., Winkelstroter, L.K. (2024). Nanotechnology innovation combined with bacteriocins as emerging strategy for the development of active and intelligent food packaging. *Sustainable Chemistry and Pharmacy*, 39, 101551. doi: 10.1016/j.scp.2024.101551.
84. Hong, S.J., Riahi, Z., Shin, G.H., Kim, J.T. (2024). Pseudomonas aeruginosa-derived carbon dots doped with sulfur as active packaging materials for fresh food preservation. *Food Bioscience,* 57, 103506. doi: 10.1016/j.fbio.2023.103506.
85. Velásquez, P., Montenegro, G., Valenzuela, L.M., Giordano, A., Cabrera-Barjas, G., Martin-Belloso, O. (2022). K-carrageenan edible films for beef: Honey and bee pollen phenolic compounds improve their antioxidant capacity. *Food Hydrocolloids*, 124, 107250. doi: 10.1016/j.foodhyd.2021.107250.
86. Ananthi, P., Hemkumar, K., Subasini, S., Pius, A. (2023). Development of biodegradable films reinforced with silver functionalized cow milk carbon dots for active food packaging applications. *Materials Today Sustainability*, 24, 100609. doi: 10.1016/j.mtsust.2023.100609.
87. Zhang, Z., Zhou, X., Wang, D., Fang, C., Zhang, W., Wang, C., Huang, Z. (2021). Lysozyme-based composite membranes and their potential application for active packaging. *Food Bioscience*, 43, 101078. doi: 10.1016/j.fbio.2021.101078.
88. Zhang, W., Rhim, J.W. (2022). Functional edible films/coatings integrated with lactoperoxidase and lysozyme and their application in food preservation. *Food Control*, 133, 108670. doi: 10.1016/j.foodcont.2021.108670.
89. Papoutsis, K., Grasso, S., Menon, A., Brunton, N.P., Lyng, J.G., Jacquier, J.C., Bhuyan, D.J. (2020). Recovery of ergosterol and vitamin D2 from mushroom waste-Potential valorization by food and pharmaceutical industries. *Trends in Food Science Technology*, 99, 351–366. doi: 10.1016/j.tifs.2020.03.005.
90. Gomaa, M., Hifney, A.F., Fawzy, M.A., Abdel-Gawad, K.M. (2018). Use of seaweed and filamentous fungus derived polysaccharides in the development of

alginate-chitosan edible films containing fucoidan: Study of moisture sorption, polyphenol release and antioxidant properties. *Food Hydrocolloids*, 82, 239–247. doi: 10.1016/j.foodhyd.2018.03.056.

91. Vicario, A.L., García, M.G., Ochoa, N.A., Quiroga, E. (2024). Assessing the impact of chañar and green tea extracts on pectin active packaging for extended food preservation. *Food Hydrocolloids*, 153, 110009. doi: 10.1016/j.foodhyd.2024.110009.
92. Mondal, K., Bhattacharjee, S.K., Mudenur, C., Ghosh, T., Goud, V.V., Katiyar, V. (2022). Development of antioxidant-rich edible active films and coatings incorporated with de-oiled ethanolic green algae extract: A candidate for prolonging the shelf life of fresh produce. *RSC Advances*, 12, 13295–13313. doi: 10.1039/d2ra00949h.
93. Severo, C., Anjos, I., Souza, V.G., Canejo, J.P., Bronze, M.R., Fernando, A.L., Ribeiro, I.A. (2021). Development of cranberry extract films for the enhancement of food packaging antimicrobial properties. *Food Packaging and Shelf life*, 28, 100646. doi: 10.1016/j.fpsl.2021.100646.
94. Zhang, X., Zhang, B., Mao, R., Huang, Z., Jing, K., Jin, C., Hu, Y. (2024). A novel multilayer film based on sodium alginate/k-carrageenan-gelatin incorporated with ZnO nanoparticles and oregano essential oil for active food packing. *Progress in Organic Coatings*, 187, 108170. doi: 10.1016/j.porgcoat.2023.108170.
95. Priya, N.V., Vinitha, U.G., Sundaram, M.M. (2021). Preparation of chitosan-based antimicrobial active food packaging film incorporated with *Plectranthus amboinicus* essential oil. *Biocatalysis and Agricultural Biotechnology*, 34, 102021. doi: 10.1016/j.bcab.2021.102021.
96. Mondal, K., Bhattacharjee, S.K., Goud, V.V., Katiyar, V. (2024). Effect of waste *Dunaliella tertiolecta* biomass ethanolic extract and turmeric essential oil on properties of guar gum-based active films. *Food Hydrocolloids*, 146, 109199. doi: 10.1016/j.foodhyd.2023.109199.
97. Abdo, S.M., Youssef, M., El Nagar, I., Mohamed, H.E., El-Kholy, S.A., Youssef, A.M. (2024). Processing and characterization of antimicrobial bioplastic films based on green microalgae *Scenedesmus obliquus* extract-loaded polyurethane. *International Journal of Biological Macromolecules*, 257, 128711. doi: 10.1016/j.ijbiomac.2023.128711.
98. Díaz-Montes, E., Castro-Muñoz, R. (2021). Edible films and coatings as food-quality preservers: An overview. *Foods*, 10, 249. doi: 10.3390/foods10020249.
99. Souabi, S., Anouzla, A. (2023). Wastewater from olive oil production. In: Souabi, S., Anouzla, A. (Eds.), *Wastewater from Olive Oil Production* (pp. 297). Springer.
100. Ben-Othman, S., Jõudu, I., Bhat, R. (2020). Bioactives from agri-food wastes: Present insights and future challenges. *Molecules*, 25, 510. doi: 10.3390/molecules25030510.
101. Gullón, B., Gullón, P., Eibes, G., Cara, C., De Torres, A., López-Linares, J.C., Castro, E. (2018). Valorisation of olive agro-industrial by-products as a source of bioactive compounds. *Science of the Total Environment*, 645, 533–542. doi: 10.1016/j.scitotenv.2018.07.155.
102. Deshmukh, R.K., Gaikwad, K.K. (2024). Natural antimicrobial and antioxidant compounds for active food packaging applications. *Biomass Conversion and Biorefinery*, 14, 4419–4440. doi: 10.1007/s13399-022-02623-w.
103. Ravindran, R., Jaiswal, A.K. (2016). Exploitation of food industry waste for high-value products. *Trends in Biotechnology*, 34, 58–69. doi: 10.1016/j.tibtech.2015.10.008.

104. Cejudo-Bastante, C., Arjona-Mudarra, P., Fernández-Ponce, M.T., Casas, L., Mantell, C., Martínez de la Ossa, E.J., Pereyra, C. (2021). Application of a natural antioxidant from grape pomace extract in the development of bioactive jute fibers for food packaging. *Antioxidants*, 10, 216. doi: 10.3390/antiox10020216.
105. Figueroa-Lopez, K.J., Vicente, A.A., Reis, M.A., Torres-Giner, S., Lagaron, J.M. (2019). Antimicrobial and antioxidant performance of various essential oils and natural extracts and their incorporation into biowaste derived poly (3-hydroxybutyrate-co-3-hydroxyvalerate) layers made from electrospun ultrathin fibers. *Nanomaterials*, 9, 144. doi: 10.3390/nano9020144.
106. de Dicastillo, C.L., Bustos, F., Guarda, A., Galotto, M.J. (2016). Cross-linked methyl cellulose films with murta fruit extract for antioxidant and antimicrobial active food packaging. *Food Hydrocolloids*, 60, 335–344. doi: 10.1016/j.foodhyd.2016.03.020.
107. Yuan, G., Lv, H., Yang, B., Chen, X., Sun, H. (2015). Physical properties, antioxidant and antimicrobial activity of chitosan films containing carvacrol and pomegranate peel extract. *Molecules*, 20, 11034–11045. doi: 10.3390/molecules200611034.
108. Moreira, D., Gullón, B., Gullón, P., Gomes, A., Tavaria, F. (2016). Bioactive packaging using antioxidant extracts for the prevention of microbial food-spoilage. *Food Function*, 7, 3273–3282. doi: 10.1039/C6FO00553E.
109. Liang, S., Wang, L. (2018). A natural antibacterial-antioxidant film from soy protein isolate incorporated with cortex phellodendron extract. *Polymers*, 10, 71. doi: 10.3390/polym10010071.
110. Zhang, X., Liu, J., Yong, H., Qin, Y., Liu, J., Jin, C. (2020). Development of antioxidant and antimicrobial packaging films based on chitosan and mangosteen (*Garcinia mangostana* L.) rind powder. *International Journal of Biological Macromolecules*, 145, 1129–1139. doi: 10.1016/j.ijbiomac.2019.10.038.
111. de Araújo, G.K.P., de Souza, S.J., da Silva, M.V., Yamashita, F., Gonçalves, O.H., Leimann, F.V., Shirai, M.A. (2015). Physical, antimicrobial and antioxidant properties of starch-based film containing ethanolic propolis extract. *International Journal of Food Science Technology*, 50, 2080–2087. doi: 10.1111/ijfs.12869.
112. Ali, A., Ahmed, S. (2021). Eco-friendly natural extract loaded antioxidative chitosan/polyvinyl alcohol based active films for food packaging. *Heliyon*, 7, e06550. doi: 10.1016/j.heliyon.2021.e06550.
113. Rambabu, K., Bharath, G., Banat, F., Show, P.L., Cocoletzi, H.H. (2019). Mango leaf extract incorporated chitosan antioxidant film for active food packaging. *International Journal of Biological Macromolecules*, 126, 1234–1243. doi: 10.1016/j.ijbiomac.2018.12.196.
114. Abbattista, R., Ventura, G., Calvano, C.D., Cataldi, T.R., Losito, I. (2021). Bioactive compounds in waste by-products from olive oil production: Applications and structural characterization by mass spectrometry techniques. *Foods*, 10, 1236. doi: 10.3390/foods10061236.
115. Maesano, G., Chinnici, G., Falcone, G., Bellia, C., Raimondo, M., D'Amico, M. (2021). Economic and environmental sustainability of olive production: A case study. *Agronomy*, 11, 1753. doi: 10.3390/agronomy11091753.
116. https://www.internationaloliveoil.org/what-we-do/economic-affairs-promotion-unit/#figures (accessed April 01, 2024).

117. Dermeche, S., Nadour, M., Larroche, C., Moulti-Mati, F., Michaud, P. (2013). Olive mill wastes: Biochemical characterizations and valorization strategies. *Process Biochemistry*, 48, 1532–1552. doi: 10.1016/j.procbio.2013.07.010.
118. Hassen, W., Hassen, B., Werhani, R., Hidri, Y., Jedidi, N., Hassen, A. (2023). Processes of valorization and management of olive by-products: The Pomace and Olive Mill Wastewater. In: (Eds.), *Wastewater from Olive Oil Production: Environmental Impacts, Treatment and Valorisation* (pp. 1–25). Springer International Publishing.
119. Fathy, S.A., Mahmoud, A.E., Rashad, M.M., Ezz, M.K., Mohammed, A.T. (2018). Improving the nutritive value of olive pomace by solid state fermentation of *Kluyveromyces marxianus* with simultaneous production of gallic acid. *International Journal of Recycling of Organic Waste in Agriculture*, 7, 135–141. doi: 10.1007/s40093-018-0199-5.
120. Ramos, P., Santos, S.A., Guerra, Â.R., Guerreiro, O., Felício, L., Jerónimo, E., Duarte, M. (2013). Valorization of olive mill residues: Antioxidant and breast cancer antiproliferative activities of hydroxytyrosol-rich extracts derived from olive oil by-products. *Industrial Crops and Products*, 46, 359–368. doi: 10.1016/j.indcrop.2013.02.020.
121. Tapia-Quirós, P., Montenegro-Landívar, M.F., Reig, M., Vecino, X., Cortina, J.L., Saurina, J., Granados, M. (2022). Recovery of polyphenols from agri-food by-products: The olive oil and winery industries cases. *Foods*, 11, 362. doi: 10.3390/foods11030362.
122. Otero, P., Garcia-Oliveira, P., Carpena, M., Barral-Martinez, M., Chamorro, F., Echave, J., Garcia-Perez, P., Cao, H., Xiao, J., Simal-Gandara, J., Prieto, M.A. (2021). Applications of by-products from the olive oil processing: Revalorization strategies based on target molecules and green extraction technologies. *Trends in Food Science Technology*, 116, 1084–1104. doi: 10.1016/j.tifs.2021.09.00712.
123. Akretche, H., Pierre, G., Moussaoui, R., Michaud, P., Delattre, C. (2019). Valorization of olive mill wastewater for the development of biobased polymer films with antioxidant properties using eco-friendly processes. *Green Chemistry*, 21, 3065-3073. doi: 10.1039/x0xx00000x.
124. Nunes, M.A., Pimentel, F.B., Costa, A.S., Alves, R.C., Oliveira, M.B.P. (2016). Olive by-products for functional and food applications: Challenging opportunities to face environmental constraints. *Innovative Food Science Emerging Technologies*, 35, 139–148. doi: 10.1016/j.ifset.2016.04.016.
125. Luzi, F., Pannucci, E., Clemente, M., Grande, E., Urciuoli, S., Romani, A., Santi, L. (2021). Hydroxytyrosol and oleuropein-enriched extracts obtained from olive oil wastes and by-products as active antioxidant ingredients for poly (vinyl alcohol)-based films. *Molecules*, 26, 2104. doi: 10.3390/molecules26072104.
126. Apicella, A., Adiletta, G., Di Matteo, M., Incarnato, L. (2019). Valorization of olive industry waste products for development of new eco-sustainable, multilayer antioxidant packaging for food preservation. *Chemical Engineering Transactions*, 75, 85–90. doi: 10.3303/CET1975015.
127. Galanakis, C.M., Tsatalas, P., Galanakis, I.M. (2018). Implementation of phenols recovered from olive mill wastewater as UV booster in cosmetics. *Industrial Crops and Products,* 111, 30–37. doi: 10.1016/j.indcrop.2017.09.058.

128. Tafesh, A., Najami, N., Jadoun, J., Halahlih, F., Riepl, H., Azaizeh, H. (2011). Synergistic antibacterial effects of polyphenolic compounds from olive mill wastewater. *Evidence-Based Complementary and Alternative Medicine*, 2011, 431021. doi: 10.1155/2011/431021.
129. Yangui, T., Sayadi, S., Rhouma, A., Dhouib, A. (2010). Potential use of hydroxytyrosol-rich extract from olive mill wastewater as a biological fungicide against Botrytis cinerea in tomato. *Journal of Pest Science*, 83, 437–445. doi: 10.1007/s10340-010-0314-5.
130. https://www.oiv.int/sites/default/files/documents/OIV_State_of_the_world_Vine_and_Wine_sector_in_2022_2.pdf (accessed April 01, 2024).
131. Ahmad, B., Yadav, V., Yadav, A., Rahman, M.U., Yuan, W.Z., Li, Z., Wang, X. (2020). Integrated biorefinery approach to valorize winery waste: A review from waste to energy perspectives. *Science of the Total Environment*, 719, 137315. doi: 10.1016/j.scitotenv.2020.137315.
132. Oliveira, M., Duarte, E. (2016). Integrated approach to winery waste: Waste generation and data consolidation. *Frontiers of Environmental Science Engineering*, 10, 168–176. doi: 10.1007/s11783-014-0693-6.
133. Ioannou, L.A., Puma, G.L., Fatta-Kassinos, D. (2015). Treatment of winery wastewater by physicochemical, biological and advanced processes: A review. *Journal of Hazardous Materials,* 286, 343–368. doi: 10.1016/j.jhazmat.2014.12.043.
134. Alañón, M.E., Castro-Vázquez, L., Díaz-Maroto, M.C., Gordon, M.H., Pérez-Coello, M.S. (2011). A study of the antioxidant capacity of oak wood used in wine ageing and the correlation with polyphenol composition. *Food Chemistry*, 128, 997–1002. doi: 10.1016/j.foodchem.2011.04.005.
135. Alañón, M.E., Alarcón, M., Marchante, L., Díaz-Maroto, M.C., Pérez-Coello, M.S. (2017). Extraction of natural flavorings with antioxidant capacity from cooperage by-products by green extraction procedure with subcritical fluids. *Industrial Crops and Products*, 103, 222–232. doi: 10.1016/j.indcrop.2017.03.050.
136. Athanasiou, P.E., Gkountela, C.I., Patila, M., Fotiadou, R., Chatzikonstantinou, A.V., Vouyiouka, S.N., Stamatis, H. (2024). Laccase-mediated oxidation of phenolic compounds from wine lees extract towards the synthesis of polymers with potential applications in food packaging. *Biomolecules*, 14, 323. doi: 10.1016/j.scitotenv.2020.137315.
137. Kaynarca, G.B., Kamer, D.D.A., Gumus, T., Sagdıc, O. (2023). Characterization of Poly (vinyl alcohol)/gelatin films made with winery solid by-product (vinasse) extract. *Food Packaging and Shelf life*, 35, 101013. doi: 10.1016/j.fpsl.2022.101013.
138. Firouz, M.S., Mohi-Alden, K., Omid, M. (2021). A critical review on intelligent and active packaging in the food industry: Research and development. *Food Research International*, 141, 110113. doi: 10.1016/j.foodres.2021.110113.
139. Teixeira, A., Baenas, N., Dominguez-Perles, R., Barros, A., Rosa, E., Moreno, D.A., Garcia-Viguera, C. (2014). Natural bioactive compounds from winery by-products as health promoters: A review. *International Journal of Molecular Sciences*, 15, 15638–15678. doi: 10.3390/ijms150915638.
140. Abd Rani, U., Ng, L.Y., Ng, C.Y., Mahmoudi, E. (2020). A review of carbon quantum dots and their applications in wastewater treatment. *Advances in Colloid and Interface Science*, 278, 102124. doi: 10.1016/j.cis.2020.102124.

141. Rubilar, M., Pinelo, M., Shene, C., Sineiro, J., Nuñez, M.J. (2007). Separation and HPLC-MS identification of phenolic antioxidants from agricultural residues: Almond hulls and grape pomace. *Journal of Agricultural and Food Chemistry*, 55, 10101–10109. doi: 10.1021/jf0721996.
142. Georgiev, V., Ananga, A., Tsolova, V. (2014). Recent advances and uses of grape flavonoids as nutraceuticals. *Nutrients*, 6, 391–415. doi: 10.3390/nu6010391.
143. Hixson, J.L., Jacobs, J.L., Wilkes, E.N., Smith, P.A. (2016). Survey of the variation in grape marc condensed tannin composition and concentration and analysis of key compositional factors. *Journal of Agricultural and Food Chemistry*, 64, 7076–7086. doi: 10.1021/acs.jafc.6b03126.
144. Pinelo, M., Arnous, A., Meyer, A.S. (2006). Upgrading of grape skins: Significance of plant cell-wall structural components and extraction techniques for phenol release. *Trends in Food Science Technology*, 17, 579–590. doi: 10.1016/j.tifs.2006.05.003.
145. Chedea, V.S., Echim, C., Braicu, C., Andjelkovic, M., Verhe, R., Socaciu, C. (2011). Composition in polyphenols and stability of the aqueous grape seed extract from the Romanian variety "Merlot Recas". *Journal of Food Biochemistry*, 35, 92–108. doi: 10.1111/j.1745-4514.2010.00368.x.
146. Monteleone, E., Condelli, N., Dinnella, C., Bertuccioli, M. (2004). Prediction of perceived astringency induced by phenolic compounds. *Food Quality and Preference*, 15, 761–769. doi: 10.1016/j.foodqual.2004.06.002.
147. Saurabh, C.K., Gupta, S., Variyar, P.S. (2018). Development of guar gum based active packaging films using grape pomace. *Journal of Food Science and Technology*, 55, 1982–1992. doi: 10.1007/s13197-018-3112-3.
148. Xu, Y., Willis, S., Jordan, K., Sismour, E. (2018). Chitosan nanocomposite films incorporating cellulose nanocrystals and grape pomace extracts. *Packaging Technology and Science*, 31, 631–638. doi: 10.1002/pts.2389.
149. Rodrigues, R.P., Gando-Ferreira, L.M., Quina, M.J. (2022). Increasing value of winery residues through integrated biorefinery processes: A review. *Molecules*, 27, 4709. doi: 10.3390/molecules27154709.
150. Engel, J.B., Ambrosi, A., Tessaro, I.C. (2019). Development of biodegradable starch-based foams incorporated with grape stalks for food packaging. *Carbohydrate Polymers*, 225, 115234. doi: 10.1016/j.carbpol.2019.115234.
151. Dobrinčić, A., Balbino, S., Zorić, Z., Pedisić, S., Bursać Kovačević, D., Elez Garofulić, I., Dragović-Uzelac, V. (2020). Advanced technologies for the extraction of marine brown algal polysaccharides. *Marine Drugs*, 18, 168. doi: 10.3390/md18030168.
152. Dutta, S., He, M., Xiong, X., Tsang, D.C. (2021). Sustainable management and recycling of food waste anaerobic digestate: A review. *Bioresource Technology*, 341, 125915. doi: 10.1016/J.BIORTECH.2021.125915.
153. Khoshnevisan, B., Duan, N., Tsapekos, P., Awasthi, M.K., Liu, Z., Mohammadi, A., Liu, H. (2021). A critical review on livestock manure biorefinery technologies: Sustainability, challenges, and future perspectives. *Renewable and Sustainable Energy Reviews*, 135, 110033. doi: 10.1016/J.RSER.2020.110033.
154. Mak, T.M., Xiong, X., Tsang, D.C., Iris, K.M., Poon, C.S. (2020). Sustainable food waste management towards circular bioeconomy: Policy review, limitations and opportunities. *Bioresource Technology*, 297, 122497. doi: 10.1016/J.BIORTECH.2019.122497.

155. Vera, J., Castro, J., Gonzalez, A., Moenne, A. (2011). Seaweed polysaccharides and derived oligosaccharides stimulate defense responses and protection against pathogens in plants. *Marine Drugs*, 9, 2514–2525. doi: 10.3390/md9122514.
156. Bojorges, H., Fabra, M.J., López-Rubio, A., Martínez-Abad, A. (2022). Alginate industrial waste streams as a promising source of value-added compounds valorization. *Science of the Total Environment*, 838, 156394. doi: 10.1016/j.scitotenv.2022.156394.
157. Fawzy, M.A., Gomaa, M., Hifney, A.F., Abdel-Gawad, K.M. (2017). Optimization of alginate alkaline extraction technology from *Sargassum latifolium* and its potential antioxidant and emulsifying properties. *Carbohydrate Polymers*, 157, 1903–1912. doi: 10.1016/j.carbpol.2016.11.077.
158. Pawar, S.N., Edgar, K.J. (2012). Alginate derivatization: A review of chemistry, properties and applications. *Biomaterials*, 33, 3279–3305. doi: 10.1016/j.biomateriales.2012.01.007.
159. Quideau, S., Deffieux, D., Douat-Casassus, C., Pouységu, L. (2011). Plant polyphenols: Chemical properties, biological activities, and synthesis. *Angewandte Chemie International Edition*, 50, 586–621. doi: 10.1002/anie.201000044.
160. Lemesheva, V., Islamova, R., Stepchenkova, E., Shenfeld, A., Birkemeyer, C., Tarakhovskaya, E. (2023). Antibacterial, antifungal and algicidal activity of phlorotannins, as principal biologically active components of ten species of brown algae. *Plants,* 12, 821. doi: 10.3390/plants12040821.
161. Negara, B.F.S.P., Sohn, J.H., Kim, J.S., Choi, J.S. (2021). Antifungal and larvicidal activities of phlorotannins from brown seaweeds. *Marine Drugs*, 19, 223. doi: 10.3390/md19040223.
162. Pradhan, B., Nayak, R., Bhuyan, P.P., Patra, S., Behera, C., Sahoo, S., Jena, M. (2022). Algal phlorotannins as novel antibacterial agents with reference to the antioxidant modulation: Current advances and future directions. *Marine Drugs*, 20, 403. doi: 10.3390/md20060403.
163. Pal, D., Raj, K. (2021). Biological activities of marine products and nutritional importance. *Bioactive Natural Products for Pharmaceutical Applications*, 140, 587–616. doi: 10.1007/978-3-030-54027-2_17.
164. Zhao, J., Jiang, H., Huang, Q., Xu, J., Duan, M., Yu, S., Zhi, Z., Pang, J., Wu, C. (2022). Carboxymethyl chitosan incorporated with gliadin/phlorotannin nanoparticles enables the formation of new active packaging films. *International Journal of Biological Macromolecules*, 203, 40–48. doi: 10.1016/j.ijbiomac.2022.01.128.
165. Wu, Z., Tong, C., Zhang, J., Sun, J., Jiang, H., Duan, M., Pang, J. (2021). Investigation of the structural and physical properties, antioxidant and antimicrobial activity of konjac glucomannan/cellulose nanocrystal bionanocomposite films incorporated with phlorotannin from Sargassum. *International Journal of Biological Macromolecules*, 192, 323–330. doi: 10.1016/j.ijbiomac.2021.09.200.
166. Zheng, H., Zhao, Y., Guo, L. (2022). A bioactive substance derived from brown seaweeds: Phlorotannins. *Marine Drugs*, 20, 742. doi: 10.3390/md20120742.
167. Yu, S., Duan, M., Sun, J., Jiang, H., Zhao, J., Tong, C., Wu, C. (2022). Immobilization of phlorotannins on nanochitin: A novel biopreservative for refrigerated sea bass (Lateolabrax japonicus) fillets. *International Journal of Biological Macromolecules*, 200, 626–634. doi: 10.1016/j.ijbiomac.2022.01.089.

168. Wang, Y., Xing, M., Cao, Q., Ji, A., Liang, H., Song, S. (2019). Biological activities of fucoidan and the factors mediating its therapeutic effects: A review of recent studies. *Marine Drugs*, 17, 183. doi: 10.3390/md17030183.
169. Luthuli, S., Wu, S., Cheng, Y., Zheng, X., Wu, M., Tong, H. (2019). Therapeutic effects of fucoidan: A review on recent studies. *Marine Drugs*, 17, 487. doi: 10.3390/md17090487.
170. Li, B., Lu, F., Wei, X., Zhao, R. (2008). Fucoidan: Structure and bioactivity. *Molecules*, 13, 1671–1695. doi: 10.3390/molecules13081671.
171. Gao, H., Duan, B., Lu, A., Deng, H., Du, Y., Shi, X., Zhang, L. (2018). Fabrication of cellulose nanofibers from waste brown algae and their potential application as milk thickeners. *Food Hydrocolloids*, 79, 473–481. doi: 10.1016/j.foodhyd.2018.01.023.
172. Pandey, A. (2021). Pharmaceutical and biomedical applications of cellulose nanofibers: A review. *Environmental Chemistry Letters*, 19, 2043–2055. doi: 10.1007/s10311-021-01182-2.
173. Cebrián-Lloret, V., Metz, M., Martínez-Abad, A., Knutsen, S.H., Ballance, S., López-Rubio, A., Martínez-Sanz, M. (2022). Valorization of alginate-extracted seaweed biomass for the development of cellulose-based packaging films. *Algal Research*, 61, 102576. doi: 10.1016/j.algal.2021.102576.
174. Álvarez-Viñas, M., Flórez-Fernández, N., Torres, M.D., Domínguez, H. (2019). Successful approaches for a red seaweed biorefinery. *Marine Drugs*, 17, 620. doi: 10.3390/md17110620.
175. Torres, M.D., Flórez-Fernández, N., Domínguez, H. (2019). Integral utilization of red seaweed for bioactive production. *Marine Drugs*, 17, 314. doi: 10.3390/md17060314.
176. Castejón, N., Adrien, A., Spitzer, L., Fernandes, S.C. (2024). Unlocking the full potential of the red seaweed *Gelidium corneum*: Beyond its use as an agar source. *Journal of Applied Phycology*, 36, 291–311. doi: 10.1007/s10811-023-03115-8.
177. Errati, H., Bencheqroun, S.K., Aboutayeb, R., Abail, Z., Lebbar, S., Dari, K., Hilali, L. (2022). Assessment of the red seaweed *Gelidium sesquipedale* by-products as an organic fertilizer and soil amendment. *Sustainability*, 14, 14217. doi: 10.3390/su142114217.
178. Kočí, V., Černý, R. (2022). Directly foamed geopolymers: A review of recent studies. *Cement and Concrete Composites*, 130, 104530. doi: 10.1016/j.cemconcomp.2022.104530.
179. Lebbar, S., Fanuel, M., Le Gall, S., Falourd, X., Ropartz, D., Bressollier, P., Faugeron-Girard, C. (2018). Agar extraction by-products from *Gelidium sesquipedale* as a source of glycerol-galactosides. *Molecules*, 23, 3364. doi: 10.3390/molecules23123364.
180. Tůma, S., Izaguirre, J.K., Bondar, M., Marques, M.M., Fernandes, P., da Fonseca, M.M.R., Cesário, M.T. (2020). Upgrading end-of-line residues of the red seaweed *Gelidium sesquipedale* to polyhydroxyalkanoates using Halomonas boliviensis. *Biotechnology Reports*, 27, e00491. doi: 10.1016/j.btre.2020.e00491.
181. Pandey, A., Pandey, S., Pathak, J., Ahmed, H., Singh, V., Singh, S.P., Sinha, R.P. (2017). Mycosporine-like amino acids (MAAs) profile of two marine red macroalgae, Gelidium sp. and *Ceramium* sp. *International Journal of Applied Sciences and Biotechnology*, 5, 12–21. doi: 10.3126/ijasbt.v5i1.16568.

182. Roberts, D.A., Paul, N.A., Dworjanyn, S.A., Bird, M.I., de Nys, R. (2015). Biochar from commercially cultivated seaweed for soil amelioration. *Scientific Reports*, 5, 9665. doi: 10.1038/srep09665.
183. Chrapusta, E., Kaminski, A., Duchnik, K., Bober, B., Adamski, M., Bialczyk, J. (2017). Mycosporine-like amino acids: Potential health and beauty ingredients. *Marine Drugs*, 15, 326. doi: 10.3390/md15100326.
184. De la Coba, F., Aguilera, J., Figueroa, F.L., De Gálvez, M.V., Herrera, E. (2009). Antioxidant activity of mycosporine-like amino acids isolated from three red macroalgae and one marine lichen. *Journal of Applied Phycology*, 21, 161–169. doi: 10.1007/s10811-008-9345-1.
185. Gan, J., Guan, C., Zhang, X., Sun, L., Zhang, Q., Pan, S., Chen, H. (2022). The preparation of anti-ultraviolet composite films based on fish gelatin and sodium alginate incorporated with mycosporine-like amino acids. *Polymers*, 14, 2980. doi: 10.3390/polym14152980.
186. Paz-Cedeno, F.R., Solórzano-Chávez, E.G., de Oliveira, L.E., Gelli, V.C., Monti, R., de Oliveira, S.C., Masarin, F. (2019). Sequential enzymatic and mild-acid hydrolysis of by-product of carrageenan process from Kappaphycus alvarezii. *BioEnergy Research*, 12, 419–432. doi: 10.1007/s12155-019-09968-7.
187. Meinita, M.D.N., Marhaeni, B., Jeong, G.T., Hong, Y.K. (2019). Sequential acid and enzymatic hydrolysis of carrageenan solid waste for bioethanol production: A biorefinery approach. *Journal of Applied Phycology*, 31, 2507–2515. doi: 10.1007/s10811-019-1755-8.
188. Ortiz-Tena, J.G., Schieder, D., Sieber, V. (2017). Carrageenan and more: Biorefinery approaches with special reference to the processing of Kappaphycus. Tropical seaweed farming trends, problems and opportunities: Focus on Kappaphycus and Eucheuma of Commerce, 155–164. doi: 10.1007/978-3-319-63498-2_10.
189. Reddy, C.R., Baghel, R.S., Trivedi, N., Kumari, P., Gupta, V., Prasad, K., Meena, R. (2015). An integrated process to recover a spectrum of bioproducts from fresh seaweeds. WO 2015/102021 A1.
190. Masarin, F., Cedeno, F.R.P., Chavez, E.G.S., De Oliveira, L.E., Gelli, V.C., Monti, R. (2016). Chemical analysis and biorefinery of red algae *Kappaphycus alvarezii* for efficient production of glucose from residue of carrageenan extraction process. *Biotechnology for Biofuels*, 9, 1–12. doi: 10.1186/s13068-016-0535-9.
191. Sedayu, B.B., Cran, M.J., Bigger, S.W. (2020). Reinforcement of refined and semi-refined carrageenan film with nanocellulose. *Polymers*, 12, 1145. doi: 10.3390/polym12051145.

Index

K

L

M

N

O

P

R

S

For Product Safety Concerns and Information please contact our EU
representative GPSR@taylorandfrancis.com
Taylor & Francis Verlag GmbH, Kaufingerstraße 24, 80331 München, Germany

www.ingramcontent.com/pod-product-compliance
Ingram Content Group UK Ltd.
Pitfield, Milton Keynes, MK11 3LW, UK
UKHW021512250926
475702UK00043B/106

* 9 7 8 1 7 7 9 6 4 1 4 2 7 *